DATE DUE

		AUG 1 0 2010
2 JUL 1984		
		AUG 2 6 2010
3 1 JUL 1984		
2 7 AUG 1984 R		
2 1 MAR 1985		
MAR - 6 1987		
FEB 2 5 1989		
JUN 15 1990		
SEP 14 1990		
DEC 2 1 1994		
JAN - 8 1997		
JUN 28 2010		Printed in USA

201-6503

Energy, The Biomass Options

Alternate Energy ─────────────
A WILEY SERIES

Series Editors:

MICHAEL E. McCORMICK

Department of Naval Systems Engineering
U.S. Naval Academy
Annapolis, Maryland

DAVID L. BOWLER

Department of Engineering
Swarthmore College
Swarthmore, Pennsylvania

Solar Selective Surfaces

O. P. Agnihotri and B. K. Gupta

Energy, The Biomass Options

Henry Bungay

Energy, The Biomass Options

HENRY R. BUNGAY

Rensselaer Polytechnic Institute
Troy, New York

A WILEY-INTERSCIENCE PUBLICATION

JOHN WILEY & SONS   New York · Brisbane · Toronto · Chichester

Library of Congress Cataloging in Publication Data:

Bungay, Henry Robert, 1928–
    Energy, the biomass options.

    (Energy alternatives series)
    "A Wiley-Interscience publication."
    Bibliography: p.
    Includes index.
    1. Biomass energy.   I. Title.   II. Series.

TP360.B86   662'.8   80-19645
ISBN 0-471-04386-9

Series Preface

During the 1970s it became clear that the world's known nonrenewable energy resources are decreasing rapidly and may be exhausted within the foreseeable future. In response to this disturbing prospect, the technologically advanced countries began to focus attention on renewable resources. As a result, there have been significant advances in such areas as solar heating and cooling, photovoltaics, wind power, bioenergy, and ocean wave energy.

The purpose of Alternate Energy: A Wiley Series is to discuss solutions to the technological and economic problems associated with the widespread use of renewable energy resources. The series is intended to introduce readers to the range and potential of these resources, to describe currently available and anticipated methods for conversion and delivery, and to consider economic aspects. The authors published in this series are well known in their fields and have made significant contributions. We have planned the series to meet the needs of all those interested in alternate energy, both practitioners and students.

MICHAEL E. MCCORMICK
DAVID L. BOWLER

Annapolis, Maryland
Swarthmore, Pennsylvania

v

Preface

Inflation, international tensions, gasoline shortages, and general feelings of frustration have sharpened our energy awareness, so there is no need for another harangue about the energy crisis. It is time to evaluate progress on developing alternate energy sources, to formulate future plans, and to consider the implications of recent progress. Nearly every scientist and engineer has some strong opinions about how best to proceed, and many are deeply enmeshed in energy programs. Objectivity has been lost—this book will certainly be criticized severely for lack of objectivity by those whose pet schemes are dealt with harshly. Although most of the conclusions in the book are explained and documented, some may view the same data and reason differently. Healthy and fruitful arguments can refine positions and provide better bases for deciding among the options. For this reason, a sincere attempt has been made to delineate these options in a critical fashion and to take a stand on each issue. Some decisions are highly debatable and subject to revision as invention and technological breakthroughs occur. Nevertheless, the intent has been to focus on each issue and the various alternatives so that strengths and weaknesses can be defined. Proponents of a given approach can marshall their arguments to be more convincing when there is merit or can desist if their plan is inferior.

Technologists are in some disrepute today. Chemical dumps, accidents at nuclear power plants, jet plane crashes, and the like prove dramatically that scientists and engineers are far from infallible. Most of us never thought we were infallible. There is a great danger that satisfactory technical answers can be found, but our nation may be eclipsed anyway. If our main goal is the continuation of a profligate energy policy, we don't deserve to succeed. Even our government officials seem to feel that anyone has a vested right to live as far from work as he desires and that it is sensible to provide fuel for the gas-guzzling vehicle of his or her choice. Legislation to force people to be reasonable is impossible, but policies that penalize energy wasting and encourage conservation are essential. Part of the technological approach to new fuels should be aimed at end uses. Thought should be given to new ways of bringing jobs and people closer together, to more efficient heating of homes, to reducing losses in transmission of energy, and to recreation that is less consumptive of gasoline. This book cannot address matters far outside of the author's competence and must be narrowed to the topic of biomass as a feedstock for producing fuels and chemicals. However, projections show U.S. energy needs to be rising much faster than can the percentage contribution from biomass. If the United States can conserve and become energy efficient, biomass can become a very significant feedstock. On the other hand, continued accelerating demand for energy will result in biomass being a minor contributor in an energy situation gone mad and headed for bankruptcy.

A recent book by Paul (1979) covers the topic of ethyl alcohol in great detail by summarizing many reports and articles. Many of the same sources reviewed by Paul are used here. Although there is some duplication of facts, my book tries for analysis and criticism of many biomass options of which ethanol is but one.

The units are those that are most readily grasped by U.S. readers, and metric units are in parentheses. Temperatures in Celsius are easily appreciated by scientists and engineers and by the U.S. public because weather reports commonly give both Fahrenheit and Celsius. Some units cause problems. For example, the bushel is the standard U.S. measure for corn, but weight instead of volume is more meaningful for energy calculations. There has been an attempt to use a common metric unit when it is less clumsy than its SIE equivalent. There is only a small difference between tons (2000 lb) and tonnes metric (2205 lb or 1000 kg); for rough estimates, a ton is used throughout the book to be the same in both systems. Whenever other units are associated with tons (e.g., tons per hectare), an appropriate conversion to the correct type of ton has been performed.

The book draws on diverse sources, and insights have been offered by present and past members of the Fuels from Biomass Group at the U.S. Department of Energy, including Roscoe Ward, Nello Del Gobbo, Beverly Berger, Ann Fege, and Robert Spicher. Conversations with George Tsao of Purdue University and Edward Lipinsky of Battelle–Columbus have been particularly helpful, as have been meetings with the staff of the Solar Energy Research Institute. However, each of these people would dispute strongly some of my statements. My wife, who disagrees with some of the conclusions, has been an invaluable editor and critic.

Acknowledgment is also given to the fermentation contractors for the Department of Energy. It was my privilege to coordinate their research for three years, to visit their laboratories, and to benefit greatly from meetings with their staffs and students. Without these associations, the book would probably not have been written.

HENRY R. BUNGAY

Troy, New York
August 1980

Contents

Energy, The Biomass Options

1

Introduction

Potential for Disaster

Rising costs and redistribution of the wealth of nations are damaging to the United States, and continuing financial deficits due to oil imports could ruin our economy. Right or wrong, cheap transportation has become essential for getting to work, for linking producers to consumers, and for business and recreation. Petrochemicals plus energy lead to fertilizer, plastics, fabrics, floor tile, and thousands of items that give comfort or pleasure. When sales resistance stiffens because of rising prices, the results are fewer goods for consumers and fewer jobs for producers.

In the past, a good year for the automobile industry could prop up the entire economy, but popularity of U.S. automobiles is unpredictable with high fuel costs and government-mandated designs to improve mileage. As the economy moves into uncharted territory, the United States faces disaster because its cheap energy, cheap feedstocks, and wide technological lead are gone. Change is upon us, and the United States must find new fuels. If these are uneconomic, we will reach our downfall just as surely as with the continuous hemorrhage of dollars for foreign oil. Without cheap energy, the United States might maintain its military posture but at prohibitive cost, and the civilian sector of the economy would suffer drastically. It is unlikely that our nation could maintain world leadership for more than a few more decades. There may be disagreement about our role in international affairs, but there is little reason to think that our citizens would enjoy the shift from a "have" to a "have-not" status with major unemployment and with goals set for us by others.

Radiant energy from the sun is inexhaustible, but care must be exercised in harnessing it. Humans may be distorting the earth's balance by increasing carbon dioxide in the atmosphere and constructing highways, cities, dams, canals, and the like that displace vegetation. Very large scale collection of solar energy may also pose hazards, but we should be able to derive energy from the sun by means that impact favorably on our environment.

Proposed solar technologies include photovoltaic devices, windmills, mirrors focused on steam boilers, solar heating and cooling for homes and buildings, schemes powered by thermal differences in the oceans, collectors in space beaming microwaves down, and photosynthetic plant biomass for conversion to fuels.

Pollard (1976a,b), in two articles on the long-range prospects for solar energy, concluded that fuels from biomass shows promise but other solar technologies appear far too costly and inefficient to become practical. However, his arguments are quite general and must be developed and scrutinized before being accepted.

When the fuel values of trees, agricultural residues, crops, or municipal wastes are compared in price to the fuel values of coal, natural gas, petroleum, or even electricity, it seems that alleviation of the energy crisis with biomass has small margins for profits. Whereas fossil fuels are found in concentrated deposits, biomass is distributed over wide areas and is costly to collect. Biomass has a high moisture content; thus combustion suffers from supplying the energy for vaporization. Nevertheless, there are already some methods for converting biomass to fuels and to chemicals at competitive prices. These are liquid fuels that command a premium for transportation and are chemicals presently made from fossil fuels. This means that there could immediately be a small but significant substitution for petroleum by commercializing biomass conversion processes. As technology improves and as by-products or coproducts such as protein for food are developed, biomass industries can become major suppliers of exactly those materials most needed to sustain and improve our life-styles.

The view is widespread that biomass will not become a major contributor to the U.S. energy picture (Landau, 1978; Weisz, 1978). Low efficiency of photosynthesis, costs of water and fertilizer, and competing uses of land to grow food are cited as overriding obstacles to establishing a biomass–energy system. However, crops must be grown for food and fiber, and only small fractions of the plants are in the final products. The stems, stalks, cuttings, and residues can be an enormous resource. Arguments about energy ratios of inputs must be reevaluated. Certainly the credits taken for biomass fractions diverted to energy production can make agriculture more economical while not requiring much more labor, capital, and operating cost at the farm.

Biomass

A little over a century ago, wood was the main fuel for homes, buildings, factories, and locomotives. Ships sometimes had wood-fired engines, but sails were most economical for ocean voyages. Improvements in engines and combustion technology and the use of coal, which is easier to handle and has higher heating value per unit mass, accelerated the industrial revolution. Eventually petroleum and natural gas became economical and much more convenient than coal. At the present time, these fuels, all of fossil origin, dominate the market. Carbon dioxide in the atmosphere or carbonate species in water were converted to organic matter by photosynthesis, and geological processes converted deposits of decaying plant material to fossil fuels. The formation of fossil fuels is not completely understood, and there seems to be no possibility of natural generation to meet future needs. While we may contest the time scales for depletion of fossil fuels, there is little doubt that petroleum and natural gas will rapidly become scarce and that coal will become much more expensive as the richest and most accessible deposits are exhausted. Wood or other biomasses now deserve consideration as fuels, but there can be no return to the wood burning characterized by belching dense smoke. Wood burning with modern combustion technology has value, but a superior use of biomass is conversion to gaseous or liquid fuels that are clean and convenient or to organic chemicals that can substitute for those presently derived from fossil fuels.

There are many analogies between coal and biomass, and these are perhaps predictable because biomass is the original form of coal. The main differences are that the sulfur content of coal is higher, which is a disadvantage for direct combustion, and that coal is found in large deposits, a very major advantage. As air pollution regulations become more restrictive, sulfur compounds must be removed from coal or from the combustion gases. This means much higher costs because the pollution control equipment must be amortized. Mixing biomass with coal is one way to lower the concentration of sulfur compounds in stack gas, but handling of mixed fuels also adds to the expense.

A later chapter on using heat to convert biomass to clean convenient fuels presents processes that were first developed for conversion of coal. With either coal or biomass as feedstocks, such plants would have to compete with existing producers of fuels and chemicals. The petroleum and petrochemical industries operate on such vast scales that great economies are realized in terms of management, labor, equipment, financing, and distribution. To compete, thermal conversion plants should also be gigantic to realize the economics of scale.

Quads

Convenient units for U.S. energy needs are quadrillion BTUs (British thermal units), or Quads for short. A quadrillion is 10^{15}, or 1,000,000,000,000,000. Presently, the United States generates roughly 76 Quads of energy annually. If this all came from oil, it would equate to almost 40 million Bbl/day (barrels per day). At the present rate of increase, our consumption could surpass 100 Quads per year before 1990. There must be conservation and a slower rate of increase because the United States cannot afford to squander energy. Our per capita consumption is over twice that of other industrial nations, and much is nonessential. Oil is the main villain because more than half comes from foreign sources, and our balance of payments is intolerably poor. Natural gas is also in limited supply and should command a premium because of its convenience; instead, it is regulated by the government at artificially low prices.

Any proposals for alleviating the energy crisis must consider the existing power grids. There is a network of gas and oil pipelines plus trucks and railroad tank cars to convey these fuels. Electric power is generated, and its distribution uses a very complicated, interconnected system of transmission lines with transformers to convert voltages. The investments in the U.S. power structures are massive and are continuing. New facilities are faced with sky-high costs of materials and labor. It is not sensible to scrap the existing system and to start over with an alternate one.

Large central power stations can use atomic energy, coal, or hydroelectric power, and those using gas or oil can convert to coal for which the United States has abundant supplies. A program is under way to force conversion from oil and gas back to coal; it is ironic that some plants have recently scrapped their coal handling facilities after completing conversion from coal.

Many of the solar technologies could be compatible with existing power systems. Generation of electricity by windmills, by photovoltaic devices, sunlight focused on boilers, ocean thermal differences, and the like could be connected directly to the electric power grid. An investment in storage would be required if intermittent solar power based on wind or light were to become major factors. An exception to storage needs would be solar collectors in space to beam down microwaves 24 hr/day, but many people believe this to be hopelessly impractical because of horrendous costs of construction in space. Centralized power stations may be the backbone of electrical power, but energy is needed as fuels for heating individual buildings and for transportation.

The U.S. Department of Energy and other groups have laid great stress on the cost of a unit of energy. This misses the point that there is a quality factor that plays a decisive role in determining the value of energy. Electricity is very high quality energy, and we expect to pay a premium for it. We are beginning to appreciate that heat is a low form of energy and that electric heating, as for homes, squanders this premium. Jet engine fuel is another elegant form of energy which must not be valued in terms of its BTU content as compared, for example, to coal, because the two cannot be equated.

The power intensity of direct sunlight on the earth's surface can exceed 1 kW/m^2, but the average for the entire continental United States is roughly 0.2 kW/m^2. Assuming that power consumption is 10 kW per capita, each person could satisfy energy needs with 540 ft^2 (50 m^2) in theory. This is about the size of the roof on a typical home. However, real efficiencies are low, and storage must be provided for periods of darkness. Perhaps an acre (4000 m^2) would suffice for each person, but this equates to all the arable land in the country if we were to rely solely on solar power.

Small, distributed systems for producing energy have been discussed at length in a popular, controversial book by Lovins (1977). The following advantages are given for changing to small, localized power sources:

1. Virtual elimination of capital costs, operating costs, and losses of the distribution system from large power sources.
2. Less likelihood of massive failures.
3. Elimination of large standby or storage systems.
4. Quick start for implementation instead of long lags for planning and approval.

5.   Opportunity for savings by mass production.

A detailed critique would not be appropriate here, but Lovins has made an important contribution by directing attention to simple systems as alternatives to large, central power stations.

Contributions by Biomass

The combustion of biomass adds 2 to 3 Quads presently to annual U.S. energy consumption mainly from the forest industries' burning of wastes for power generation. Home fireplaces and wood-burning furnaces are very important in Vermont but of minor consequence elsewhere. In a few locations, bagasse, the sugarcane residue after extraction of sugar, is burned to generate steam. If total U.S. annual energy consumption does not increase rapidly from the current 76 Quads, biomass can play a significant role. In the last year or two, energy usage has leveled off, thus indicating a possibility of no more sharp increases. Of course, biomass can only expand to the limits of available lands for growing, with further increases based on improvements in genetics and cultivation techniques. The rates of increase in biomass production are probably less than the increasing energy usage to which we have become accustomed; thus the importance of biomass depends on sensible management of our total energy demands.

Published estimates of the potential annual contribution of biomass range from 1 to 20 Quads. The low range can be dismissed as propaganda by advocates of other technologies because we already are doing better than their projections. A century ago, the Eastern forests alone supplied 2.4 Quads of fuel per year. New annual growth on the 2 billion acres (8 million km^2) of the United States equates to about 30 Quads (14 Quads of cultivated crops, 10 Quads of forests, and 3.5 Quads of grasses). It does not seem unrealistic to strive for 8 Quads of biomass energy by 1990 and 15 Quads by the year 2000, especially if new agricultural practices are introduced.

Gaddy (1977) estimates that 25 percent of the total U.S. energy requirement could be satisfied by extending conventional agriculture to idle lands to produce biomass. Anaerobic digestion to methane would be the conversion procedure. The cost of a plant for 50 million ft^3 (1.4 million m^3) per day of methane is $75.4 million, with about $\frac{2}{3}$ of this allocated for purchase of the digesters. One such factory produces roughly 0.015 Quad/yr, thus implementation of this approach is not appealing in terms of effort and capital expenditure.

Goldemberg (1978) has reviewed the energy picture in Brazil. Biomass accounts for about 30 percent of Brazil's total energy needs, and the government is encouraging increased use of wood by industry and of the crops-to-ethanol programs. The average intensity of solar radiation on Brazil is approximately double that on the United States. Warm temperatures, copious water, and high humidity lead to rapid growth of many types of vegetation.

Factories That Produce Starting Materials

Present-day fuels can be categorized as raw materials. They are relatively cheap and have alternate uses as feedstocks for the chemical industry. Conversion of biomass to fuels raises the prospect of operating factories whose products must compete with cheap raw materials. Direct combustion of biomass is a different economic matter, and areas in the United States remote from coal and close to forests have long used wood to heat homes, factories, and public buildings. Economics and logistics will determine when the best energy use of biomass is direct combustion. Other fuels from biomass will be analogous to petroleum refining: a cheap feedstock will be upgraded to more useful and valuable products. These products cannot compete *on a BTU basis* with wood, oil, coal, or natural gas any more than can high-octane jet fuel be expected to compete with crude oil.

Once we accept the concept of converting biomass to valuable fuels and chemicals instead of cheap BTUs, there are some important implications for profitability. In all private ventures, return on investment is an overriding factor. Very cheap fuels fit into the utility–monopoly segment of the economy where 10-year payoff is acceptable. The investor taking risks in the private industrial sector needs a shorter time scale, and people managing a factory look to 2- or 3-year payback as the upper limit on capital expenditures. In other words, no group in their right minds would proceed without government-backed loans and profit guarantees to construct and operate a colossal plant for conversion of coal or biomass to cheap fuels. A company might very well invest its own money in a chemical plant to produce premium fuels or valuable chemicals if the investment were reasonable and a profit likely.

Biomass Technologies

Conversion of biomass to products may be thermochemical or biochemical. Heat energy can rearrange the molecules in biomass to yield compounds such as methanol, acetic acid, or acetone. A reductive atmosphere with high temperature and pressure produces oil from biomass. Reacting biomass with water gives a mixture of carbon monoxide and hydrogen known as synthesis gas. The main routes from fossil fuels to useful chemicals pass through synthesis gas or olefins. Synthesis gas can yield such products as methanol, or the hydrogen can be separated and reacted with nitrogen to form ammonia. Olefins can lead to alcohols, oxo compounds, and the like. The heat energy needed to drive any of these processes probably should be derived from a portion of the biomass and from combustible, low-value by-products.

Thermochemical conversion of biomass was once a commercial venture in the United States. A division of Ford Motor Company heated wood to produce volatile organic chemicals, tar, and coke until the 1930s, when products from petroleum made the operation uneconomic. Now that petroleum is not so cheap, the old biomass process might be resurrected and improved. As biomass was once a chemical feedstock, the economics should not be too far out of line today.

Whereas thermochemical processes put intact biomass in the pot, bioconversion processes are directed at certain fractions. Most of biomass consists of the polymers cellulose, hemicellulose, and lignin. Lignin, which is very resistant to hydrolysis, is discussed later. None of these polymers can as yet be transformed directly by biochemical processes to useful chemicals other than their respective monomers. Cellulose is composed only of glucose; hemicellulose is mostly five-carbon sugars with some six-carbon sugars and uronic acids. Fermentation processes are known to convert these materials to a large number of relatively valuable products.

Complete biomass can be fermented by an anaerobic process termed *digestion.* There is no attempt to inoculate with desired organisms, and a mixed culture develops. When operated correctly in terms of physical factors, pH, and detention time, the main product can be methane, which, when purified, is suitable as pipeline gas. Lignin is little affected by anaerobic digestion and is found in the waste stream along with the spent culture and other recalcitrant organic compounds. Solids from the waste are called *sludge;* sludge is poor fertilizer but has value for conditioning soils high in clay.

While thermochemical conversion is a more brute-force approach, bioconversion takes biochemicals through sophisticated fermentation pathways to more valuable biochemicals. Water is an essential ingredient in fermentation but is undesirable in thermochemical processes because of its high heat of vaporization. It may turn out that relatively dry biomass is best converted to fuels by thermochemical means while bioconversion is well suited for wet biomass.

Another challenge for research and development of biomass conversion processes is to find an optimal product mix. Thermochemical steps tend to produce several products, and sales for some may so overshadow sales of others that markets may be saturated and by-product value may be very small. Fermentation by pure cultures of microorganisms gives one or two products, but separation from the medium is often costly. Better purification technology is needed.

Tong (1978) has noted that the past 10 years have seen a quadrupling of petroleum prices while sugar and starch have not doubled. Based on carbon content, the costs in Table 1.1 were derived. Actually, on a fermentable sugar basis, molasses was recently depressed with a temporary price of 6¢/lb (13.2¢/kg). Tong feels that carbohydrate feedstocks could supply 4 million tons/yr of C_3 and C_4 chemicals by fermentation processes if grain production were increased by 10 percent. Improved fermentation technology would assure biomass a position competitive with petroleum feedstocks.

Flickinger and Tsao (1978) have presented the status of fermentation of cellulosic substrates and pointed out the very important factor of price stability. A chemical plant using one major raw material has only limited bargaining power with its suppliers. Shortages can quickly drive prices up. On the other hand, cellulosic substrates for hydrolysis to sugars are plentiful, varied, and pretty much

Table 1.1   Comparison of Costs (1977 Basis)

Feedstock	Unit Cost, ¢/kg	Cost per Kilogram of Carbon, ¢
n-Paraffin	33	38
Ethylene	24	28
Molasses sugar	11	27.5
Cane sugar	13.2	33
Cornstarch	22	55

interchangeable. If one source encounters shortages, competing crops could substitute with little upset of the price structure. The message is the need for use of all fractions of the biomass. Ethanol from cellulose might barely pay for the cost of raw materials and processing, but there are very significant values to the hemicellulose, lignin, and alcohol stillage residues that can result in sizable profits.

Handling of large amounts of biomass may resemble the operations in pulping of wood. After many years with an old and little-changed technology, innovations in paper pulping are appearing. Elmore (1978) has presented some of the newer ideas and equipment designs. Very large, agitated reactors several hundred feet tall are used for countercurrent contacting of wood chips with alkaline pulping solutions. Similar equipment might be used for extracting hemicellulose from wood with hot, dilute acid or for pretreatment to weaken the cellulose for hydrolysis. Another interesting feature of newer pulping technology is more concern for recovering heat from process streams. Borrowing from pulping technology can save time and substantial research and development (R&D) costs in the commercialization of fuels from biomass.

Renewable Resources

Biomass must not be oversold as a renewable resource. To achieve the high productivities of biomass required for energy farming to be profitable, the entire plant above the ground would be collected. This differs greatly from current practices of both forestry and food cropping where trimmings and residues are left to rot. Most of the earth's plant biomass is a standing crop based on return of the minerals and organic constituents from dead and dying species. With no return and no residues, erosion and depletion can ruin the soil. Land is eroded less by water or wind when it is protected by residues, and federal and state directives can insist on specific materials being left on the site. Alternatives such as terracing to prevent erosion by water or constructing windbreaks are expensive. Although not insurmountable, erosion problems are serious and can impact unfavorably on the economies of biomass production.

Farming, as we know it today, is not an operation of merely taking from the

land; large amounts of fertilizer are added. With no residues, even greater applications of fertilizer will be required. The trace constituents of fertilizer are of little
consequence, but nitrogen and phosphorus could become overly expensive.
Chemical fixation of nitrogen is energy intensive and thus it grows more expensive each year. Biological fixation of nitrogen by microorganisms in the root nodules of legumes or by other microorganisms will someday be controlled effectively. In the meantime, current means of replenishing nitrogen in the soil must
suffice for establishing energy farming. The key nonrenewable ingredient in fertilizer is phosphate. Phosphate rock deposits should be mined out in 400 to 500
years, which is a shorter life span than that predicted for coal mines. Phosphate
availability can, therefore, be a very real restraint on biomass production.

Kemp and Szego (1976) have advocated "energy plantations" and have
pointed out how a forest could provide wood for a power plant with the stack
gases furnishing CO_2 to algal ponds to yield additional products. Spent streams
from the ponds would irrigate and fertilize the forest. Dubinsky, Aaronson, *et al.*
(1978) have proposed a highly integrated system based on algae using seawater,
sunlight, flue gases, sewage, and wastewater from industry and from municipali-
ties. The algae would produce food, fuel, and a lower grade of water for cooling
or processing. Szego's group (Intertechnology/Solar, 1978) found that it is very
difficult to match the various operations, so that it may be impractical to have a
forest, power plant, and algae unit with each sized and located for optimal
performance. In other words, an algal unit that is limited to the CO_2 from the
power plant may not be significant economically. Very elaborate systems could
well founder on the logistical and mismatch problems, especially since acceptance of the products and side streams is uncertain. Furthermore, if a major user
of process water were to discontinue its use, disposal could be costly and could
seriously perturb the overall economics.

Soil, Food, and Energy

It is estimated that 3 million years were required for human society to propagate
to a population of 1 billion in the 1830s. One hundred years later, there were 2
billion with only 30 more years to reach 3 billion. Four billion was reached in 15
more years, and a world population of 5 billion is projected 10 years from now.
As population advances logarithmically, fertile soil is being destroyed. In the
past, new lands were found to replace worn-out soil, but there is an end in sight
for developing new agricultural areas. Of all the nations that once exported
grains, only the United States and Canada remain as major suppliers.

There is little hope of avoiding mass starvation if the world's population keeps
expanding at the present rate. Even if undeveloped nations install drastic birth
control measures now, their population increases have so much momentum that
food supplies could not keep up. The United States and Canada will be expected
to provide food for the starving. However, there are no longer large surpluses and

idle land. The United States and Canada will be unable to feed the multitudes, and world commerce will be distorted by beggar nations whose plight cannot be ignored by more fortunate humans.

These considerations were raised by Brink, Densmore, et al. (1977), who reported on soil changes on selected U.S. farms. Corn, now the dominant annual crop, exhausts the soil. Strains of corn have been created and selected that are resistant to insects, are highly productive, and are responsive to fertilizers; however, without heavy application of fertilizer, yields would be meager. Erosion causes major losses in fertility except for crops of grasslike character, the soil is exposed to abrasion and scouring by wind and water for much of the season. Some U.S. farmers practice crop rotation where corn is followed by alfalfa, a legume that replenishes soil, but unfortunately, erosion and poor crop management are prevalent.

Lockeretz (1978) presents a grim history of the dust bowl in the United States where years of good rainfall encouraged agricultural practices that led to disaster in times of drought. Lessons learned during previous dry periods were quickly forgotten when times were good. Improper attention to preparing for dry years and overconcern with high yields might lead to new dust bowls that would be blamed on fuels from biomass. Marginal lands are just that: marginal. An imprudent attempt to cultivate energy crops could ruin those lands for many years to come and create dust that would impair many adjacent acres of better lands.

Recycling of nitrogen or phosphorus to the soil with wastes from the biomass-to-fuel conversion process must be considered. Ash from a thermal conversion process could present difficulties if the needed elements are in chemical forms that have poor solubility. This is likely to be of less importance, however, than the inefficiency of application whereby much of the fertilizer is washed off or is percolated through the soil. Recycling will not be very helpful if most of it is gone before being incorporated into the plants. Marine farming has the major advantage of using ocean waters that can be fertilized from land runoff or from bottom sediments. However, ocean surface waters are almost devoid of nutrients; an ocean farm without fertilization has poor prospects.

Jensen (1978) takes a gloomy view of the chance for agricultural gains matching the increases in population. Although crop productivities have shown remarkable increases due to genetics, fertilization, irrigation, and management, yield ceilings must result in smaller steps in improvement. In other words, when the ceiling is distant, a steep increase can be sustained. Today, many of the major modifications have already been made, so further refinements are likely to be gradual. Even if yield were to increase at present roughly linear rates, population growth tends to be logarithmic. Agriculture in the United States competes for resources that may be diverted to other uses. Western U.S. irrigated agriculture faces gradual elimination as water becomes too costly because of other demands. The ultimate in productivity may never be practical because farmers must judge when harvesting is more economical rather than seeking additional gain at the expense of labor and resources while risking losses to weather and insects.

With a world food crisis and with soil depletion, it seems heartless to strive for fuels from biomass. Using some of the fuels for production of food can justify diversion of land from food crops to energy crops, but any significant energy contributions will require vast lands. There would almost certainly be an impact on either conventional forestry or food or both. The only approach consistent with clear conscience is to produce energy as a coproduct of food or fiber. Residues are available now if other means of erosion control and return of organic material to the soil are practiced. Conversion of biomass to fuels will provide water, minerals, and nutrients for recycle.

Economics may force changes in our diet. With human starvation driving up the price of grain, it will be costly and will present a poor international image to feed most of our grain to cattle. Beef is likely to become a luxury affordable by few, and protein directly from plants will take its place. Single-cell protein from microbial fermentations can be processed to resemble hamburger or other meat fractions and can provide adequate nutrition. We may find that vegetable protein and single-cell protein are to our liking when steak costs too much of our incomes. Diversion of grain from cattle to people would free massive amounts of stalks and leaves now used as fodder to be converted to fuels and chemicals.

We are perhaps decades away from the time when we can reason calmly about how best to preserve and improve our civilization. There are limits to living space and food. The atmosphere and the oceans are not endless sinks for pollution. Our planet must be shared, but Hardin (1968), in a classic paper about the problem of common resources, has emphasized that human nature leads to inequitable distribution. Furthermore, allotting each person a fair share is complicated by religious views, national policies, customs, and attitudes about birth control. Expedient solutions must be found quickly before accelerating population growth and rampant pollution cause irreparable damage to our civilization. Rules for distribution must be developed for mankind, but there are no guidelines for what would be just. It is unlikely that vested interests can compromise or that the conflicts of management versus labor, capitalism versus socialism, and haves versus have-nots can be resolved. Humanitarian instincts demand protection for the deprived and the handicapped, physically or mentally. However, mankind becomes genetically enfeebled by propagating defects.

Managers of research and development must constantly be aware of the "moving target." This means that goals must be updated and that the competition is not standing still. For example, Kodak would be the brilliant innovators in instant color photography if Polarold didn't keep announcing technological breakthroughs. Fuels from biomass has as its central dogma the supposition that fossil fuel prices will continue to escalate. However, Organization of Petroleum Exporting Countries (OPEC) nations could manipulate prices just as biomass factories were completed and undercut their profits. A major change in transportation such as more efficient use of trains and buses could weaken demand for fuels and undermine OPEC solidarity. Just as the automobile made buggy whips obsolete, a new technology or a change in travel preferences could greatly reduce the number of automobiles. Success in finding other sources of cheap energy could reduce demand for petroleum and greatly extend its lifetime.

Summary

All things considered, biomass is a very plausible energy feedstock. There is already a strong technological base, and products suitable as fuels or petrochemical substitutes can be made. Concentrated efforts on the most cost sensitive steps could lead rapidly to new industries that supply premium fuels at attractive prices.

The salient points for discussing fuels from biomass are:

1. Are there sufficient quantities to add appreciably to our energy needs?
2. Can affordable products be made from biomass?
3. With what will products from biomass compete?
4. Are there trends in technology that auger for or against more profitable products from biomass?

2

Biomass Feedstocks

The many permutations of resources, technologies, and products present a confusing picture of fuels from biomass. Furthermore, advances continue on all fronts: crop yields are increasing, old conversion technologies are being improved, and exciting new processes are being announced. Some biomass in the form of wastes costs very little but is too small a resource to support a fuel industry; other materials that would be excellent for producing fuels are presently more valuable for food or fiber. Economics must dictate the uses for biomass and the choice of conversion processes. Undoubtedly, there will be several commercial ventures for biomass fuels with a variety of feedstocks and processes. Some will succeed because of superior technology; others will survive because of good regional matches between feedstock availability and ready markets. An attempt is made to sort out the main threads of biomass processes, but with full knowledge that a technological breakthrough can quickly alter priorities.

The economic framework for fuels from biomass allows so little margin for transporting biomass that a nation must be prepared to produce its own feedstocks. It matters little what ideal plants can do in terms of productivity, for species must be chosen that thrive on the soil available and in the climate of the country. This does not mean that some tropical countries with lush vegetation cannot play leadership roles in valuable chemicals from biomass. However, cheap fuels are incompatible with large collection and transportation costs for biomass. In fact, the energy expenditure for harvesting, hauling, and shipping must not exceed the energy content of the products.

The U.S. climate ranges from subtropical to cold-temperate, with a spectrum from rain forests to deserts. Some mountainous regions have abundant trees but are so rugged that harvesting them is impractical. Arid lands have been converted to rich productive farms by irrigation, but water costs and availability are threatened by sinking water tables. Low rainfall in recent years has jeopardized many western farmlands.

Native vegetation may prove not very productive for energy needs. Adaptation over decades of wet-dry cycles often emphasizes survival rather than optimum use of the good years. Planting a selected crop with attention to its cultivation can give markedly higher yields. For example, sorghums over a 35-yr period in Oklahoma on dry soils have yielded 3 tons of biomass per acre per year, whereas native buffalo grass seldom achieves 0.6 tons.

Table 2.1   Energy Potential of Existing Sources of Biomass

Resource	10^6 dry tons/yr	Quads/yr
Crop residues	278.0	4.15
Animal manures	26.5	0.33
Unused mill residues[a]	24.1	0.41
Logging residues	83.2	1.41
Municipal solid wastes	130.0	1.63
Standing Forests[b]	384.0	6.51
Total	925.8	14.44

[a] Does not include unused bark from wood pulp mills.
[b] Surplus, noncommercial components. Total forest wood equals 40 Quads (Reed, 1979).

Table 2.1 shows clearly that wastes or residues are small resources, and less than half could be collected economically. Trees or crops grown specifically for energy could provide many more Quads. Of course, a great amount becomes more important in the long range. It may take several years to decide to embark on a massive program for biomass fuels and to prepare and implement a plan. In the meantime, as more modest ventures are made, wastes and residues are eminently logical feedstocks. Municipal and industrial solid wastes are an environmental hazard for which disposal credits can be taken, but only 1.6 Quads/yr could be produced. The first factories for biomass fuels should use wastes and residues whenever possible and claim these credits. As sales increase and additional feedstocks are required, other sources must dominate over wastes. Process improvements and larger factories with the economies of scale will allow profitability as feedstock prices increase. The cheap feedstocks allow for a wider margin of error during the crucial period of inaugurating a new industry.

The quality of biomass must not be overlooked. For direct burning, moisture content, heat of combustion, and cost of subdivision are indices of biomass quality. Other thermochemical conversion processes are concerned with the effects of biomass composition on yield and relative proportions of products. Bioconversion also requires subdivision, but in this case the moisture is beneficial. The major difference with bioconversion is the overriding importance of the biochemical composition. Plants rich in sugars, starch, or easily hydrolyzed carbohydrate polymers provide readily fermentable substrates. The most abundant plant polymer, cellulose, requires hydrolysis to glucose to achieve rapid fermentation. Even plants with high concentrations of other carbohydrates have large amounts of cellulose, thus a means for processing cellulose to additional products is important. Some aquatic plants contain structural polymers other than cellulose, but little is known about hydrolysis and fermentation of these materials.

General categories of feedstocks and the possible conversion technologies are shown in Table 2.2. In general, very wet biomass is poorly suited for combustion or thermochemical conversion because energy is wasted in drying or in vaporiz-

Table 2.2 Feedstocks and Conversion Technologies

Feedstock	Combustion	Thermochemical Conversion	Bioconversion
Wastes			
Municipal solids	Very good	Very good	Good
Plant trimmings	Good	Good	Good
Manures	Poor	Fair	Excellent
Terrestrial growth			
Trees	Excellent	Very good	Fair
Sugary plants	Poor	Poor	Excellent
Stalky plants	Very good	Good	Good
Aquatic growth	Poor	Poor	Very good

ing the excess moisture. All feedstocks are acceptable for bioconversion, but those that require little subdivision or pretreatment are best.

Leaving quality aside, sheer productivity per unit of growing area provides a good index for selecting a plant species. Obviously land is expensive, and thus crop productivity per unit area is important. Even in open ocean waters where acquisition costs are nil, harvesting is expensive, and it is desirable to reach the maximum crop per time increment. With a given velocity of the harvesting vessel, higher productivity means more crop collected per hour. Some estimates of productivity are presented in Table 2.3.

Note from Table 2.3 that some tree species are fairly productive. Sugarcane is excellent but is suited to the climate of only a very few of the United States. Seaweeds have great promise, but major engineering problems exist with respect to their culture and harvest.

Residues

Residues are field crop trimmings, stalks, and foliage left behind on the ground; forest residues; low-grade mill fractions; animal manures; and municipal wastes. Biomass residues currently have some value for animal feeds, fiberboard products, fertilizer, and erosion control.

Costs of residues range widely. Wheat straw and other small-grain straws are less expensive than corn stover, which is worth $20 to $40/ton. Manure may range from no value up to $17/ton for rich poultry manure. Forest residues delivered to the user can cost from $25 to $60/ton depending on location.

Alich, Schooley, et al. (1977) performed a study of 10 selected sites in various sections of the United States for potential of processing residues to fuels. Evaluations were made of:

- Characteristics of residues, including the value or cost of current disposition, requirements for collection and transportation, and the energy content of residue as a feedstock.

- Condition of residues, seasonality of production and location, and current disposition.
- The type of equipment required to collect and transport residues.
- Total production and consumption of energy in the area.
- Specific energy needs of processing plants in the area.
- The feasibility of using the residues for the production of energy.
- Institutional factors and recommended technological development.

Table 2.3   Productivity of Agricultural Crops, Forests, and Seaweeds

Climate	Area	Yield	
		tons/ha · yr	tons/acre · yr
Temperate			
Rye grass	U.K.	23	10
Kale	U.K.	21	9
Sorghum	U.S., Illinois	16	7
Maize	Japan	26	11
	Iowa	16	7
	U.S., Kentucky	22	9
Potatoes	The Netherlands	22	9
Sugar beets	U.S., Washington	32	14
Wheat (spring)	U.S., Washington	30	13
Sycamore	U.S. (SE—NC)	12	5
Loblolly pine	U.S. (SE)	10	4
Hybrid popular	U.S. (NC)	20	9
Aspen	U.S. (NE)	4	2
Seaweeds	U.S., Massachusetts	33	15
Subtropical			
Alfalfa	U.S., California	33	15
Sorghum	U.S., California	47	21
Bermuda grass	U.S., Georgia	27	12
Sugar beets	U.S., California	42	19
Wheat	Mexico	18	8
Rice	U.S., California	22	10
Eucalyptus	U.S., California, Louisiana, Florida	26	11
Seaweeds	U.S., Florida	73	32
Tropical			
Napier Grass	Puerto Rico	85	37
Sugarcane	Hawaii	64	28
Sugar beets	Hawaii (2 crops)	31	14
Maize	Peru	26	11
Rice	Peru	22	10

It was concluded that energy from residues would be practical at most of the sites that use existing conversion technology. With rapid advances being made on conversion processes, residues appear to be the basis for immediate commercialization of fuels from biomass, and crops grown specifically for energy purposes can be phased in later.

Forest Residues

The heating value of mill residues depends on species, wood: bark ratio, and moisture content. On a dry basis, 7900 to 9700 BTU/lb (4400 to 5400 cal/g) is a reasonable estimate for residues in North America, with higher values if considerable bark is present. Moisture also affects combustion efficiency; typical 50 percent moisture gives a combustion efficiency of 68 percent, compared with 82 percent for dry wood.

There is practically no sulfur oxide emission problem when burning wood or residues. Wood is sometimes mixed with high-sulfur coal to meet pollution standards for the stack gases. Coal has ash content ranging from 5 to 25 percent, but wood is less than 1 percent ash, and bark may reach 5 percent ash. Wood ash may be better suited than coal ash for spreading on growing areas because the toxic metal content is low. During handling and storage, however, dirt or sand may adhere to wood fractions and adversely affect combustion and increase wear on equipment.

Howlett and Gamache (1977b) have reviewed wood residues as fuel. They point out that the bulk density of wood and bark residues means costly transport and storage. Some recent prices for residues are given in Table 2.4.

Of the 86 million tons of forest residue produced annually in the United States about 67 million is wood and the rest is bark. Approximately 75 percent of residues were used for some purpose, with 56 percent for nonenergy products such as lower grades of pulp. Most of the bark is burned. The main unused residues are those from logging where trimmings, thin branches, foliage, and defective trees are left. In the Pacific Northwest there is up to 24 dry tons of such residue per acre (59 tons/ha) of logging. For rapid commercialization of a fuels from biomass process, these residues present a readily available feedstock.

Table 2.4 Approximate Value of Residues

Material	Cost per Dry Ton, $
Chips	
Softwood	27 to 40
Hardwood	20
Shavings	7.50
Sawdust	1.50 to 5.00
Bark	1.00 to 5.80

Manure

Animal manures total 5 million tons daily (0.8 million tons, dry basis) containing 32,000 tons of nitrogen and 10,000 tons of phosphorus. Human wastes constitute less than 8 percent of the animal wastes and thus are a minor biomass resource. Livestock in pens produce about 40.5 million tons of manure per year (dry basis), but dairy cattle and poultry represent many small, widely dispersed resources that cannot capture the economies of large-scale operation. Manure resources are shown in Table 2.5.

Cattle Manure

Cattle feedlots are fairly similar with respect to diet of the cattle, and manure from one lot might be expected to be much the same as that from another lot. This is far from the truth, however, because handling practices have profound effects. There are two very different types of lot: dirt and environmental. In the dirt lots, animal wastes fall directly onto the ground so that feces, urine, rain water, and dirt can be mixed by animals treading through the mess. At intervals, this material is bulldozed into piles. The moisture content then depends on ambient drying prior to entering the pile, pile configuration, and duration of storage. Percentages of dirt and stones will depend on how the scraping was performed. Microorganisms are active in the piles; thus organic compounds are metabolized. An old manure pile may contain very little easily digested material.

The environmental lot has a permanent floor, and animal wastes are frequently flushed to receivers. This fresh, wet manure is very amenable to anerobic digestion. Manure as deposited usually has about 80 percent moisture. A slurry of manure is present in the receivers of an environmental lot, so that a solid–liquid separation would be desirable if a thermochemical conversion process were contemplated.

Garrett (1977) presented typical cattle manure composition (Table 2.6). Although fresh samples are quite variable in composition, the values are consistent on a dry basis. The heat of combustion of dry manure is in the range of 5750 to 6730 BTU/lb (3195 to 3740 cal/g).

Table 2.5   Manure in the United States

Animals	Numbers (Millions)		Potential Contribution (Percentage of Total)
Cattle	110	Feedlot	39
		Dairy	35
Sheep and goats	22		Very small
Swine	54		8
Poultry	825		18
Horses	7		Very small
(Humans)	(210)		(Not counted as manure)

Table 2.6 Typical Manure Composition

Constituent, Weight Percent	As Received	Dry	Dry—Ash Free
Carbon	18.	37.4	50.7
Hydrogen	2.8	5.6	7.6
Nitrogen	1.4	2.8	3.8
Sulfur	0.2	0.5	0.7
Oxygen	13.7	27.4	37.2
Ash	13.2	26.3	—
Moisture	50.00	—	—
Totals	100.0	100.0	100.0

Beck, Halligan, et al. (1977) have commented on the impact of manures on U.S. energy needs. The larger cattle feed lots in the United States are listed in Table 2.7. If all the manure on these feedlots were collected, the energy of this feedstock would be 0.09 Quad, which is a tiny amount compared to 76 Quads/yr used annually. Furthermore, the 42 plants for converting manure would produce about as much fuel as one standard-sized gasification plant using coal. The perspective for ammonia is slightly different, however, because thermochemical conversion can yield ammonia from manure in significant amounts. Since ammonia production represents about 3 percent of the annual natural gas consumed in the United States, manure from large feedlots has potential to save about 0.6 percent of this natural gas.

Food-Processing Wastes

On a national basis, there appear to be enough food wastes to constitute a resource for conversion to energy. However, these materials are so widely distributed that only a few isolated sources have economic potential. The largest potato chip factories could each supply sufficient wastes for a small alcohol plant, and other wastes rich in starch or sugars could support fermentation facilities. Whey, a cheese-processing waste available in many states, provides good perspective on the energy prospects of food wastes. Mueller Associates (1979) surveyed whey in the United States. The approximate composition is 4.8 percent lactose and 0.19 percent protein, which is well suited for fermentation to alcohol. All the whey in the United States would equal 85 million gallons (320 million l) per year, which is about the output of three or four large factories for alcohol from corn. The logistics of whey collection are such that the alcohol factories within a reasonable collection distance of whey sources would be very small.

Although each type of food-processing waste is too small a resource for energy purposes, combinations of several types could support a few large alcohol plants in localities with favorable logistics. For example, potato wastes plus whey could provide a fermentation medium with better nutritional balance and with more

Table 2.7 Capacity and Locations of Major Centers of Feedlot Cattle in the United States

State and City	Number of Cattle	Assessment and Comments
Arizona		
Casa Grande	415,000	Practical site; most within 25-mi radius
Phoenix	349,000	Practical site; most within 25-mi radius
Yuma	(41,000)	Combined with Niland, California site
Total, practical sites	764,000	
California		
Bakersfield–San Margarita	209,000	Practical site; two-thirds around Bakersfield
Dos Palos	243,000	Practical site; somewhat decentralized
Niland	681,000	Practical site (Imperial Valley); most within 25-mi radius; 41,000 from Yuma, Arizona added
Tulare	(179,000)	Potential site; may be possible to combine with part of Bakersfield–San Margarita–Dos Palos site
Total, practical sites	1,133,000	
Colorado		
Greeley	691,000	Practical site; most within 25-mi radius; includes 250,000 capacity Monfort feedlot
Sterling	353,000	Practical site; most within 25-mi radius
Total, practical sites	1,044,000	
Iowa		
Atlantic–Audubon	270,000	Practical site; as with all Iowa and Nebraska sites, most of individual feedlots are less than 10,000
Cherokee	495,000	Practical site
Total, practical sites	765,000	
Kansas		
Albert	318,000	Practical site
Garden City–Dodge City	600,000	Practical site; almost two distinct locations
Scott City	220,000	Practical site; most within 25-mi radius
Total, practical sites	1,138,000	

Nebraska		
Grand Island	430,000	Practical site
Kearney	386,000	Practical site; most within 30-mi radius
Omaha	436,000	Practical site; possible combination with Atlantic–Audubon, Iowa site
Sioux City	200,000	Practical site; could be optimized with Omaha
Scottsbluff	(185,000)	Potential site; possible combination with Sterling, Colorado site
Total, practical sites	1,452,000	
New Mexico		
Carlsbad	(173,000)	Potential site
Texas		
Abernathy	464,000	Practical site; most within 25-mi radius
Amarillo	226,000	Practical site
Bovina	858,000	Practical site; includes 122,000 from immediate locations in New Mexico; most within 25-mi radius
Stratford	668,000	Practical site; includes Western Oklahoma
Total, practical sites	2,216,000	
Grand total, practical sites	8,512,000	
Other possible sites		
Crystal City, Texas	106,000	
Blackwell, Texas	97,000	
Iowa Falls–Webster City, Iowa	120,000	
Atkinson, Nebraska	109,000	

fermentable carbohydrate than from either alone. It is not established that the alcohol plant wastes would not present a pollution problem, but the total strength should be greatly reduced by converting most of the organic material to ethanol. Crediting the alcohol plants for treatment of food-processing wastes would, of course, help the economics. Rather than build factories tailored to specific wastes or to combinations of wastes that could be unreliable because of business conditions or process modifications, it would probably be wise to accept rich wastes such as whey as supplemental feedstocks into processes based on major feedstocks.

Terrestrial Energy Crops

The cultivation of energy crops will not differ greatly from existing practices–the most noticeable feature will be the vast size of energy farms or plantations. Compared to food, lumber, or paper, energy is very cheap. This dictates very careful attention to all possible economies in producing feedstocks, especially in conserving labor. Steps in an integrated farm and energy factory are shown in Table 2.8. Considerable savings can be realized by circumventing steps; for this reason perennial species are more desirable than annuals that must be replanted. Pine trees grow very rapidly but do not grow back from the roots after harvest. Certain hardwoods do grow back, or coppice, and the well-developed root structure gives very rapid growth. After several crops are taken from the same roots, vigor declines, and the stumps must be pulled prior to a new planting. Coniferous trees that require replanting each time may have compensating advantage in the production of oleoresins, which are highly desirable co-products.

Site preparation includes the initial clearing, grading, and placement of roads, ditches, pipes, and channels, in addition to the plowing, disking, or harrowing prior to each planting. Biomass cleared from the land probably would be used for startup of the conversion facility. Because grading is expensive, it should be avoided except where essential for roads or proper irrigation. Rocks and steep inclines are serious impediments to many farm operations but are often overlooked by those advocating growth of biomass for energy on marginal lands. Sowing or planting have sound bases, and no departures from present practice would be expected.

Cultivation incorporates fertilization, thinning, irrigation, pest control, and other details necessary for good growth of healthy plants. Biomass farms would require little thinning or weed control since foreign biomass can be tolerated by the conversion processes—even some pests can be accepted if total productivity is not greatly impaired.

High productivity means that nutrients in the soil will be depleted rapidly and must be replaced by fertilization. Nitrogen and phosphorus are most critical, and both are expensive. Leguminous plants fix their own nitrogen from air by means of the bacteria in root nodules, but cost analyses show that biomass from legumes is more expensive than that from trees by a factor of 2 or 3. This is a result of only modest productivity from legumes and the costs of harvesting.

Table 2.8   Operations at a Biomass Energy Facility

1.	Site preparation
2.	Sowing or planting
3.	Cultivation
4.	Cutting or harvesting
5.	Collection
6.	Transportation to conversion facility
7.	Storage
8.	Shredding or grinding
9.	Conversion
10.	Packaging and distribution of product

Phosphorus can be viewed as a conserved element because the processes of growth do not produce volatile phosphorus compounds. Neither thermochemical nor biological conversion of biomass leads to phosphorus in forms that are useless for fertilization; thus recycle of spent streams to the growing area could be worthwhile. There is also a possibility of recycle of nitrogen compounds; however, by-products such as protein would take nitrogen from the system. Furthermore, some pathways lead to nitrogen gas or to ammonia that could escape. Several reductive conversion processes solubilize nitrogen as its ammonium salts; these are excellent for fertilization if not mixed with undesirable materials.

Water is a prime restraint on land selection. All the highly productive crops require plentiful water; thus rainfall or irrigation is mandatory. Where water is expensive, biomass for energy has poor prospects. A saving feature is the opportunity to use water polluted by human or animal wastes. With suitable precautions for the health of the energy farmers, waste waters can supplement other water and also provide valuable nutrients.

Harvesting, collection, and transport of biomass must be cheap and efficient. It may be necessary to use expensive equipment on an 18- to 24-hr/day basis, and automation is desirable to avoid long labor shifts and the lighting required for night operation. Hauling distances will determine the most economic size and location of energy farms.

Storage, too, can add significant costs, especially for crops that are harvested for a brief period of the year. Silos or sheds to provide protection from the elements would have to be enormous to house 10 or 11 months reserve of biomass for a conversion factory. An important development is very tight bales that resist water penetration; unless these bales are submerged in water, the interior is stable for prolonged times. Stacks of such bales would provide compact and relatively inexpensive storage.

The remaining steps shown in Table 2.8 cover the processing of biomass as discussed in subsequent chapters.

Symbiotic microorganisms that colonize the root structure of plants not only fix nitrogen but assist in cycling of soil nutrients. For example, plants grow poorly in soils deficient in phosphorus but do much better when the roots are associated with mycorrhizal fungi. Ruehle and Marx (1979) feel that research on fungal symbionts needs much more emphasis in view of the need for plant material for fiber, food, and fuel. An example of a missed opportunity is growth of tree seedlings on rich nursery soils that favor the wrong symbionts. When transferred to the field, these root fungi adapt poorly to harsh sites and tree survival is threatened because new fungal species are slow to develop. It should be quite cost-effective to determine which fungal symbionts are best for given species and soils and to be sure that they are present.

Site Selection

Climate, variations in soil, topography, erosion, and irrigation are vital considerations in site selection and the choice of plant species for energy crops. Salo, Inman, et al. (1977) have reported potential plantations on ten sites in the United States. Their criteria for suitable sites were (1) at least 25 in. of precipitation annually, (2) arable land, and (3) slope not more than 30 percent (17°).

It is likely that mechanical harvesting systems can be developed for use on biomass farms with steep slopes. Local topography, nevertheless, is an important factor in determining the feasibility of establishing a farm in a given location. Huge areas are effectively eliminated because of rugged terrain and rocky soils. Much of the West Coast and the eastern regions of the country that are climatically suitable have high hills and mountains. Large areas in states such as Washington, Oregon, California, West Virginia, Tennessee, and much of New England have topographies that will probably prevent complete utilization of many parcels in otherwise suitable locations.

Working estimates of productivities are presented in Table 2.9.

Variations in soil, erosion, and irrigation are important considerations in site selection and choice of plant species. Similar rainfall and temperature patterns do not necessarily assure equal biomass productivity, because soils may differ. The prairie regions of the Great Plains tend to a hard, compacted, very fine silt loam that favors shallow-rooted plants such as short grasses. In contrast, sandy loam has much less runoff of water and favors tall, more deeply rooted species. Very sandy soil has almost no runoff of water, and grasses grow in bunches. Soil with high runoff stores little water, thus reducing the growing season. For example, winter wheat matures early and gives a good yield on packed soil whereas corn does poorly because it matures late when this soil is dry. Fertile, hard soils favor plants that grow rapidly in the rainy season; drought can devastate such areas because the normal plants are adapted to a wet season.

In the sand-loam soils, almost all the rainfall penetrates, and surface evaporation is greatly reduced. Fertility is still sufficiently high so that crops grown during years of favorable rainfall are almost as good as those on the hard lands. During dry years crops are produced because the moisture is distributed to a considerable depth; when drought threatens, plants are able to draw on the reserves found in the deeper layers of the soil.

Table 2.9 Projected Yields for Energy Farms

		Yearly Biomass Yield			
		Current		Future	
Site	Biomass Candidate	dry tons/acre	tons/ha	dry tons/acre	tons/ha
Washington	Red alder	10	23	20	45
California	*Eucalyptus*	13	29	22	50
Wisconsin	Hybrid poplar	5	11	10	23
Illinois	Hybrid poplar	8	18	15	34

Crop failures occur most frequently on the hard or silt-loam soils and least often on the sand-textured soils. However, during favorable years, yields are highest on the former and lowest on the latter. The sandy-loam soils represent a safe intermediate condition—during favorable years crops are almost as good as on the silt-loam soils, and during dry years a fair crop can be produced.

Estimates for the North Central states in 1971 were that 6 percent of all cropland needed conservation treatment (van Bavel, 1977). More and more, highly erosive and sloping soils are being used to produce crops for export, with ever-increasing losses of good topsoil. A new survey of U.S. croplands found, however, that erosion losses were about half of the previous estimates (*New York Times,* May 20, 1979, p. 48). The national average is 4.8 tons/acre (11 tons/ha). Soil is carried away by excess water to streams and rivers where it deposits as sediment. A total of 2 billion tons of soil were eroded in 1977, enough to deposit 1.3 in. (3.3 cm) on an area equal to that of the state of Rhode Island. According to the U.S. Department of Agriculture, erosion at a rate of 4 to 5 tons/acre·yr (11 tons/ha), is acceptable because new topsoil is formed at about that rate. The state-by-state totals are given in Table 2.10. Several tons of soil could be washed away for each ton of crops harvested.

The problems of fertilization have already been discussed; erosion is a serious compounding complication. Harvesting the entire plant, leaving no trimming or residues on the ground, removes some of the crude erosion control now practiced. But if residues are sufficiently dense, they will break the fall of raindrops, reduce splash, and slow runoff of water. Residues also harbor pests and disease microorganisms, thus an inert material might be a better erosion deterrent than plant debris.

Conventional wisdom states that crop residues must be returned to the soil to maintain productivity and to control erosion. Of course, soil type, slope of the terrain, and exposure to the wind are some of the factors that determine how much residue is needed. Shrader (1978) has reviewed the topic and found that 0.3 to 0.8 tons/acre (0.8 to 1.9 tons/ha) of residues is recommended for controlling wind erosion. Some authors feel that residues are essential for supplying nutrients and organic matter, but this notion seems to be based on the idea that residues are not good for much else. The ancient and reliable practice of terracing crop land to greatly reduce erosion would be costly. Irrigation, drainage, diversion, soil conditioning, and physical retention of soil are also expensive. Although erosion control based on hydraulics is cheaper than terracing, it would be highly site specific and beyond the ken of many farmers. On a huge energy farm, it would be cost-effective to design a system for erosion control.

Torpy, Barisas, et al. (1979) estimate 25 percent greater erosion with large-scale removal or residues. There is no foundation for blaming high productivity of biomass for causing greater erosion, but thoughtless pursuit of biomass yields without considering the consequences could be disastrous. If the initial energy farms are based on short-term profits and erosion controls seem uneconomic, one outstanding crop may portend success while erosion damages the soil so badly that many years will be required before another good crop can be harv-

Table 2.10   United States Soil Erosion Losses in 1977

State	Per Acre Loss, tons	Total for State, tons
Alabama	8.9	39,904,000
Arizona	0.45	593,000
Arkansas	6.48	51,809,000
California	0.76	7,626,000
Colorado	2.5	27,687,000
Connecticut	4.13	855,000
Delaware	3.49	1,854,000
Florida	4.17	13,298,000
Georgia	6.58	42,645,000
Hawaii	13.71	3,975,000
Idaho	2.36	14,843,000
Illinois	6.72	160,056,000
Indiana	5.15	68,607,000
Iowa	9.91	261,719,000
Kansas	3.78	108,797,000
Kentucky	9.4	51,163,000
Louisiana	7.9	46,637,000
Maine	2.99	2,724,000
Maryland	6.08	10,238,000
Massachusetts	2.11	598,000
Michigan	2.28	21,604,000
Minnesota	2.31	52,906,000
Mississippi	10.92	79,623,000
Missouri	11.38	165,701,000
Montana	1.08	16,523,000
Nebraska	5.74	118,623,000
Nevada	0.04	43,000
New Hampshire	1.14	308,000
New Jersey	6.54	5,105,000
New Mexico	2.0	4,552,000
New York	4.85	28,895,000
North Carolina	7.64	47,204,000
North Dakota	2.01	54,215,000
Ohio	3.59	42,199,000
Oklahoma	3.68	43,435,000
Oregon	1.09	5,617,000
Pennsylvania	5.49	31,015,000
Rhode Island	2.96	77,000
South Carolina	4.77	15,948,000
South Dakota	2.45	44,416,000
Tennessee	14.12	69,461,000
Texas	3.47	105,525,000
Utah	0.55	991,000
Vermont	2.04	1,231,000
Virginia	6.61	21,211,000
Washington	1.52	12,116,000

Table 2.10 (*Contd.*)

State	Per Acre Loss, tons	Total for State, tons
West Virginia	3.65	3,632,000
Wisconsin	3.6	42,336,000
Wyoming	1.57	4,651,000
Caribbean area	40.64	14,753,000

ested. Credibility of the fuels from biomass concept might never recover from such a blow.

Crop residues are only one means for covering soil. Gravel, pebbles, and stones have been found more effective than crop residues in decreasing evaporation losses by reflecting heat and in reducing impact and soil erosion (Adams, 1966).

The rapid movement of water through sandy soils leaches away soil nutrients, and water is not retained near plant roots where it is needed. Fertilization and irrigation must be increased for highly porous soils to maintain sufficient amounts for good plant growth, so costs can be relatively high. Drip irrigation is the frequent, slow application of water to the plant root system. Capillary action then distributes the water around the roots. Soil remote from a distribution point is not wetted, and as little as 10 percent of the soil in a field may be moist for drip irrigation of newly planted crops. As the crop matures, there are indications that $\frac{1}{3}$ of the soil at the depth of the roots should be wet and that even more moisture is beneficial. Nevertheless, significantly less water is required for drip irrigation than for normal irrigation, and this is very important in nations such as Israel where water supplies are critically limited.

With normal irrigation, labor costs are high because the systems are tended. For example, valves must be turned and streams diverted with dams that may be changed by shoveling dirt from one channel to another. By contrast, drip irrigation is regulated; in this way a few central valves or automatic timers can control flow, so that labor costs are low. Because much of the soil surface is never touched by irrigation water, weed growth is not troublesome with drip irrigation. Another advantage is that soluble fertilizer can be added to drip irrigation water, focusing application directly to the plant. This not only reduces the amount of fertilizer used, but also contributes to labor saving.

There are many advantages to drip irrigation for porous soils or for arid lands where water is costly. Drip irrigation is finding increasing applications, but mainly for food crops. The initial investment cost for the distribution system is high ($600 to $1200/acre or $1500 to $3000/ha), and the revenues generated by biomass feedstocks for making cheap fuels may be insufficient to justify drip irrigation. Over and over again, we see that lands are marginal because of deficiencies of soil, water, or accessibility. It will be costly to get high productivity of cheap biomass from marginal lands.

Land Acquisition

The forest industry in the United States procures its raw timber material through a variety of land tenure arrangements. These range from fee ownership of lands to the one-time purchase of the right to harvest standing timber, or "stumpage," on private or public forest lands. The timberlands may constitute one large, contiguous tract. More commonly, however, there are a number of tracts at widely separated locations, and probably under a variety of tenure arrangements.

The assembling of an aggregation of land large enough to support a biomass farm for energy production purposes could present considerable difficulties. Total land requirements for a biomass farm large enough to fuel a 150-MW thermal power plant would be, for example, on the order of 60,000 acres (24,300 ha) assuming an average biomass productivity of 10 dry tons/acre·yr (25 tons/ ha). There are very few contiguous tracts of this size in private ownership that are not already committed to agriculture, range and pasture land, or traditional forestry and that at the same time would be suitable for intensively managed short-rotation forest crops.

Large tracts of land owned and managed by federal and state agencies do exist in various parts of the country, particularly on the West Coast. In the Pacific Northwest, for example, there are vast areas managed by the USDA (United States Department of Agriculture) Forest Service. This agency is under considerable pressure to use its productive forest lands to produce raw material for the local timber industries. Furthermore, in recent years much national forest land has been withdrawn from timber production and devoted to other uses, such as wilderness areas.

Black (1978) has raised questions about prices for biomass such that farmers will participate in energy programs. He suggests that a system based on spot prices in an open market for biomass might be unworkable and that contracts and commitments would better provide predictable supplies. Acreage contracts rather than tonnage might be more attractive to the farmer because risk would be reduced. Cooperatives might be very desirable for commercializing biomass fuels because of logistics, dependability of the market, elimination of the middlemen, and tax advantages.

Silviculture

Kemp and Szego (1976) were early advocates of energy plantations. With direct burning as the conversion technology and with the goal of minimizing dollars per BTU, they prepared preliminary estimates of suitable tree species, culturing practices, and harvest cycles. Many of the ideas of Kemp and Szego are developed in a massive study of silviculture covered in seven reports by the MITRE Corporation. Most of this section draws heavily on these reports.

Howlett and Garmache (1977a) list as advantages of short-rotation forestry:

1. High yields per unit land area.
2. Shorter time from initial investment to a profitable crop.
3. Increased mechanization to lower labor cost.
4. Adoption of agricultural harvesting techniques for lowered collection costs.
5. Very low planting costs through use of coppicing species that grow back from stumps.
6. Quick advantage of genetic improvement compared to present forestry when a generation of trees may be 30 to 50 yr.

Of the disadvantages, the most serious is cultivation of one or a few species that could be highly susceptible to epidemic diseases or insect infestation.

Average annual yields will be much greater for short-rotation forestry. Whereas conventional forestry produces only 1 to 3 dry tons/acre (2.3 to 7 tons/ha) per year, short rotation should give 5 to 10 tons (11 to 23 tons/ha). Comparisons are shown in Table 2.11. Of course, management and field operations will be much more costly because of the frequent attention.

Essential to rapid establishment of short-rotation silviculture is thorough site preparation to remove competing vegetation and to condition the soil. This will resemble practices for field crops more closely than will those of present-day forestry. Much higher densities will be employed. Conventional planting of 200 to 700 stems per acre (490 to 2000/ha) is a sharp contrast with short-rotation plots with 1200 to 77,000 stems per acre (3450 to 221,000/ha).

Softwoods currently account for a far greater proportion of timber harvests in the United States than do hardwoods. Of the 12.5 billion cubic feet (3.54×10^8 m^3) harvested in 1972, about 9.5 billion (2.69×10^8 m^3) consisted of various softwood species. The reasons are to be found in the general versatility of the softwoods, which are ideal for structural building materials and as components in most types of woodpulp. Hardwoods, on the other hand, are useful primarily as decorative building materials and as components, usually along with softwoods, in certain types of woodpulp. For short-rotation biomass production, however, softwoods generally appear to offer less promise than hardwoods. Conifers in general do not fit the short-rotation criteria very well; most conifers do not exhibit rapid juvenile growth, starting somewhat slowly compared to hardwoods. In addition, true coppice reproduction of conifers is generally not possible.

Eucalyptus trees are of special interest because of the outstanding growth rates reported and their widespread commercial use in Argentina as a source of coke for their steel industry. The status of eucalyptus has been thoroughly reviewed by Mariani, Wood, et al. (1978). Eucalyptus are evergreen trees belonging to the myrtle family. There are about 450 distinct species and numerous variants within the genus *Eucalyptus*. They dominate 95 percent of Australian forests and are

**Table 2.11 Productivities of Various Tree Species on Conventional and
Short Rotations**

Rotation Type and Species	Rotation Length, yr	Mean Annual Biomass Increment, dry tons/acre · yr	Tonnes/ha
Conventional rotation			
Douglas fir	73	1.4	3.2
Loblolly pine	30	2.7	6.1
White pine	55	2.5	5.7
Balsam fir	40	1.7	3.9
Aspen	18	1.2	2.7
Red alder	45	2.5	5.7
Short rotation			
Eastern cottonwood	9	4.9	11.1
Black cottonwood	4	5.8	13.1
Sycamore	4	5.6	12.7
Red alder	4	8.5	19.3
Hybrid poplar	8 to 10	8.9 to 10.3	20.2 to 23.3
Loblolly pine	10	4.5	10.2

found in many other countries as well. Some species produce tall, thick trees whereas others are shrubs or shrubby trees. Australia has a wide range of soils and climate but is generally relatively dry. Its coastal areas have sufficient rain to support forests, whereas grasses and scrub growth are common in the interior. A lumbering industry is based on eucalyptus.

Eucalyptus regenerates primarily from seeds that are produced in great numbers. Following damage by fire, drought, grazing, or cutting, there is vigorous vegetative rejuvenation. Many species put down subterranean lignotubers capable of producing new shoots if the aerial portion of the plant is destroyed. The plant's capacity for rapid growth depends on "indefinite shoots" and "naked buds," structures that increase in height and length as long as favorable conditions exist.

Despite its wide adaptability, eucalyptus does poorly when there are sudden temperature drops to below freezing. The trees may survive, but a growing season may be lost or severely impaired. In the species that are most important for lumbering in Australia, a soil fungus has a beneficial symbiosis with eucalyptus. Planting seeds in soils without this fungus may account for erratic results in attempts to transplant the trees to other continents.

There is an entertaining history of eucalyptus in the United States with frequent tragedies such as cattle grazing on young plants, floods, and disastrous winters. Introduced in 1856 from Australia, eucalyptus was planted extensively in California as a source of lumber. Other trees were more successful, and interest in eucalyptus waned by 1912. Irrigation turned many of the potential forest lands to very rich croplands in the 1930s and further reduced incentives for developing new trees. More recently there has been research on species for ornamentation and for recovery of marginal lands.

The U.S. Forest Service has conducted research on eucalyptus in several states. An extensive program in Florida suffered a serious setback because the unusually severe winter of 1976/77 killed or impaired many trees. The Florida Division of Forestry has planted almost 3 million eucalyptus seedlings to establish about 6 thousand plantation acres.

It is very likely that eucalyptus can grow well in California and in the southern states with warm climates. Further research may find species better adapted to sudden frosts. However, for energy purposes, these trees can only be important to the United States in a few favorable locations.

The roads used in a short-rotation silviculture plantation may be of several different types, depending on the frequency of use and the type of use. The number of roads to be built and their spacing will depend on the harvesting system to be used, site conditions, and the topography of the plantation. Construction costs for forest roads generally range from $4000 to $8000/mi ($2500 to $5000/km). Total annual maintenance costs about $300/mi ($190/km). Should winter maintenance be required, the figure would be increased to about $1100/mi ($680/km).

Harvesting costs depend in part on the type of harvesting system used. If trees are large enough, perhaps 4 in. (10 cm) in diameter or more, a conventional mechanized harvesting system could be used, at an estimated cost of $200 to $560/acre ($500 to $1400/ha) (DeBell and Harms, 1976). If crop trees are small enough to permit the use of a harvesting system patterned after a corn silage harvester, costs would probably be considerably lower, about $30 to $80/acre ($74 to $200/ha).

It is a serious misconception that trees can be cut almost any time of the year. Coppicing is dependent on season. Sprouting is best when trees are harvested during the dormant season; poor results are obtained when trees are cut soon after the appearance of leaves. Sprouting vigor relates to carbohydrate reserves in the root system that are high during the dormant season and depleted at the time of leaf formation. Sycamore stumps coppice if cut at various seasons, but tender sprouts from midsummer cutting are killed during the following winter. Sprouts formed in the dry season start slowly.

Erosion of the forest can be serious if people and heavy machinery disrupt its floor during the rainy season. There will be times when too much rain or snow impedes harvesting. When heavy machinery is in the woods for harvesting, it might be expedient to conduct other operations, such as fertilization or soil conditioning, at the same time. A noncoppicing species could prove quite practical if cutting *and* planting could be performed.

Ideally, the harvesting and processing of biomass should be integrated processes. If, following harvesting, there are several months of poor growth, costs will be impacted unfavorably by idle workers and equipment and the need for increased storage capacity at the conversion facility.

Harvesting of Trees

Whole-tree harvesting has become quite common in some areas of the country. Cut trees with branches and foilage intact are moved a short distance and fed into a chipper that pneumatically conveys "whole-tree chips" directly into vans for highway transportation. The chips are used for making grades of pulp and paper that are not seriously affected by the bark and foilage content. Chippers that are integrated with cutting equipment are also being developed. These units mow down trees and feed chips to a trailer. A prototype machine for rapid harvesting of small trees and residue has been tested (Koch and Nicholson, 1978). A 575-hp (horsepower), tracked chipper cuts at 6 in. (15 cm) above the ground while taking a 9-ft (2.7-m) swathe. Chips are blown by air into a trailing vehicle of 10-ton holding capacity; another trailer takes its place when the hopper is full and must be taken for dumping. The design speed is 1 mi (1.7 km) per hour on land free of large rocks and with slopes of less than 30°. This translates to cutting about one acre per hour. A view of the system is shown in Fig. 2.1.

Plantation Costs

Inman, Salo, et al. (1977) reported on the potential tree plantation sites in greater detail. Estimated cost of the energy of the wood was $1.20 to $2.00 per million BTU at well-located tree plantations. Less productive sites could not achieve these costs. Capital costs were relatively low, comprising only about 10 percent of the total. However, irrigation and fertilization came to almost 40 percent of the total, (and these operations are not common in traditional forest management). The South and Southeast were identified as most promising for biomass production, mainly because of good rainfall and warm climate. It would be uneconomic to divert prime farm lands to silviculture because land acquisition costs would be prohibitive. Contrary to some unfavorable projections about energy balances for farms, solar energy captured as biomass would be 10 to 15 times the conventional energy input to the plantations.

Certain details of the proposed plantations are quite interesting. Work roads 20 ft (6.8 m) wide must interlace the farms to allow wagons to move chips from the harvesters to storage areas. Plantation sizes would be from 25,000 acres to over 50,000 acres (10,000 to 20,000 ha) depending on productivity, and 100 acres (40 ha) would be set aside for storage of chips in piles 10 ft (3.4 m) high. Such large amounts are difficult to visualize, but roughly 100 football fields piled 10 ft high for chip storage alone conveys the image of a very large operation.

The unit farm could produce about 250,000 dry tons of biomass per year with a rotation period of 6 years. Biomass on the sites would be cleared and sold, and seedlings or cuttings planted in rows at 4 ft by 4 ft spacing, a density of 2725 plants per acre (6733/ha). Irrigation would be used for the first 3 yr with reliance on rainfall after that. For the 4-month period during the year when irrigation is required, an automatic "traveler" system has been selected. This method uses a

Figure 2.1 Prototype unit for mowing trees. Courtesy P. Koch.

pump, a power unit, a main pipe, and hoses. The sprinkler is a water cannon with a delivery of up to 1000 gal/ min (3700 l/ min). A traveling unit can drag the hose behind it. Such a unit can be operated by one person, and it has the flexibility for use on different terrains.

Heavy fertilization would be practiced during the first year of each rotation. Light application by aircraft will be most economical for subsequent years. Table 2.12 shows fertilizer costs. Cost estimates for land preparation are given in Table 2.13. Prospects for tree plantations for energy are attractive, and if converted to clean fuels or chemical feedstocks of much higher value than wood for burning, an excellent return on investment is possible.

Table 2.12   Fertilizer Costs for Hypothetical Plantations at Various Sites

| | | Dollars per Ton Formulated | |
Site	Urea	Superphosphate	KCl
Wisconsin	104.71	100.31	74.21
Missouri	105.08	93.14	73.52
Louisiana	102.18	86.19	70.69
Georgia	108.66	88.54	77.59
Illinois	100.17	94.16	76.05
New England	115.99	93.82	81.92
California	100.35	104.71	91.51
Washington	124.95	99.05	76.37
Mississippi	98.21	79.56	73.42
Florida	108.31	81.16	80.08

Carbohydrate Crops

The main commercial sources of sugars and starch at present are sugarcane, sugar beets, and corn. Sweet sorghum shows promise as an energy crop, and other plants such as manioc or kudzu deserve consideration. The stalks, trimmings, and residues from crops after the starch or sweet juices have been removed are presently of low value. They are used for animal feed, burned for power generation, incinerated, or dumped. If used for producing fuels, the starches or sugars are high-quality biomass easily fermented to organic chemicals, and the rest of the plant is cellulosic biomass that requires treatment to become a source of fermentable sugars. An integrated factory utilizing both high- and low-quality biomass could produce fuels and petrochemical substitutes at considerably lower costs than present factories that utilize only the easily extracted carbohydrates.

Table 2.13   Estimated Costs for Land Preparation

| | Cost in Dollars per | |
Operation	Acre	Hectare
Clearing	250	618
Raking	12	30
Burning	4	10
Drainage, leveling, and minimum roading	100	247
Plowing	15	37
Disking	8	20
Subsoiling	12	30
Herbicide application	4	10
Herbicide	20	49
Total	$425	$1051

A detailed analysis of fuels obtained from carbohydrate crops was conducted by Lipinsky and colleagues (1977) at Battelle Memorial Institute. Some of their findings were:

- Total production is, obviously, the yield times the area. Sugarcane has the highest yield but the smallest potentially cultivatable area of the crops considered. Sugar beets provide a modest yield but can grow in a wide geographical range. Sweet sorghum has a vast range, but yields are uncertain and more field research is needed. Corn grows well in many states and gives yields approaching those of sugarcane.
- Sugarcane and corn have the greatest short-term potential. Sweet sorghum may be best over the long term.
- Costs of fermentable solids are roughly the same for each crop. Logistics may be the deciding selection factor.

Corn

Corn is a highly attractive candidate for production of fuels because (1) the yield is relatively high; (2) its range covers the entire continental United States; (3) its photosynthetic mechanism is the efficient C-4 type (see Chapter 5); (4) there exists a well-developed system for seeds, fertilizers, and equipment; and (5) methods are available for insect and pest control. At present, the food value of corn exceeds its potential value for fuel. However, 90 percent of corn is fed to cattle, not directly to people. In time, development of new protein sources and new eating habits may reduce the demand for meat and, consequently, for corn. It would be foolhardy to predict that Americans will abandon their strong desire for steak, although citizens bled to near poverty by an outflow of oil dollars to OPEC could eventually seek less expensive foods. There is, however, an immediate opportunity for using spoiled corn and corn stover for energy, and a long-range possibility of growing the whole plant for conversion to fuels.

Lipinsky, Sheppard, et al. (1977b) report that farmers in the United States are capable of growing 150 to 200 million tons of corn grain, with 122 to 159 million tons of dry residues generated in the process. Corn for grain is normally profitable, and at high prices (e.g., at least $3 per bushel) farmers can be encouraged to increase production. Average grain yields have been increasing about 3.3 percent annually over the past 25 years and are expected to continue at about 2 percent per year. There are at least 10 locations in the midwestern corn belt, where sufficient corn grain could be brought together for a large complex to manufacture ethanol. Residues in these locations are sufficiently concentrated geographically so that processing plants for conversion of corn stover would be assured sufficient feedstock. Grain corn is shelled in the field and the residues—cobs, stalks, and husks—are discarded. It should not be difficult to design a new harvester to gather grain while collecting the remaining biomass; however, whole-plant harvesting raises problems with regard to soil erosion and fertilization to replace nutrients that would otherwise have been derived from the rotting biomass.

Conventional corn has a valuable but minor by-product in corn oil. Experimental strains show an increase from the usual 4.5 percent to 17 percent corn oil, but at the expense of other constituents. Although food or industrial value of the oil mitigates against its use as fuel, the credits taken for oil could give the other biomass a more attractive cost. Multiple cropping shows promise for corn. For example, winter rye would give a significant increase in total biomass yield with better distribution of fixed costs.

In the United States, corn is grown on 72 million acres (29 million ha) with an average yield of 86.5 bushels/acre for harvest. Some corn is not harvested but used for livestock forage. Silage yields are about 16 tons/acre (5.9 tons/ha). Higher yields can be obtained with close planting. The product is termed "green manure" and has a high moisture content.

Silage corn is worth roughly $50/ton. Grain is worth over $100/ton. When the grain is used for fermentation to alcohol, protein-rich fractions are removed for use as feed; by-product credits reduce the fermentation substrate costs to $45 to $65/ton depending on the region. "Distress" corn is damaged by mold, insects, disease, or malformation. If severely damaged, it may sell at half the price of healthy corn. Corn residue is the aerial portion minus the grain. The farmer might accept $15/ton for residue, but transportation and handling would bring the total to about $40/ton.

The leading grain-producing states (Iowa, Illinois, Indiana, and Nebraska) accounted for almost 60 percent of total U.S. production from 1973 through 1975. Illinois, the highest yielding of the four major corn-producing states, averaged 101 bushels/acre between 1973 and 1975.

Iowa, Wisconsin, Minnesota, and New York are the leading corn-silage-producing states. These four states averaged approximately 11.4 tons/acre (26 tons/ha) of silage (fresh weight, containing 35 percent dry matter) and produced about 36 percent of all U.S. silage production from 1973 to 1975. California, Washington, Texas, and Colorado have the highest silage yields, averaging approximately 18 tons/acre (40 tons/ha) (fresh weight) on irrigated land. On a dry-weight basis, the aerial portion of the mature corn plant is 53 percent grain, 10 percent cob, 25 percent stalk, 6 percent leaf, and 6 percent husk.

One billion bushels of corn (less than 20 percent of an annual crop) would provide grain for conversion to 2.8 billion gal of ethanol with an energy value of 0.2 Quad. This equates to 2 to 3 percent of U.S. needs for gasoline. Using the corn stover and waste corn could more than double the amount of ethanol.

The use of corn as the raw material for making alcohol has advantages over sugarcane in that corn is easily stored and is an article of commerce. A distillery can purchase grain as needed. This reduces storage requirements and reduces the working capital, in addition to reducing the risk arising from poor crop yields in the immediate vicinity of the plant.

Starch, which constitutes about 75 percent of the content of the grain, can be converted into glucose. This is done with an enzyme-containing material such as malted barley or amylases from a fungal source.

Sugarcane

The U.S. imports approximately half of its sugar. Of the 6 million tons produced each year in the United States, almost half comes from sugarcane and the remainder from sugar beets. Cane is grown in Florida, Hawaii, Louisiana, and Texas. Puerto Rico grows sugarcane, but with excessive manual labor; automation would greatly exacerbate unemployment on the island. Government subsidies are required to make U.S. sugarcane a profitable product, except in times when the price cycle nears a peak.

Polack and West (1978) estimate that over 1 million tons/yr of alcohol could be made from all the cane sugar produced in the United States. This would be only 0.4 percent of our auto fuel requirements. With present-day technology and current prices, U.S. cane sugar may be too valuable to fit into the picture of cheap fuels. However, new technology and the concept of using the whole sugarcane plant may alter this situation.

Agricultural research is leading to increases in the yield of sugarcane per acre, better ripening characteristics, and to varieties resistant to pests, diseases, freezes, and the like. Recent work in the United States and Australia has been aimed at increasing the productivity, in effect, by crowding more cane onto the land by multiple stalk planting and close row spacing. By these techniques, cane yields have been effectively doubled, with little apparent loss in sucrose production. Some of the techniques will require new methods of mechanical harvesting. Commercialization of these ideas will increase available biomass and decrease unit cost.

Chemical ripeners that may extend the season and increase yields are coming into substantial use. These materials (glyphosine is the most widely used) cause earlier maturation (i.e., rise in sucrose content). Cost:benefit ratios are under continuing study at many locations, and expanded commercialization can be expected.

Sugarcane forms a tight canopy of leaves that are sharp and can injure workers. Although there are mechanical cutters for whole cane, much of the world's cane is harvested manually after the leaves are burned. The author was in one of the sugarcane regions of Brazil when some fields were harvested. There was striking beauty of the glow of burning fields on the horizon at night, and interesting patterns of flames could be seen from an aerial view of the district. Occasionally, unexpected high winds spread the flames, and harvesting cannot keep up; a burned field that is not promptly harvested loses sugar rapidly. The stalks, because of their high moisture content, resist burning and remain erect but blackened. In the United States leaves are sometimes burned when the stalks are cut and on the ground; leaves are considered bulky and useless. In many countries where labor is cheap, workers with machettes hack the erect cane stalks and load them on wagons or trucks.

Much energy is wasted in burning, and the leaves could constitute a significant source of fuel or lignocellulosic feedstock. The nighttime spectacle of burning is accompanied by a daytime pollution problem. Some of the pollution is papery ash from burning bagasse in the power plants at the sugar mills; this probably could be collected at the source. Other ash from burned fields floats in the air, is inhaled, and forms spots on clothing. Although burning of cane fields is permitted in the United States, there is an obvious need for demanding only mechanical harvesting and making use of the leaves. Since there is also a loss of roughly 5 percent of the sucrose during burning, the potential saving could help pay for processing whole cane.

At the sugar mill, intact stalks are washed with water, but inevitably some dirt and mud are carried into the extraction step. Extraction is carried out in a large machine that crushes the cane and has several stages for contacting with water and squeezing the extract from the solid residue. Wet bagasse could be dried to get a better fuel, but its heat of combustion when still containing 40 percent moisture is sufficient to supply the steam needs of the sugar mill. Auxiliary fuel (oil, coal, or wood) is used for startup of the steam boilers.

Nathan (1978), in a survey of fuels from sugar crops, states that U.S. sugarcane in 1975 totaled 25 million tons from 770,000 acres (312,900 ha) of cropland. Yields in Florida averaged 46 tons/acre (105 tons/ha), but experimental plots indicated a distinct potential of 130 tons/acre (294 tons/ha). Typical fertilization was

Kilograms per Hectare	Pounds per Acre
27 as P_2O_5	24
333 as K_2O	297
166 as N	148

Studies in Texas showed that sugarcane depletes the soil by only 0.8 to 1 kg of nitrogen per ton of cane. Table 2.14 summarizes date for important U.S. sugarcane regions.

Lipinsky, Kresovich, et al. (1978c) report that narrow spacing of the plants may not give the expected increase in total biomass yield. In Louisiana, narrow spacing gave 12 to 14 dry tons/acre (28 to 31 tons/ha) whereas controls with normal spacing gave 9 tons/acre (21 tons/ha). At another Louisiana site, narrow spacing yielded 18 to 22 (40 to 50) versus 14 tons/acre (31 tons/ha) for the controls. However, experiments in Florida showed an initial advantage for close spacing, but the final yields were not significantly different from those of the controls. A possible explanation is that muck soils used in Florida affected the response of sugarcane varieties that were better suited to other soils.

Sugar recovery uses various combinations and sequences of evaporation, crystallization, and carbon treatment for color removal to get several crops of crystals. Crops are collected until the final syrup becomes too high in contaminants and in heavy-metal ions. This syrup is termed "black strap molasses" and is treated to remove heavy metal ions to make it suitable as a fermentation substrate.

Any or all of the extracts or concentrates from sugarcane can be considered

Table 2.14 Sugarcane in the United States

	Florida	Louisiana	Texas	Hawaii	Puerto Rico
Area harvested					
Acres	279,000	334,000	27,000	110,000	132,000
Hectares	112,000	136,000	10,400	46,000	57,000
Yield, wet					
Tons/acre	46	33 + 22	40	58	28
Tons/ha	105	74 + 50	91	131	63
			From second crop		
Growing cycle, months	15	5	12	24	12 to 18
Soil type	Muck and peat	Sand and clay	Loam and clay	Variable	Variable
Disease problems	Minor	Serious	Serious	Some	Serious
Harvesting	Heavily mechanized	Mostly mechanized	Mostly mechanized	Heavily mechanized	Mostly hand labor

as fermentation substrates. Several crops of high-grade table sugar could be taken, and side fractions could be used for energy fermentations. Bagasse, the cane stalks minus sugar, can presently be burned for generating steam, and when cellulose hydrolysis is commercialized, it will be possible to mix fermentable sugars from cellulose with cane sugar. The pentose fraction from biomass hydrolysis, however, will probably be handled separately because of its greater acceptability to organisms other than the usual alcohol-producing yeasts.

Polack and West (1978) have reviewed the state of the art of bioconversion of sugarcane and have emphasized that the 2.7 to 3.7 percent efficiency of capture of sunlight by sugarcane is much higher than the 1 to 2 percent averaged by most crops. Sugarcane is grown in most of the tropical nations of the world that have good rainfall. Typical composition is shown in Table 2.15.

Sugar factories are generally located near the growing sites because the juices degrade after harvest. Raw sugar from these plants is 96 percent pure and is a commodity for world trade. It can be refined to almost pure sugar by dissolving, adsorbing impurities, and recrystallizing.

Cane molasses is valuable for manufacture of rum, for other fermentations, and for supplementing cattle feed. Price is highly variable because sugar prices are extremely sensitive to very small undersupplies or surpluses with respect to world demand.

Bagasse is the principal ingredient of the wallboard, Celotex, and has also been used in papermaking. Other specialty applications have included the manufacture of furfural, hexoses, and pentoses. Although most American sugar mills have used liberal quantities of fossil fuels, rising costs and the threat of allocations of natural gas make this no longer feasible. Enough bagasse is produced in processing to supply all the fuel needs of the factory. However, old factories may have inefficient power plants that must supplement the bagasse with other fuel. Since the value of bagasse is tied to fuel replacement cost, bagasse is no longer the cheap "give away" by-product that it once was. Its attractiveness as a feedstock for a variety of processes is, therefore, much decreased.

Lipinsky, Birkett, et al. (1978a) point out that sugarcane processing uses *cold* water for extraction because hot water dissolves substances that are inhibitory to crystallization. Production of fuel alcohol by fermentation sidesteps crystallization; thus the more complete and more rapid extraction with

Table 2.15 Typical Composition of Sugarcane

	Percent by Weight		Percentage of Soluble Solids
Water	73 to 76	Sucrose	70 to 88
Dry fiber	11 to 16	Reducing sugar	4 to 8
Soluble solids	10 to 16	Nonsugars	(Difference)

hot water is advantageous. The sugar concentration in hot extracts is only slightly below that in the sugarcane; therefore, some of the energy required for concentrating more dilute cold water extracts is saved. Furthermore, heat kills most of the organisms in the extract and reduces contamination problems in the fermentation.

Practical mechanical harvesting of sugarcane was developed in the 1920s (Cochran and Ricaud, 1979). A machine for erect cane was known as the "soldier" or "whole cane harvester." After World War II, when mechanical harvesting became widespread, there was interest in removal of leaves. However, burning before the cane is cut or burning of piles or heaped rows is still the predominant method. A British unit, the McConnel harvester, collects cane, removes leaves, and bundles the cane; an additional feature of importance is the ability to operate on sloping land. Another device, the combine harvester, has extractor fans to remove leaves. The cane is cut into short billets of 12 to 16 in. (30 to 40 cm) in length. For rugged terrain and dense fields, a device called the "push rake" and "V-cutter" is suitable, but it collects soil and extraneous material.

A major technological breakthrough may be the Tilby separator marketed by Intercane Systems, Inc. and patented by the Canadian government (Tilby, 1971). One such unit is in commercial operation in Florida producing animal feed and fiber. The old extraction process greatly weakens fibers in the stalk, but the Tilby separator provides an excellent fiber suitable for many paper products. Switching from water extraction to a dry, mechanical separation process puts cane processing in a brand new perspective.

The Tilby separator divides a stalk of cane into three components: the juicy pith from the center; the shell or rind with high-quality fiber; and the thin, cover layer of wax, which is also of high quality. Remarkable speed and efficiency in accomplishing this separation have been observed. As shown in Fig. 2.2, cane stalks are aligned and forced over a knife blade for longitudinal splitting. Each half of the stalk is abraded by a special wheel that rubs off the wax layer. Rapidly rotating blades then scrape out the pith, discharging it to a conveyor. Long rind fibers are collected at the side. Stalks move through the machine at 20 ft/sec, and several stalks are handled simultaneously. The process appears to have numerous advantages: (1) it is a dry process (no water is added); (2) its energy consumption is apparently lower than that of a conventional grinding mill; (3) it produces a long, strong, fiber that is a premium product (especially compared to bagasse), useful in making paper or special pressed boards; (4) the pith and entrapped juice are clean and quite pure in contrast to muddy cane juice from a conventional mill; and (5) it prefractionates the cane plant into its component parts, each of which can then be processed optimally. In the long run, prefractionation would seem to be the best way to handle cane (just as it is in the cases of oil refining or beef production). The high-grade pith produced by this machine is currently being sold as animal feed on a small scale. However, it could be ideal for fermentation, either with or without initial separation of the liquid from the solid.

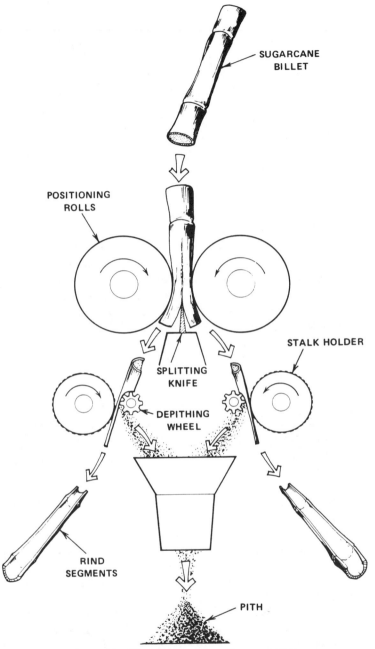

Figure 2.2 The Tilby separator. Courtesy NTIS.

During trials with sugarcane in Florida, the Tilby separator did not function well over extended periods (Lipinsky, 1979) because of dirt and debris that jammed the mechanism. Some of this trouble might have been anticipated because shortcuts were taken in the design of the cleaning step. With better cleaning and with some improvements of the separator, it is very likely that this method can supplant conventional extraction of sugarcane and can be applied to sweet sorghum as well.

Other research on sugarcane processing also shows promise. A German process reported by Rudolph, Owsianowski, et al. (1977) uses steam treatment of cane at slightly elevated pressure to improve extraction. The juice and the wet pith are sterile, and either can be fermented directly.

Sugarcane is not immune to pests. Populations of 30 to 100 rats/acre (75 to 250/ha) have been commonly observed. Rats not only eat the cane, but are vectors for transmitting insect and microbial infestations. The usual means of rat control is application from airplanes of poisoned baits. The moth borer is a serious insect pest that tunnels through stalks, causing losses of about $5 million per year, in addition to costs for control using insecticides. The cane weevil in Hawaii is controlled by burning cane residues and by care in planting. Nematodes that attach to the root system are controlled with parasitic fungi. Some fungi actually have loop-shaped snares that close on the nematode worms, enabling digestive enzymes to act. The most important cane diseases are caused by fungi, but resistant strains of sugarcane have been developed. Weeds are controlled by application of herbicides.

Nathan (1978) reports that costs per ton of undried millable sugarcane including seed, fertilizer, labor, fuel, harvest, and machinery are $15.61 in Texas, $16.81 to $21.49 depending on soil type in Florida, and $19.32 in Louisiana. Tops and leaves are about 30 percent of the millable cane.

Sweet Sorghum

Sweet sorghum is an attractive crop for energy. In the past it has been grown only for syrup because too much starch (about 2 percent of juice solids) is present for easy crystallization of sucrose. Farmers sometimes grow an acre or so of sorghum to make table syrup, but the practice has been declining because of little demand. However, the starch is easily hydrolyzed to glucose for fermentable sugar in addition to the sucrose; thus sorghum juices are quite suitable for production of ethanol or other fermentation products.

Sorghum already grows in at least 19 states. It is tougher and more adaptable than sugarcane but is easily damaged by drought. Sorghum matures in 100 to 120 days and could be harvested from July to December. Lipinsky, Kresovich, et al. (1978c) have considered integration of sweet sorghum with sugarcane processing or with sugar beet production. Because sorghum has a short season, it is sensible to supplement it with a longer-season crop. There is no commercial extraction process for sweet sorghum, but it now appears that the Tilby separator can do a satisfactory job. This could permit an excellent match with sugarcane processing or provide a pith fraction that is probably compatible with sugar beet extracting equipment.

Sweet sorghum fresh yields range from 20 to 49 tons/acre (45 to 112 tons/ha). Close spacing of plants gives yields higher than those with conventional spacing (Lipinsky, Kresovitch, et al. 1978c). Tests in Texas, Mississippi, Louisiana, and Ohio showed differences in yield among varieties of sweet sorghum and yields somewhat below those obtained with sugarcane. However, sweet sorghum is a 4-month crop whereas the sugarcane season is 9 months in Louisiana and about the same elsewhere in the United States. It is also easier to plant and requires less fertilizer than sugarcane. Varietal yields from the Ohio planting are shown in Table 2.16. The Sart variety experimental plantings in Texas were infested by the sugarcane borer and were destroyed. In general, however, pests and diseases seldom cause serious damage. Possibly this is because sorghum crops are fairly uncommon and widely separated so that transmission is minimized by distance. With vast, contiguous acreages, costs for control of pests and diseases could prove high.

Sugar Beets

Sugar beets are grown in 17 states with over 60 percent of the total coming from California, Colorado, Minnesota, North Dakota, and Idaho. The total acres harvested is 1,300,000 (525,000 ha). As is discussed in Chapter 5, many plants use a "C_3" path instead of the highly productive "C_4" metabolic scheme of sugarcane and corn. Sugar beets belong to this lower yielding group, and fresh yields average 16 tons/acre (43 tons/ha) with a range up to 21 tons/acre (58 tons/ha). Sugar concentrations are highest in the roots. Although resistant to cold, heat, and drought, biomass yields and sugar content are low for sugar beets.

If a factory were to be built right now for the fermentation of sugars to fuel-grade alcohol, a reasonable plan would be use of sugarcane juices or beet sugar during the harvest season. When juice supplies became exhausted, a switch should be made to molasses that is an article of commerce and usually readily available but in short supply in 1979.

Cassava

Cassava (tapioca) is a tropical tuberous plant used as a stable food by more than 300 million people throughout the world. Research has led to improved yields in

Table 2.16 Comparison of Sweet Sorghum Varieties

Variety	tons/ha Yield, Dry Weight	tons/acre
Sart	25.6	11.3
Ramada	18.1	8.0
Rio	15.4	7.0
MN 1202	9.4	4.1

the range of 15 to 18 tons/acre (40 to 50 tons/ha), and there are indications that these yields could be doubled. Cassava can be grown in soils too poor for sugarcane, and its water and fertilizer requirements are quite modest.

Cassava root tubers are rich in starch that is readily fermentable, however, the high fiber content of the roots makes it more difficult to swell the starch and to hydrolyze it. Cookers at high temperature are used to prepare the starch gel even though the gelatinization temperature is only 65°C. DeMenezes, Arakaki, et al. (1978) have found that addition of fungal cellulases increases the rate of sugar formation by about 15 percent and greatly reduces the viscosity of the reaction mixture.

McCann and Prince (1978) have considered alcohol–fuel production for Australia using various crops including cassava. They envision an agroindustrial complex with the alcohol factory located conveniently within the farming area to minimize transportation costs. In addition to ethanol, products would be obtained from wastes and sidestreams by anaerobic digestion to yield methane for fuel and from microbial solids for animal feed. Wastewater would be used to irrigate the fields. Cassava may soon become the focus of ethanol production in Brazil, for even if sugarcane retains its dominant position, it could be unwise to depend on a single crop.

Jerusalem Artichoke

This plant has no relation to Jerusalem; the name is a corruption of a name for sunflower. The most useful portion is the tuber, which is somewhat similar to a potato. The storage polymer is inulin, which is unusual because it is a polymer of fructose, the sweetest sugar. Adult plants resemble wild sunflower plants except for being quite tall. When grown from seed tubers or sections of tubers, a bushy plant 6 to 10 ft tall (2 to 3 m) can pack the field. Jerusalem artichoke is native to the United States, grows in all states even on rather poor soils, and has good resistance to frosts, drought, and pests. Although this crop has potentially good yields and might be a source of cheap sugar, it would be wise to use the fructose for the more valuable application as a sweetener rather than as a substitute for glucose in ethanol fermentation. However, there are excellent prospects for using the tops for energy while processing the inulin to sweetener.

Pineapple

Pineapple can be cultivated at latitudes up to 60°, but the growing period in colder climates can be 3 yr whereas 14 months is adequate in equatorial areas. A second fruit crop is often harvested from the same plant after one-half the time for the first crop. The water requirement for pineapple is about two-thirds that for sugarcane, and pineapple can survive periods of drought. Marzola and Bartholomew (1979) have compared pineapple with other potential biomass–energy feedstocks and concluded that its prospects are bright. Their estimates are shown in Table 2.17.

Table 2.17 Alcohol Production and Water Use for Some Energy Crops

Crop	Ethanol Production		Water Use	
	gal/acre·mo	l/ha·mo	in/mo	mm/mo
Sugarcane	98.5	921	7.1	180
Pineapple	103.1	964	3.3	83
Cassava	66.4	611	4.9	125

There is a high proportion of the pineapple plant in leaves instead of the sugar-bearing fruit. Depending on age, the leaves can be 90 percent of young plants and 45 to 65 percent of the total mass of mature plants. Leaves and bagasse from sugarcane can be burned to provide process energy, but pineapple leaves are too wet. Dried pineapple leaves might be burned if an inexpensive drying operation can be devised. Otherwise, energy from pineapple may depend on using the leaves for hydrolysis to fermentable sugars that could be combined with sugars extracted from the fruit. Although pineapple appears to be a reasonable candidate for producing alcohol, the fruit is such a desirable food that many people would be offended by its use for fuels to power automobiles.

Grains and Grasses

Benson, Allen, et al. (1978) have been evaluating grains and grasses as feedstocks for fuels. Possible land for such plants is shown in Table 2.18. In 1975 there were 347 million acres (140 million hectares) devoted to grain crops and hays. Soil, temperature, and water strongly influence yields. About 25 percent of the land not already in farms is in the eastern U.S. states with annual rainfall exceeding 25 in. (64 cm). Areas with 16 to 25 in. of rain make up 50 percent of the total, and less falls on the remaining 25 percent of the available land. If rainfall were the only limiting factor, the better land would yield 3 to 5 tons/acre (7 to 11 tons/ha) and the worst would yield less than 1 ton (0.4 tonnes). These plants thus are relatively unproductive as compared to the most promising candidates for energy feedstocks. This is reflected in high prices per dry ton of biomass.

Table 2.18 Potential Land for Grains and Grasses

Type	Acres (millions)	Hectares (millions)
Range	365	148
Permanent pasture	106	43
Cropland in tillage	450	182
Federal range	300	121
Other	30	12
Total	1251	506

The United States, with its diverse topography, soil, rainfall, and temperature, has agriculture that has adapted remarkably well. The North Central states are blessed with generally adequate rainfall, fertile well-drained soil on fairly level ground, and a favorable climate. This region is well suited to many crops such as wheat, and the current predominance of corn and soybeans is a consequence of crop economics as well as crop suitability. The High Plains area from the Dakotas through Texas generally has good soil and topography but less rain. A higher proportion of pasture and rangeland places emphasis on the more drought-resistant crops such as wheat and sorghum. To support a major livestock industry, forage crops and native or cultivated hays and grasses are important.

The more western states vary considerably in rainfall and thus tend to be heavily dependent on irrigation except in certain areas. Irrigation is practiced on high-value crops such as vegetables, fruit, and rice, but crops in these areas still generally are a fairly representative cross section of crops grown throughout the United States.

The Southwest has a long growing season with sparse rainfall and extensive acreages with very little productivity in the absence of irrigation. The East has many rich farms, ample rain, but some states are so rugged that their farms are few and small.

The South–Southeast also has good annual rainfall but is generally less fortunate than the Midwest in the availability of rich soils with reasonable topography. There are greater problems with regard to maintaining soil fertility and some of the problems inherent for agriculture in warm, humid areas. This region grows a considerable variety of crops, including tobacco, cotton, and peanuts.

Benson, Allen, et al. (1978) also cataloged many of the grains, grasses, and exotic plants. Several of the following subsections are based on their report.

Kenaf

Kenaf (*Hibiscus cannabinis*) appears to have been cultivated originally as a fiber plant in western Africa, and this practice has spread through the tropics. Next to cotton, it is the most widely cultivated fiber plant in the open country from Senegal to Nigeria. The leaves and flowers are used as a vegetable, and plant parts are used in medicines and in various native rites.

Kenaf is a fast-growing annual crop that generally grows on soils that also can produce good yields of cotton, soybeans, and corn. The climatic adaptation is best in the southeast portion of the United States; under favorable conditions, yields reach 7.5 to 8.0 tons/acre (17 to 18.0 tons/ha). In the midwest, yields appear to be highly dependent on climatic factors, especially temperature and soil moisture. Kenaf will produce a fiber crop quickly (90 to 120 days) and requires little care during growth.

Seeding rates will vary mainly with location, row width, and the germination percentage of the seed. However, a plant population of 75,000 to 100,000 plants per acre is desirable, which requires 6 to 8 lb (3 to 4 kg) of seed per acre. Seed should be planted after the danger of a killing frost is over and when there is sufficient soil moisture. Kenaf is relatively immune to disease and insects, although it is highly susceptible to root-knot nematodes. Response to fertilization varies with the soil; in certain soils added fertilizer has given no significant increase in yield. It will not tolerate much standing water, and production would thus be restricted to regions with relatively good drainage characteristics.

The most satisfactory method of harvesting kenaf uses forage choppers for either the green (high moisture content) or air-dry (standing crop killed by frost or chemicals) plants. Green material may be stored as is silage or sugarcane bagasse, using bulk methods. Air-dried materials can be stored in large piles or stacks.

Babassu

The Brazilian alcohol program also has directed attention to the potential of a particular palm tree—babassu, which covers about 33.4 million acres (13.5 million ha in nine states) of that country. One acre yields about 6.9 tons (15.6 tons/ha) of coconuts per year. The epicarp (external layer) is fibrous and makes up 12 percent of the coconut. The mesocarp (intermediate layer) is starchy and is 23 percent by weight of the coconut. The endocarp (inner core) is 58 percent of the weight and is rich in oily almonds, which are presently the only commercial product. Yields from coconuts are 15 percent charcoal, 3.5 percent vegetable oil, 2.5 percent animal feed, 0.7 percent of ethanol, and some other products (Carioca, Scares, et al. (1978). The production of ethanol from babassu flour works well, as expected.

Guayule

Guayule (*Parthenium argentatum* Gray) is a shrubby plant that has been extensively investigated for rubber production. The plant is desert oriented, as it is acclimated to north central Mexico and the Great Bend area of Texas. It usually attains a height of 2 to 3 ft (0.7 to 1 m) and has crooked, brittle branches. Its slender leaves are grayish green, and it produces inconspicuous flowers on short slender stems. The North American Indians discovered that the guayule plant contained rubber and extracted it by chewing. Because of the plant's resin content (which made it burn with a fierce, hot flame), it was used to fuel adobe smelters in mining areas of northern Mexico. Until extensive use depleted much of the guayule, it was also used as a fuel for the bread ovens of Mexican women in the northern provinces.

Rainfall requirements for reasonable nonirrigated production are 15 to 16 in. (38 to 40 cm) per year. Yields vary with maturity of the plant and with climatic and other conditions. With a 4-yr cycle, "rubber" yields are of the order of 1 ton/ acre or 500 lb/acre·yr (0.57 ton/ha·yr). The total biomass yield is about 1.5 tons/acre·year (3.4 ton/ha) (estimated from sparse data). The plants, including roots, are sacrificed at harvest after a several-year growth period.

Yokoyama, Hayman, et al. (1977) found that up to sixfold increases in rubber production by guayule plants was obtained by application of 2-(3,4-dichloro-phenoxy)-triethylamine. Although of great potential value in itself, this also shows the potential of further elucidation of the synthetic pathway and of genetic improvements in this and other plants.

The usual procedure in guayule rubber production has been to sow seeds thickly in nursery beds, grow to an arbitrary size, clip the plants within a few inches of the crown, and transplant to fields. It is usually more than a year with the most favorable conditions and sometimes several years before the plants spread fully through the fields.

Guayule may be harvested by undercutting the plants to a depth of 7 to 9 in. (18 to 23 cm), drying for several days, and then collecting with a conventional baler. Another method is to harvest the aerial portion of the plant and leave the roots. The roots of older plants will die while those of 2 to 3-year plants will sprout. This process is called *pollarding*.

Guayule produces toxic substances, and hybrid species that produce greater amounts of rubber also have more toxins. Mears (1979) feels that the toxic side effects can seriously endanger workers handling the plants. Skin rashes have been widespread in regions of India where botanical relatives of guayule are abundant. Mechanical harvesting would minimize the danger to humans, but some underdeveloped nations are hoping to create jobs by harvesting by hand. Breeding to obtain good yields of rubber and low levels of toxic materials could slow development by many years.

Giant Reed

Giant reed (*Arundo donax*) is a tall, erect perennial canelike or reedlike grass that grows 6.7 to 26.3 ft (2.0 to 8.0 m) in height. It is one of the largest of the herbace-ous grasses. In English-speaking countries it has several names—bamboo reed, Danubian reed, donax cane, Italian reed, Spanish reed, or Provence cane; most commonly, it is called giant reed.

The fleshy creeping rootstocks form compact masses from which arise fibrous roots that penetrate deeply into the soil. The outer tissue of the stem is of a silice-ous nature, very hard and brittle with a smooth glossy surface that turns pale yellow when fully mature. Rapid growth, a characteristic of giant reed, may occur at a rate of 12 to 28 in./week (0.3 to 0.7 m/week) over a period of several months. New growth is soft, and very high in moisture and has little wind resistance.

Giant reed is native to the countries surrounding the Mediterranean Sea and has been introduced into almost all the subtropical and warm-temperature areas of the world. In the United States the plant has been cultivated successfully as far north as Virginia and Missouri. It has been used for ornamentation throughout the South and along ditches for erosion control in the Northwest. Abundant wild growths have spread along the Rio Grande River and through other areas of the Southwest.

Giant reed will survive extended periods of severe drought accompanied by low atmospheric humidity or periods of excessive moisture. The plant's ability to tolerate or grow well under conditions of apparent extreme drought results from the development of resistant rhizomes and penetrating roots that reach deep sources of moisture. Giant reed can be seriously retarded by lack of moisture during its first year, but drought does not cause damage to stands 2 to 3 yr old. The plant produces the most vigorous growth in well-drained soils where abundant moisture is available.

Giant reed is a perennial crop. New plantings are commonly started by root cuttings. Either chemicals or cultivation may be necessary for weed control. Established plantings require little attention other than periodic removal of large weeds.

Annual yields of dry matter from giant reed vary widely. In India, 3.2 tons/ acre (7.2 tons/ha) have been reported for wild stands. In the United States along the Rio Grande, fairly good wild stands yield 8 tons of oven-dried reed per acre (18 tons/ha). In Italy, average annual production under cultivation yielded 35 tons/acre (79 tons/ha) of green reed per acre and 13 tons (29 tons/ha) of dry-cleaned cane for pulp production.

The bamboos are largely underexploited in the United States. Research on bamboo in this country has been directed at the kraft pulp industry and has shown that per acre yields are better than yields from comparable slash pine yields.

Jojoba (Simmondsia Chinensis)

The jojoba plant is another desert-oriented plant. It has been studied as a producer of oils. The plant is a long-lived perennial that requires several years to come into production. Annual yields of the beans are indicated to be in the $\frac{1}{4}$ to $\frac{3}{4}$-ton/acre (0.6- to 1.7-ton/ha) range, which tentatively places it in the marginal category for biomass production.

Sunflower

The sunflower is widely adapted throughout the United States. It usually is considered to be a weed, but significant acreages are under cultivation for seed production. It appears to merit consideration, although yield figures have not yet been developed. It differs from the majority of the so-called underexploited plants in that it is adapted to northern climes.

Tropical Grasses

Tropical grasses could be alternatives to sugarcane for obtaining peak yields of biomass. Several crops could be collected per year, and the soil types, irrigation, fertilization, and management practices could provide economic advantages over those for sugarcane. Alexander and Molina (1978) have experimented in Puerto Rico with sugarcane and with tropical grasses. A hybrid Napier grass,

Sordan 7A, is the outstanding short-rotation grass tested so far. Yields were better with a 6-month harvest than with three 2-month harvests, even though growth appears greatest in the early weeks. From 3 months onward, new growth translates to an accumulation of solids in tissues that first contained mostly water.

The estimated cost with the average yield expected at present (12 dry tons/acre·yr, 27 tons/ha·yr) is roughly $90/ton. However, there are prospects for increasing the yields to 40 dry tons/acre (90 tons/ha), and this would lower the cost to slightly over $20/ton.

It must be questioned whether energy crops proposed as alternatives to sugarcane can be competitive unless their costs are very cheap. Sugarcane has a high value for its juices, and bioconversion costs are very low for juices and syrups compared to conversion costs for cellulose because no expensive pretreatment is required.

Salt-Tolerant Land Plants

There are no fundamental barriers to growth of plants on saline waters. Many, many marine algae are known, and at least one species, *Dunaliella,* grows in concentrated salt solutions. Terrestrial plants grow in estuaries, at the seashore, and on saline desert soils. There are also wild, salt-tolerant relatives of crop plants. The use of saline waters of low value for purposes other than growing crops for energy or for food could have profound effects on biomass costs and on diversion of fresh waters to other needs.

Epstein and Norlyn (1977) have investigated tapping the gene pool of barley to select strains that thrive on seawater. Several thousand seeds from genetic crosses were germinated with standard nutrient solutions and switched to salt water after a few days of growth. About 9 percent of the plants completed their life cycle. Seeds from the survivors were planted in salt water, and a few individuals emerged. Only about 0.3 percent survival was noted for the overall screen because of the severity of the selection pressures. Field testing was conducted on sandy soil to allow seawater to flow with no salt build up due to evaporation. Fertilizer and superphosphate were used to supply nitrogen and phosphorus. The seeds germinated, matured, and flowered. Heads of grain appeared normal, and analysis indicated satisfactory grain quality. Yields ranged from 0.04 to 0.6 ton/acre (0.1 to 1.2 ton/ha), and the three best entries averaged 0.5 tons/acre (1 ton/ha). For comparison, the U.S. average for barley in 1975 was 1 ton/acre (2.4 tons/ha). These results from early in the project show promise, and other plants more suitable for energy farming may, through careful selection, be conditioned to survive on saline waters.

Garrett (1979) has described cultivation of tamarisk plants on very poor soils in California. Water of low salinity or of up to 5 percent salt can be used for irrigation, and the yields with very high salt are about one-half those with fresh water. The best yields were about 14 dry tons/acre (32 tons/ha); heavy application of nitrogen fertilizer could give about 40 percent higher yield. Soils that were very alkaline and contained high concentrations of zinc or boron gave reasonably good yields of tamarisk; thus this plant could be grown in areas with no other agricultural value. The leaves exude excess salt, which discourages insects and disease organisms. The roots penetrate deep into the soil to search for water. When saline or brackish water is available, tamarisk seems to be an excellent choice for growing biomass on soils now considered to be almost hopeless for agriculture.

Engineering Considerations

There are some interesting engineering considerations for optimizing photosynthetic efficiency. Effect of light intensity on rate of photosynthesis is shown in Fig. 2.3. Note that the rate is reasonably good in dim light and that saturation and decline occur in bright light. Evolution has led to accommodation so that most plants encounter relatively dim light and function poorly at high light intensities.

Terrestrial plants thrust upward on rich soil such as a jungle or tropical rain forest to compete. There is some photosynthesis near the ground, but most of the activity is in a relatively thick layer of leaves with bright light at the top and dim light in the low levels of the canopy. Not only trees can achieve a "climax community," a term for established dominance by the species best adapted to the soil, climate, and conditions; grasses do well by establishing a large amount of green surface, and transport of photosynthesized molecules has a short transport path through the plant's vascular system.

Agricultural crops of high productivity, such as corn and sugarcane, also create a large amount of green surface. As the sun moves through the sky, some plants that are termed *phototrophic* turn to face the sun. Other plants have sufficient surface so that some are well oriented with respect to the sun whereas the rest use diffuse light from the stratosphere.

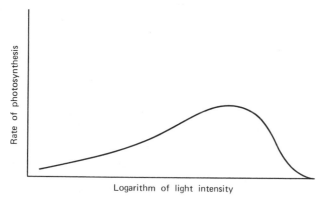

Figure 2.3 Effect of light intensity on photosynthesis.

There is not much that engineers can do to optimize photosynthesis for terrestrial plants except to ensure that water and nutrients are available when needed. On the other hand, understanding how nature has engineered land plants may provide leads for increasing productivity of aquatic plants.

Aquatic Biomass

Aquatic plant growth has problems different from those for terrestrial plants. Microalgae and their relatives such as kelp have been grown at very high rates in experimental units. Although these rates are unlikely to be reached in the field, potential productivities are much higher than those of most other plants. Freshwater plants such as hyacinth and duckweed are also candidate crops for energy, but public lakes and streams would be unlikely areas for growing plants considered as very objectional weeds. Artificial ponds would be expensive to construct, especially in view of the vast areas that would be required. Acquisition costs of land would be high except where the terrain is rugged or water is scarce—conditions unsuitable for aquaculture.

Salt marshes or coastal lagoons could furnish inexpensive areas. Extensive wetlands border some of the coastlines of the United States or are part of river delta systems. Smaller marshes occur near the Great Lakes and in the Northeast and Northwest. Louisiana has 3,750,000 acres (1,518,000 ha) of coastal marshes; Florida, Georgia, and Virginia also have large marsh lands. Estimates of biomass productivity in marshes range from 8.5 to 28.5 tons/acre·year (19 to 64 tons/ha) fresh weight. Wetlands are important in food chains and as wild refuges. In terms of energy needs, there are not enough wetlands for a major contribution, and environmental impacts would require careful study before a decision could be made that might upset these important ecological systems. The open oceans represent the greatest underutilized areas of the earth's surface. These waters are low in fertility and support relatively little plant growth. Bottom sediments are rich in nutrients; upwelling of bottom waters results in abundant growth as exemplified by the anchovy industry off the coast of Peru.

Aquatic Farms

Two types of aquatic farming are possible: plants floating or anchored in open waters and plants in natural or artificial impoundments of water. A very simple farm might consist of a lake on which duckweed or water hyacinth is grown. These floating plants can be collected by automatic equipment, easily shredded if desired, and fed wet to a bioconversion process. The drawback is availability of lakes, streams, or ponds. Vast areas are needed, and the public has competing, popular uses for bodies of water. Man-made ponds are expensive. The cost of an unlined pond (dirt bottom) built by a bulldozer on relatively flat land is estimated at $2000 to $5000/acre ($5000 to $12,400/ha). If the crop grosses about

$5000/acre·yr, even an unlined pond is a marginal investment. Many soils would be too permeable to water, and soil conditioning or a bottom liner would add much to the cost.

The future of aquatic farming seems to lie in using seawater because it is free and not restricted to other uses. Much research and engineering will be required to devise structures for the open oceans, but coastal ponds are well within our present technological capabilities. Some cheap land is available near our shores, particularly in Texas, and other flat, thinly populated areas are near brackish water or not prohibitively distant from the seas.

Whereas land plants thrust upward for light and can exchange directly with air, submerged plants may be limited by sunlight and carbon dioxide. Turbid waters allow little light penetration, and most submerged plants are associated with microbial communities that cause turbidity. In addition to carbon dioxide, its relatives, carbonates and bicarbonate ions, can contribute to photosynthesis. Microbial metabolism produces carbon dioxide; algae themselves evolve carbon dioxide when in darkness. For fresh waters, light and carbon dioxide availability permits a theoretical yield of about 40 to 50 tons of dry biomass per acre (90 to 112 ton/ha·yr) per year. Of course, insolation depends on the position of the sun and is seasonal. Intense light can saturate the photosystems of organisms near the surface for low efficiency of radiant energy utilization. Types of loss are shown in Fig. 2.4.

In nature there are seldom such rich supplies of nutrients that light becomes a problem. Plants attached to the bottom grow in fairly deep water as long as some light filters down. However, in fertile waters growth can be limited by light, carbon dioxide, nitrogen, phosphorus, or other elements in chemically suitable forms. Algae in the laboratory grow at rates that would extrapolate to over 1000 tons/acre (2240 ton/ha) per year but in nature rarely achieve 30 tons/acre (67 ton/ha) per year. This latter figure approaches the productivity resulting from

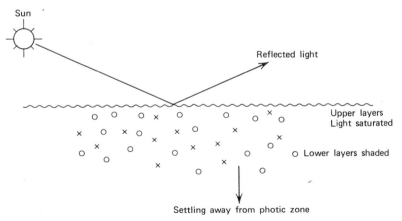

Figure 2.4 Low photosynthetic efficiency of submerged plants. (O = green plants; X = particulate matter).

light limitation due to self-masking as cell density shades organisms not near the surface.

Some freshwater floating plants that are very productive thrust above the surface to obtain more sunlight. Examples are hyacinth and duckweed. The wind can drive these plants from a patch of water, and they will slowly drift back or form new growth in open areas. No marine plants are known that project above the surface. Perhaps the violence of winds, waves, currents, and surf has discouraged evolution of projecting, floating, saltwater plants.

The engineering problems are readily defined for submerged plants: increase light availability, and ensure adequate supply of carbon dioxide and nutrients. In tropical climates or intense sunlight there is an additional problem of temperature control. Shallow algal ponds may overheat; thus deep ponds are used to buffer the temperature over the day's heating and the night's cooling.

There is at least one way to get more light through the air–water interface. Solar collectors could focus light on fiber optic conduits into the body of water. In other words, light from a wide, dry area could be brought to a volume of water. If fiber optic strands are scratched, light escapes; thus diffuse illumination is possible along a submerged fiber instead of bright light at its tip. Although feasible, a fiber optic approach seems impractically expensive.

Instead of bringing light through a limited surface, it makes more sense to increase the surface. Waves provide more surface but not enough relative to the energy input. Mechanical supports that mimic the projection of land plants could greatly increase surface. The simplest support might be a corrugated surface at a slight incline to facilitate flow. While cheap and uncomplicated, the increase in area is only two- or threefold, not sufficient to repay the investment. Structures have two possible means for wetting: rise from the pool by capillary action or supply by pumped flow or spray. Capillary action avoids power costs but has no means of dislodging new growth. As growth builds up, capillary action may be ineffective in bringing water to the sites of photosynthesis. Flow or spray would tend to carry new growth away.

Several types of support structure are shown in Fig. 2.5. A free-floating device for open waters as shown in Fig. 2.5b would require a containing section because growth that escaped would be lost to the depths. Harvesting would require dumping contained growth into some type of collector. Operation in open waters could present fiendish problems. On protected fresh waters, duckweed or water hyacinth appear better than the use of algae and special structures.

Carbon dioxide availability, a major concern in fresh water, may pose no problem in seawater because carbonate species are present in high concentrations. However, one or two replacements of seawater per day might be required if really high biomass productivity can be achieved. Total carbonate in the oceans is about 0.002 moles/l, 1 percent as carbon dioxide, 89 percent as bicarbonate, and 10 percent as carbonate. The increasing concentration of carbon dioxide in our atmosphere is worrying climatologists (see Chapter 11). Conversion of carbonates from the oceans to biomass and thence to fuels that release carbon dioxide by combustion may further disturb the world's balance. While burning of fos-

Rotating Discs for Growing Algae

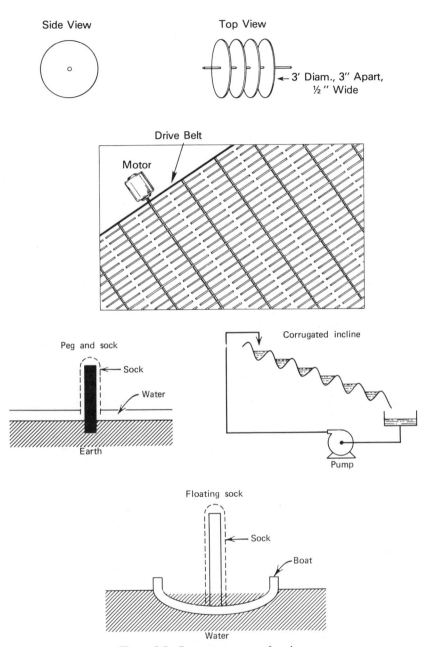

Figure 2.5 Support structures for algae.

sil fuels or release of oceanic CO_2 affect the atmosphere, the latter more directly aids readsorption because the driving force to the seas will increase because of carbon dioxide depletion. Nevertheless, photosynthetic uptake of atmospheric carbon dioxide should benefit its balance, whereas use of carbonates from seawater is little better than burning of coal, oil, or natural gas.

Freshwater Aquatic Plants

Aquatic plants that have received serious consideration as energy crops are *Eichhornia crassipes* (water hyacinth), species of the genus *Lemna* (duckweed), *Hydrilla,* and *Typha latifolia* (cattail). Water hyacinth is a floating plant with large leaves that extend upward from the water's surface from clusters of roots beneath the surface. It thrives in warm, sluggish waters in the southern United States and Central America but is not winter hardy in temperate regions. For over a century, water hyacinth has created problems for industry and recreation in Florida, where millions are spent annually for its control. Figure 2.6 shows water hyacinth.

Duckweed is a minute floating plant with a wide natural range. The fronds attain a maximum of 0.04 in. in length and width; several elongated roots hang from the underside of each frond.

The slender leaves of *Hydrilla* are entirely submerged and grow on an elongated axis with roots extending into bottom sand or mud. *Hydrilla* is typically found in warm climates.

Cattail (*Typha latifolia*) abounds in and near shallow ponds and marshes throughout the United States. The plant is rooted to bottom mud, and a portion of the foliage may be submerged. Dense clusters of tall spikes rise above the surface.

Freshwater plants have several advantages over algae. Ponds for growth of floating plants do not require a level bottom, and construction of an impoundment can be far cheaper than scraping and leveling for an algae pond. Algae thrive best with agitation and frequent change of water. Unlike the seaweeds, growth of the freshwater plants appears to be nearly independent of flow and residence time, as they grow well at relatively slow exchange rates of 2 weeks or more. Algae should have pH control whereas floating plants resist pH upsets, probably because growth sites are not immersed in the medium. Harvesting of floating plants should be easy by raking or coarse screening.

Productivity of Aquatic Plants

Although excellent yields of floating plants have been demonstrated, there is poor documentation for some of the extravagant claims that have been published. Very meager data are available, and high yields seem based not on annual production in the field, but on extrapolations from short-term observations during highly favorably conditions. Comparisons of marine plants with freshwater

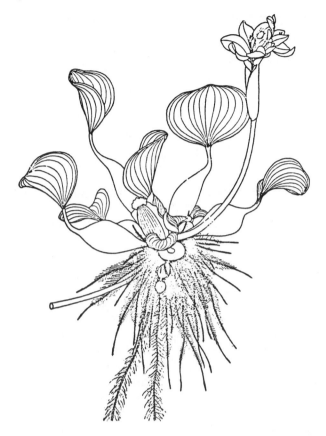

Figure 2.6   Water hyacinth. Courtesy NTIS.

macrophytes are being conducted at the Harbor Branch Foundation of the Wood Hole Oceanographic Institute, Ft. Pierce, Florida (Ryther, 1978; Ryther, Williams, et al., 1979). Water hyacinth, duckweed, and *Hydrilla* were grown in separate shallow ponds, each with a volume of 4000 to 5000 gal (15,000 to 20,000 l). The medium, consisting of well water with varying concentrations of sodium nitrate and sodium phosphate, was passed through open concrete burial vaults. To determine the effect of water exchange rates on growth, residence times were varied from 0.5 to 16.5 days. Table 2.19 summarizes yields for various residence times and concentrations of nitrogen (N) and phosphorus (P). Differences in yields at varying residence times were attributed to nutritional deficiencies other than N and P.

Some data for aquatic plants are presented in Table 2.20. The ponds were harvested weekly with the amount collected set equal to the new growth for that interval. Note that with the conditions for the experiment, water hyacinth was far more productive than the other entries. Water hyacinth had high yields during the spring with a drop in June and July. However, crop management was being

Table 2.19   Yield Comparisons for Floating Plants

Residence time, days	Influent N, μmoles/l	Influent P, μmoles/l	Mean Yield (g dry wt/m² · day)		
			E. crassipes	L. minor	H. verticellata
16.5	1500	150	11.1	3.8	—
5.0	1000	100	15.0	—	—
	500	50	—	4.6	5.1
0.5	100	10	16.7	—	—
	50	5	—	4.5	8.5

developed during this first year of research and development, and better yields are expected for the summer months. An important point was the four- to sixfold difference between winter and summer. As long as impoundment is not overly expensive, it would be wise to construct an oversized pond and to harvest at less than the maximum rate during the good months. Excess growth would be in-pond storage that would be harvested during leaner months. An oversized pond is not a viable option for algae because of the very high construction cost and high operating costs.

Hillman and Culley (1978) have reviewed the practical uses of duckweed, but mostly from the point of view of waste treatment. In warm climates, ponds of duckweed grow well if fed farm wastes. The plant growth would be harvested for anaerobic digestion to methane. The authors present a conceptual design for a dairy plant with integrated waste treatment by duckweed and recycle of the plants as cattle feed in addition to anaerobic digestion of sludges.

Obviously, a pond for growing an aquatic plant should have sufficient nutrients to allow efficient use of the available light. The system should be operated with only slight excess of expensive nutrients so that they do not limit growth but are not wasted. For algae, a yield of 20 tons/acre (44 tons/ha) would require 1600 lb (730 kg) of N and 160 lb (73 kg) of P. If light is limiting, excesses of N and P should be present in the ponds unless nitrogen fixation can supplement the supply of N.

Table 2.20   Productivity of Floating Plants in Artificial Ponds

Organism	Mean Yield (g/dry wt · m² · day)
Lemna minor (common duckweed) 5/18/77–5/17/78	3.8
Spirodela polyrhiza (giant duckweed) 1/17/78–4/24/78	3.4
Hydrilla verticellata 8/24/77–4/24/78	3.9
Eichhornia crassipes (water hyacinth) 5/16/77–5/16/78	17.9

A pond design for algae has been proposed and analyzed for costs (CSO International, 1978). Capital cost was estimated at $3474 per acre ($8,584/ha) for an unlined pond, and a very inexpensive plastic liner would add $870 ($2150/ha) for material and about an equal amount for installation. An unlined pond has a great hazard of excessive loss of water by seepage, but liners are also problematic. If leaks develop, gas can form under the liner and cause it to bulge and balloon. Lining is probably a last resort if seepage cannot be controlled or if biological sediments do not coat and seal the bottom. In any event, the capital cost is excessive because a yield of only 20 tons/acre is expected. The algae will be worth from $20 to $80 per ton, and even neglecting operating cost the payback on the capital would take too long.

Pond Design

The conceptual design of a process using floating plants should consider primary sedimentation. Over a period of years there could be so much sludge on the bottom of the pond that drainage and removal of sludge is required. The interval can be extended at low cost by making the pond deeper initially. A pond may someday need cleaning because of accumulation of settled organic matter and silt carried in from natural streams. A well-designed pond for floating plants might last for several decades without cleaning, even with heavy loading. Unclarified sewage might be fed if the additional nutrient value of the suspended solids offsets the cleaning cost. Anaerobic digestion of sediments would add CO_2 and nutrients to the overlying waters.

Harvesting

Heavy infestations of water hyacinth waters of central Florida have been nuisances for many years. The growth mats impede small boats, make swimming and aquatic sports impossible, are unsightly, and create odors when decomposing. Removal is currently accomplished by special boats with conveyors that lift the mats into receiving boats or barges that are towed away for land disposal of their contents in dumps. Such a system is costly, but the flexibility of the operation results in little interference with other uses of the lake. With a pond intended solely for water hyacinth, a more effective, cheaper means of harvest is needed.

Strong winds move floating plants to one side of a body of water and create piles of biomass. Although this might seem to present opportunity for inexpensive harvesting, there is a serious impairment to growth. Large open areas of water are devoid of water hyacinth or have plant densities far below those for high productivity. Thus, rather than designs that encourage wind transport, it is essential to fix the plants so that the water surface is evenly and densely covered.

A pond for water hyacinth might have ropes strung like shoelaces from one side to the other with a spacing of about 5 ft. By turning pulleys the growth attached to one set of strands is brought to one shore while the growth on the other strands goes to the opposite shore. Scrapers would deposit the growth in conduits where flowing water would take it to the digesters. Plants torn off the ropes and migration from the areas not harvested should reseed the open areas very quickly.

Although the rope–pulley–scraper–conduit system would be expensive, other pond costs would be very low. Slow turning of the pulleys would consume little energy. Note that rope is cheaper than motors and pulleys; thus it is more economical to position the ropes across the larger dimension of the pond—in other words, longer but fewer strands.

Limitations

Water hyacinth is a semitropical plant. During the severe winter of 1976/77, cold temperatures in Florida killed plants and led to a period of very low productivity for Ryther's experimental ponds. Elevated temperatures approaching 40°C do not seem harmful; thus the ponds could serve as heat sinks for a combustion unit. Passing flue gases directly through pond water scrubs CO_2 and pollutants from the gas and heats the water. Traces of sulfur oxides and nitrogen oxides will react to form acids that not only lower pH to a more desirable level but provide sulfur and nitrogen compounds as potential nutrients. The predominant pH effect comes from the CO_2, which also serves for photosynthesis. Of great importance is the heating of the pond, because this should extend the geographical range for water hyacinth to most of the southern United States. Winter storms with low temperatures and high winds do occur in these states, but heated ponds will be somewhat resistant. Probably a stock of water hyacinth should be maintained in a greenhouse so that the pond can be reinoculated to speed recovery from a disastrous freeze.

The other floating plant for which there is sufficient basis for assessment is duckweed. Its geographical range is at least twice the U.S. range of water hyacinth. Ryther has had numerous problems with test ponds of duckweed. Winds drive the plants to one end of the pond. Severe winds have lifted the plants from the water and dispersed them over the countryside. Plants that are overturned usually do not recover and then die. Even when conditions are favorable, yields are far inferior to those of water hyacinth. Ryther speculates that crowding is a minor problem for water hyacinth because it can grow by projecting upward from a strong base. Duckweed is very short and has no place to grow if closely packed. Ryther feels that weekly harvesting may be too infrequent for duckweed and plans to try shorter times for less dense packing.

Hydrilla seems to be of no interest. During the hot season it is inferior to water hyacinth, and in October it flowers and ceases growing. Unless new varieties are found or means are developed to suppress flowering, this aquatic plant can be dismissed from consideration.

Cattails

Cattail is adapted to live partly in water and partly in air. This species of amphibious plant has extensive underground or creeping stems that are rooted in the mud and spread rapidly. While the roots and portions of the stems and some leaves are under water, often most of the shoot is aerial. Because of its frequent occurrence at the water's edge, cattail has a wide range of adjustment and may grow partially submerged. Since both mechanical and conductive tissues are well developed, the plants are able to grow erect without being supported by the water.

Cattails have excellent geographic range, growing in all states, even in Minnesota, which has terribly severe winters. Very high productivity has been claimed for cattails in northern states, but the experimental data are scant. Even if the optimistic estimates are off by an order of magnitude, cattails may have yields as good or better than those of duckweed. As an alternating crop with water hyacinth, cattails may be a poor choice because they attach to the bottom. The need for a fairly level bottom obviates the cheap impoundments that would have sufficed for water hyacinth alone. Furthermore, there is no compatibility of harvesting methods, and the ropes for water hyacinth would have to be removed to prevent their damage by the cutters needed for cattails.

Impressive biomass yields have been reported for managed crops of cattails in Minnesota. These yields are summarized in Table 2.21. High productivity places cattails on the list of serious candidates for fuel feedstocks, but it shares with aquatic plants the problems of finding or constructing suitable ponds or marshes for growth.

A conventional pond, even if very shallow, would not be a particularly good growth area and would constitute an unneeded expense. An artificial wetland could be created by flooding flatland, but suitable terrain may be difficult to find. With terracing, a gently sloped area could be converted to a series of ponds in which the water cascades from one level to the next.

Seaweeds

For centuries seaweeds that are mostly algae have been choice foods in the Orient, and phycocolloids (alginates, agar, and carrageenan) from seaweeds have become quite important to industry as suspending, thickening, emulsifying, and stabilizing agents. Ryther (1978) has studied marine algae with two objectives: aquaculture on sewage effluents and massive culturing to provide biomass for energy purposes.

Goldman, Ryther, et al. (1977) and Goldman (1978) have reviewed in detail the various algae of possible interest for mass cultivation. Included are a discussion of algal culture throughout the world and comments on personal visits to

Table 2.21 Yield of Managed Cattails (Dry Basis)

Year	Tons per Acre	Tons per Hectare
1974	16.7	37.5
1975	19.6	44.0
1976	21.6	48.5

many of the sites. These reports are highly recommended for completeness. Goldman feels that the problems with algae will restrict their energy applications to specific sites for small-scale biomass production. Although there is merit to this view, there is a possibility of devising better pond designs or open ocean systems that could make algae production economically attractive for fuels.

The principal source of agar has been *Gelidium* sp., which is harvested in Japan for export. Seaweeds for carrageenan are harvested from natural populations throughout the world, and phycocolloids are extracted and refined. The prime source is *Chondrus crispus* (Irish moss). Such resources are limited and must be expanded to meet increasing demand. Phycocolloids from various species can have markedly different properties, and commercial blends are made. World-wide surveys of natural seaweeds and screening of species have identified those best suited to production of desirable products. The trend is to cultivation of seaweed crops because natural stands are inadequate to meet marketing needs and can be damaged by pollution, overharvesting, and storms.

The Japanese initiated a seaweed cultivation program several decades ago. Scattering fragments of seaweed in natural waters may establish beds; however, many species attach to the bottom. In Japan, the Philippines, Taiwan, and other countries, seaweeds are fastened to ropes or nets strung in shallow bays. Weeds or epiphytes are removed by hand. *Gracilaria* plants are now grown in shallow ponds formerly used for fish culture because algae are more valuable. Commercial seaweed cultivation is not yet practiced in North America because established methods are highly labor intensive; thus it is more practical to import from countries with very low labor costs. Japan can justify high labor costs because the dried product sells for over $11/lb ($25/kg).

The Canadian National Research Council has sponsored a detailed investigation by Neish (1976) of seaweed culture. Plants of different origin were found to have widely different growth rates and can have greater resistance to epiphytization (growth of competing or contaminating species). Phycocolloid content depends on nutrition and operating parameters.

Ryther's group at Wood Hole, Massachusetts and in Florida have screened many species with emphasis on red algae. Tanks and ponds have been used for suspended culture; open baskets have been held at the surface of ponds or have been sprayed. Starting in about 1970, systems were investigated for treating municipal sewage in ponds of marine algae to obtain commercially valuable crops. This shows promise, and the very high productivities of algae have generated interest in these organisms for energy crops. If cultivated for energy, some algae could be diverted to phycocolloid production, but these markets are miniscule compared to the enormous needs for energy.

Most of Ryther's work with marine algae for energy has been with *Gracilaria* sp. and *Neoagardhiella*. Each is a good producer of phycocolloids and is highly productive in total biomass. Methane can be made from this easily digestible biomass, but other and better energy-related products are likely from the interesting biochemicals of algae. Productivity has been highly seasonal because of the angle of the sun and the effects of temperature. The red algae, *Gracilaria foliifera* and

Neoagardhiella baileyi, are sensitive to the type of nitrogen sources. On the basis of equal total nitrogen, the ranking for best to poorest growth was ammonia, sewage effluent, nitrate, and urea. Growth of these plants can occur at very low levels of nitrogen, concentrations that are found in an unpolluted environment, provided, of course, that there is rapid enough water exchange to satisfy the absolute nitrogen requirement of the plants.

The availability of nitrogen was also found to influence the wet: dry weight ratio, the percentage of ash, and the caloric content of *Gracilaria*. Nitrogen-enriched cultures ($C:N \leq 10$) were about 90 percent water, 5 percent ash, 5 percent volatile solids and had a heat of combustion of 4500 cal/g gram of volatile solids. Nitrogen-starved cultures ($C:N \geq 30$), on the other hand, contained about 94 percent water, 2 percent ash, 4 percent volatile solids, and 4100 cal/g volatile solids. In other words, seaweeds grown in artificial culture with adequate concentrations of nutrients may contain over one-third as much additional energy per unit of fresh weight as nutrient-deficient plants.

Forty-three species, including representatives of the brown, green, and red algae of seaweeds, were screened for their growth potential. Over half of these, including all of the brown algae, failed to grow or subsist in culture. Some of the green algae (*Ulva, Enteromorpha*, and *Chaetomorpha*) grew well for short periods of time, but periodically disintegrated into reproductive spores. The most successful were the red algae, particularly several species of the large genus *Gracilaria*; however, the culture system employed had been developed over some considerable time, specifically for the growing of the commercially valuable red seaweeds such as *Gracilaria, Neogardhiella,* and *Hypnea*.

Spray-irrigated, out-of-water cultures appear promising. Growth proceeded upward toward the light, resulting in a thick mat of algae that assumed a strikingly different habit and morphology from the normal, water-grown plants. However, these cultures became heavily epiphytized after several weeks, usually with the green algae *Entermorpha*, and eventually had to be discarded and replaced.

Kelp Farming

Oceans could provide the enormous areas required to furnish sufficient biomass to meet U.S. energy needs. The giant kelp *Macrocystis pyrifera* grows anchored to rocks in coastal waters and receives nutrients from dissolved and particulate matter in currents. To grow kelp in the oceans, a support structure must exist, and fertilization is essential because nutrients are scarce in oceanic surface waters. As a basis for discussion, a unit farm of 100 square miles (259 km^2) will be considered.

The Dynatech R/D Company (1978a) performed a preliminary economic assessment of aquatic biomass systems. There were protests from some reviewers who felt that the assumed yields were too low and the design bases were too pessimistic resulting in very high costs (Dynatech, 1978b). Benemann, Weissman, et al. (1978a) have pointed out weaknesses in the Dynatech assumptions about growing microalgae that may have inflated the costs somewhat. Furthermore, advocates of land-based systems for aquatic plants were highly critical of using 100 mi^2 as the design size because the ponds or impoundments would be difficult to locate. It must be pointed out that 100 mi^2 for biomass cultivation is relatively tiny in terms of U.S. energy needs. The designers of large coal gasifiers talk in terms of plants that each produce 250 million ft^3 (7 million m^3) of gas per day from 33,000 tons (30,000 tonnes) of coal per day. This is larger than any oil refinery yet built but equates to only about $\frac{1}{10}$ of a Quad. Obviously 200 such plants could supply the U.S. present demand of 20 Quads of pipeline gas. A biomass farm with annual productivity of 10 dry tons/acre (22 tons/ha) at a size of 100 mi^2 miles is only $\frac{1}{20}$ the capacity of one large coal gasifier; thus 4000 farms could supply 20 Quads. If biomass advocates really think they are in the energy business, it is essential to deal with very large areas. Conversely, if land-based systems for aquatic biomass are not compatible with a 100-mi^2 farm, there is a strong argument for reducing their share of research funds.

Natural beds of giant kelp seem to be limited by dissolved nutrients, especially nitrogenous compounds. The plant can translocate nutrients to the growing tips from lower levels where the concentrations are significantly higher. For example, the surface and waters down to about 30 ft (10 m) usually have about 4 μg-atoms/l, of fixed nitrogen, which compares unfavorably with 9.4 μg-atoms/l, the value for growth at one-half the maximum rate. At depths of 300 to 900 ft (100 to 300 m), nitrate ranges from 15 to 30 μg-atoms/l, which would provide an excess. In other words, upwelled water blended to about 20 to 50 percent of the total would supply adequate nitrogen. Kelp strains presently off the California coast prefer temperatures below 20° C, so temperature is a factor in selecting water for upwelling. The undersea currents provide layers of varying temperature and chemical composition; thus each site has an optimum depth for upwelling. This may change during the year. Strains that grow best at a particular temperature could be selected.

Currents are a major factor in designing support structures. Strong currents or high probability of storms mean extra reinforcing of structures. Very strong currents can string out an anchored kelp plant and move it further below the surface and away from needed light.

Proposed support structures for kelp in ocean waters are nets in tension anchored to the bottom (James and Murphy, 1976). Continental shelves appear much more practical than do very deep locations in terms of lengths of cables. The Pacific continental shelf is fairly narrow for North America. There are sufficient areas on the Atlantic continental shelf and Gulf of Mexico for 4000 of the 100-m^2 farms, but there may not be enough sites with suitable currents, lower waters with good nutrients, and freedom from frequent, violent storms. One possible arrangement is shown in Fig. 2.7. Note the pump and distribution system for upwelled water. Both the net and the pumping system must withstand currents and storms. Design engineers should consider the frequency of unusually severe storms and should decide whether to employ especially strong cables and nets. In the deep open oceans, positioning with a propelled system is probably more cost-effective than sending cables to the bottom.

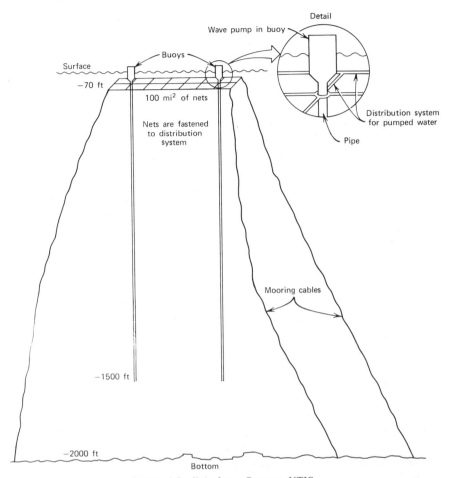

Figure 2.7 Kelp farm. Courtesy NTIS.

Fertilization with upwelled water also presents some engineering challenges. Pumps fueled with petroleum would consume the equivalent of a significant fraction of the solar energy incident on the farm. Some estimates show fuel for pumping to be almost twice the fuel value of the kelp based on a yield of 10 tons/acre·yr. In any event, it is not reasonable to use large amounts of petroleum in pumping to produce cheap fuel from kelp. It may be practical to power the pumps with wave action, but relatively calm seas would lead to periods of inadequate fertilization. Designers have assumed 6-ft average waves, but this may be overly optimistic.

Methods for pumping water to the surface have been analyzed by Hoffman, Strickland, et al. (1976). Wave pumps, windmills, and turbines driven by wave action have been considered. The designs deemed most worthy of testing were a wave turbine–propellor pump (Fig. 2.8) and a bellows pump (Fig. 2.9). The turbine pump lets wave action compress air; the check valve admits new air as the wave recedes. This low-pressure air can drive a turbine that turns a pump. Any

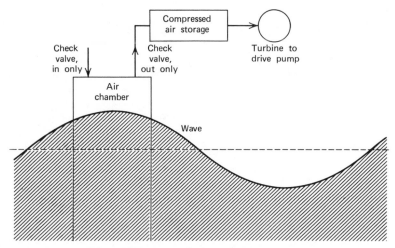

Figure 2.8   Wave turbine pump. Courtesy NTIS.

structure at the surface must withstand storms; thus the bellows pump beneath the surface has obvious attraction. Each arm of the distribution network would have a bellows and check valves so that variations in head due to wave height would compress and release the bellows to create pumping action.

It is interesting to compare wave energy with the fuel energy of the kelp. For a given area of the ocean, wave energy is roughly 190 percent of the fuel value of the kelp. Instead of harnessing wave energy for upwelling to supply nutrients, it might be better to generate electricity that could go to the shore through cables or, if transmission losses are too great, to electrolysis factories at sea for producing hydrogen and oxygen. In any event, such schemes are unattractive when compared to land-based hydroelectric power.

Great uncertainty exists about the residence time of the upwelled water in the vicinity of the kelp. Temperatures are lower for deep waters than for the surface; thus density will cause pumped water to sink. Desirable sediments that might dissolve to replenish nutrients will also sink. However, currents are of more consequence than sinking. If the currents carry water from one farm to another or to other sections of the same farm, no harm is done. The main losses would occur at the boundaries of a farm to open waters. Detention time, uptake rate by kelp, nutrient concentrations, head losses in pumping, wave heights, pipe diameter, materials of construction, anchoring, distribution systems, and trade-off between costs and growth rates present a complicated design problem, and much of the needed information is not yet available. Cooling large areas of the ocean's surfaces may adversely affect climate and make kelp farming hazardous to civilization.

Natural upwelling such as those responsible for the anchovy industry off the coast of Peru would circumvent the need for pumping. As commercial fishing exploits these areas, kelp farming might be precluded. Natural upwelling occurs

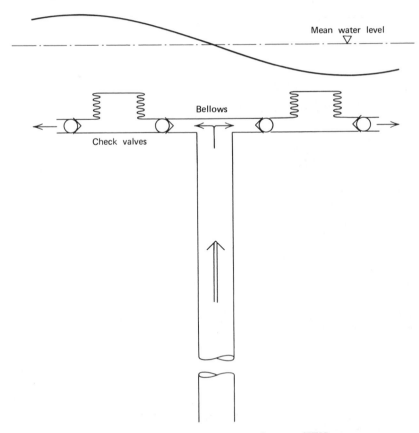

Figure 2.9 Wave bellows pump. Courtesy NTIS.

to a small extent on the East Coast (of the United States) but is important on the West Coast from April through September. Upwelling is usually greater on the West Coast of a continental mass, and the steeper slope of the western slope of the United States accentuates the effect. Even when upwelling is strong, there may be several days when currents are very weak or when flow is reversed. Fishermen report marked changes in aquatic life during these periods, so it is reasonable to expect effects on kelp growth. Artificial upwelling could augment natural upwelling during favorable periods and could act alone the remainder of times. It seems highly doubtful that solely natural upwelling could sustain continuous high kelp productivity.

North (1979) at the California Institute of Technology has been investigating kelp nutrition; trace elements influence vigor and growth rate of kelp. It may be necessary to complex undesired ions or to supplement certain trace elements in the streams pumped from the depths. If upwelled waters require treatment or supplementation, kelp farming will probably be too costly.

A kelp plant is illustrated in Fig. 2.10. Natural plants have juvenile forms that

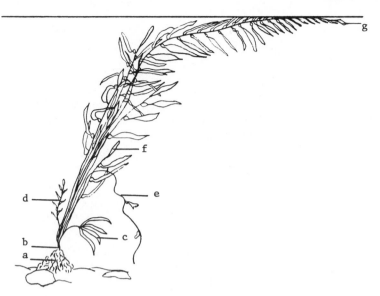

Figure 2.10 A kelp plant (*Macrocystis*): (*a*) holdfast; (*b*) primary stipe; (*c*) sporophylls; (*d*) developing young frond; (*e*) deteriorating senile frond; (*f*) terminal blade, or growing point of young frond; (*g*) terminal blade of mature frond in the canopy. Courtesy NTIS.

attach to rocks and grow into adults with lengths that may exceed 60 ft (18 m). In kelp farms, nets would be about 20 ft (6 m) beneath the surface. Initial seeding would require human skin divers to tie juvenile plants at intervals along the net. As the plants grow, fragments of reproductive tissue would encounter the net to start new plants. Twice per year, ships would cut the kelp at about 10 ft above the net and bring the harvest on board. Unfortunately, loss of the tip prevents further growth of the frond; thus it ages and eventually dies without contributing to the crop (North, 1979). This represents a material and energy loss from the system.

It is conceivable that floating factories could process kelp at sea and return wastes as fertilizer. Products would be transported to shore periodically. Factories anchored at sea, as are oil drilling towers, would not be practical because their costs would be astronomical, final products would be conveyed to land, and kelp would still require transport to the factories from more remote areas. Transport of kelp to shore appears best in preliminary cost estimates, and chopped kelp in slurries could be conveyed in pipelines to the factories from the unloading piers.

Oil drilling platforms are regulated by the U.S. Coast Guard as ships because they are floated to their eventual site. Kelp farms might be classified similarly if constructed on land. A 100-mi^2 farm might have nylon nets weighing 15,000 tons, and construction on land or at sea seems formidable. The U.S. Coast Guard regulations add to the cost because extra mooring cables or stouter lines might be demanded to ensure structural integrity. Liability insurance might be expensive if a net could break loose and foul a ship's propellers or drift into a harbor or a bridge. Navigational warnings to keep ships away from farms would be costly, and sites in normal shipping lanes would be prohibited. Environmental impact statements are probably going to be required to satisfy state and federal agencies. If opposed by a vocal and politically powerful group, these statements could be held up in the courts for years.

The composition of giant kelp is shown in Table 2.22. The structures of the various biochemicals are not particularly relevant because they are mostly carbohydrate polymers that can be digested anaerobically or hydrolyzed to sugars fairly readily. There are uncommon sugars and sugar derivatives of un-determined fermentability in the hydrolysates. Use of kelp as a feedstock for fermentation would have to await testing and development of microbial cultures.

The very high ash content of kelp may require a washing step. This will add cost for fresh water and should be carried out near the sea so that the washings do not pollute streams or lakes. Osmosis will cause cells to swell and leak some of their organic content; this is loss of feedstock and could pollute the receiving bay. Should washing be omitted, the residue from anaerobic digestion of saline mate-rials would probably be poorly proportioned for fertilizer. The liquor separated from the sludge would present a very annoying waste treatment problem. All things considered, a washing step seems very sensible.

Although kelp is presently a profitable source of specialty biochemicals, the markets are too small to justify large farms. Several fuels might be made from kelp, but the only bioconversion process as yet proven is methane by anaerobic digestion. Advocates for kelp for energy use two very debatable assumptions: yields of 40 tons/acre·yr (90 tons/ha·yr) and sizable credits for using digester sludge for fertilizer. This yield, about 20 times the best yield for natural kelp beds, seems to have come from extrapolation of growth data over a short period of time for laboratory plants grown under ideal conditions (Wilcox, 1979). Credits for fertilizer overlooks the fact that a closely related operation, anaerobic diges-tion of municipal sewage, produces sludges that are poor fertilizers and must be given away or sold as very cheap soil conditioners. Anaerobic digestion con-

Table 2.22 Composition (in Percent) of *Macrocystis pyrifera*

Fresh plant	12.5 solids, 87.5 water
Solids	44.5 ash (28.7 KCl, 7.5 NaCl, 4.3 Na$_2$SO$_4$)
Organic portion	15 mannitol (of dry material)
	18 alginic acid
	1 fucoidan
	1 laminarin
	5 cellulose
	7 protein
	6 to 10 unidentified

Source: Flowers and Bryce (1977).

sumes most of the rich organic matter; thus it is unlikely that sludge from kelp will be prime fertilizer.

Jones, Fong, et al. (1978) reviewed the prospects of methane from giant kelp and designed a possible plant. A base case and a speculative case with highly optimistic assumptions were developed. By-product credits for digester residue were considered. The base case showed that the methane cost would have to be from $17 to $30 per million BTU depending on kelp yield and credits. For the speculative case, the range was $9 to $22 per million BTU. Discouraging as these costs are, it must still be pointed out that they are based on yields substantially higher than any kelp yields ever demonstrated in nature or in cultivated beds.

By early 1980, the results of growing kelp in the oceans were poor. Juvenile kelp plants attached to the support cables grew poorly, and few, if any, survived. Storms, failure of upwelling systems, and ill-defined nutritional requirements may all have contributed to the failure. News reporters found documents from the U.S. Department of Energy files showing that expert reviewers and the department's program managers had recommended declination of proposals to grow kelp in open waters. Nevertheless, upper management overruled declination and funded a project at over $1 million per year. It is interesting that the program managers are no longer with the department, but some of the upper management is still there to blame the scandal on those who are gone.

Kelp for energy is the same sort of concept that stimulates the imagination as the concept that put men on the moon. It would require similar dedication of scientists and engineers and the expenditure of much money. Using the oceans is attractive, and the news media have featured the proposed kelp farms. The catch is the very low value of methane gas, and this must be cheap because the motivation of fuels from biomass is cheap energy. Only by making questionable assumptions and using optimistic cost estimates can kelp farming appear promising. While kelp farming cannot be dismissed as hopelessly impractical, other biomass projects seem more worthwhile. Major research on the engineering problems of growing kelp in the oceans cannot deserve high priority until high productivity can be demonstrated in small kelp beds.

Floating Marine Plants

Floating macro algae such as *Sargassum* are held at the surface by air sacs or pneumatocysts. No nets or cables would be required, and the plants grow from small pieces. As with any plant growing in the ocean, fertilization is required to get high yields. If upwelled water were used, there is a serious question about retention of nutrients near the plants because cold water brought from the depths could sink quickly past plants located close to the surface. Ryther (1980) feels that there are vast shallow areas of the Gulf of Mexico that could retain the nutrients for *Sargassum* sufficiently near the plants, and currents could be used to assist harvesting. Productivity is potentially high, and costs would be much lower than those for kelp.

Microalgae

Large areas on the earth have high solar insolation, cheap arrid land, elevated temperatures, and large amounts of seawater or brackish water. Although conventional agriculture is totally impractical, these conditions are very well suited to culturing algae. Dubinsky, Aaronson, et al. (1978) studied algal species with respect to production of lipids and hydrocarbons. Some green algae had over 50 percent lipids (dry weight basis) and 12 species were found with over 20 percent. High levels of hydrocarbons were very unusual, but one species, *Botryococcus braunii*, can have up to 20 percent. Alternate sources such as desert plants that contain hydrocarbons are very low in productivity per unit area compared to usual agricultural crops. Algae with 25 percent "oil" would have commercial prospects, and by-product credits for food protein could be quite attractive. It must be pointed out that the technology for growing, managing, and harvesting mass cultures of algae is far from being ready for commercial trials.

Microalgae have long proved practical for municipal waste treatment in units called *sewage lagoons* or *oxidation ponds*. The algae must settle to the bottom of the pond or be collected for disposal because their release to streams provides nutrients for microorganisms thus depleting oxygen and killing fish and other desirable aquatic life. Although many of the large ponds are for sewage treatment, there is increasing interest in growing algae as a commercial product. Goldman, Ryther, et al. (1977) have discussed a wide variety of ponds used throughout the world for growing algae. Many of these designs are used in commercial operations in the Orient to produce high-value food. Table 2.23 lists some of the large ponds in various countries. No algae system has yet achieved consistent high yields and low operating costs.

Some microalgae have doubling times of less than one day under ideal conditions and derive most of their energy from incident light. In ponds, algae may be limited by light because of self-masking by dense concentrations, by settling to below the zone of good light penetration, by turbidity of the medium, by shading of the pond, and by insufficient sunlight. Many areas of the United States are too far north to have adequate sunlight for culturing algae in high yields except for a few months in the summer.

Microalgae are of interest as food because over 50 percent of the weight can be protein. *Chlorella* is grown commercially in the Orient as a specialty food that retails for about $4.50/lb ($10/kg) dry weight. *Spirulina,* filamentous blue–green alagae, is a traditional food in Africa around certain alkaline lakes where it grows naturally. It is harvested with screens and sun dried before consumption. Part of a large evaporation pond near Mexico City is used to grow *Spirulina.* In bicarbonate manufacturing, alkaline solutions of high salinity were found to be ideal for *Spirulina.* Production is about 1 ton/day of a product selling at about $2.25/lb ($5/kg). The high pigment concentration makes *Spirulina* of value for poultry feed to impart color to eggs.

Pirt (1979) has proposed growing algae in solar panels that could cover the

Table 2.23   Large Ponds for Mass Culturing

Year	Location	Usual Species	Total Size Meters^2	Best Yields, g dry wt/$m^2 \cdot$ day Maximum	Average	Description[a]
1951	Cambridge, Mass.	*Chlorella*	56	11	2	N, F, SC
1951	Essen, Germany	*Chlorella*	6		4	N, F, SC
1953	Tokyo, Japan	*Chlorella*	15		3.5	
1954–1955	Tokyo, Japan	*Chlorella*	2.8–5.5	28	16	N, F, SC
1957	Tokyo, Japan	*Chlorella*	7.1–78.5	14	8.6	N, F, SC
1957	Tokyo, Japan	*Scenedesmus*		14		
1958	Tokyo, Japan	*Tolypothrix*	5		6.4	N, F, SC
1958	Tokyo, Japan	*Chlorella*	314			N, F, SC
1959–1960	Plymouth, England	*Phaeodactylum*	2.6	—	10	N, M, SC
1961	Richmond, Calif.	Green Algae	2700			W, F, C
1962	Jerusalem, Israel	*Chlorella*	4	16	12	N, F, SC
1960–1977	Dortmund, Germany	*Scenedesmus*	80	28	10	N, F, SC
1960–1962	Trebon, Czechoslovakia	*Scenedesmus*	12		15	N, F, SC
1967	Trebon, Czechoslovakia	*Scenedesmus*	50	25	16	N, F, SC
1966	Trebon, Czechoslovakia	*Scenedesmus*	900	19	12	N, F, SC
1966	Tylitz, Poland	*Scenedesmus*	50	16	12	N, F, SC
1968	Rupite, Romania	*Scenedesmus*	50	30	23	N, F, SC
1969	Firebaugh, Calif.	*Scenedesmus*	1000	35	10	TD, F, C
1970–1971	Jerusalem, Israel	*Chlorella*	300	60	27	W, F, C
1973	Woods Hole, Mass.	Diatoms	4	13	—	WSW, M, C
1974	Fort Pierce, Fl.	Diatoms	4	25	—	WSW, M, C
1975–1976	Woods Hole, Mass.	Diatoms	180	10	—	WSW, M, C
1967–1977	Mexico City, Mexico	*Spirulina*		20	10	B, F, SC
1976	Haifa, Israel	Green Algae	120	35	15	W, F, SC
1977	Yagur, Israel	Green Algae	1000			W, F, SC
1974	Bangkok, Thailand	*Scenedesmus*	87	35	15	N, F, SC
1974	Bangkok, Thailand	*Spirulina*	87	18	15	N, F, SC
1976	Japan	*Chlorella*			21	N, F, SC
1976	Taiwan	*Chlorella*			22	N, F, SC
1977	Taiwan	*Chlorella*	250–500			N, F, SC
1976	Richmond, Calif.	*Micractinium*	3	11	—	W, F, SC
1977	Richmond, Calif.	*Micractinium*	12	23	—	W, F, SC
1977	Richmond, Calif.	*Micractinium*	2700	7.3	—	W, F, SC

Source: Goldman (1978).

[a]   Key: N = artificial nutrients; W = wastewater; WSW = wastewater-seawater mixture; B = brine; TD = agricultural tile drainage; F = freshwater algae; M = marine algae; SC = semi-continuous harvest; C = continuous harvest.

land or the sea. If provided with 100 percent CO_2, the algae would produce a product gas with 80 percent O_2 and 20 percent CO_2. The theoretical output of biomass is 110 tons/acre·year (250 tons/ha·yr) of dry weight in temperate zones and twice as much in the tropics.

Märkl (1977) found the effects of light intensity and CO_2 tension to be interdependent in continuous cultures of the autotrophic alga *Chlorella vulgaris*. Agitation was excellent in the apparatus, so CO_2 was used efficiently. At higher light intensities, a higher concentration of CO_2 was required to achieve the maximum growth rate, and the rate of transport of CO_2 from gas to liquid was greater.

Brown (1977) has proposed an interesting scheme for using algae as fuel in power plants. Combustion gas would be scrubbed for CO_2 to be recycled to the algae ponds to overcome CO_2 limitations on growth rate. He envisions a two-layer lake with a natural bottom layer separated by a plastic sheet from a nutrient-rich top layer. The lower layer would serve for heat stabilization and for anaerobic digestion of sewage sludge to methane. Algae in the upper layer would be collected by sedimentation to ducts in the plastic sheet. Schemes such as this provide a stimulus to imagination but are faced with engineering problems and capital costs that probably are incompatible with the potential profits of a factory that has cheap power as its major product.

Kollman and Kollman (undated a,b) have claimed very high yields of algae grown in ponds with a transparent covering, good mixing, and addition of carbon dioxide gas. The stated yield of 360 dry tons/acre·yr corrects to over 300 ash-free tons (674 ton/ha·yr), which is far greater than the reported yields of most other investigators. A proposed plant for anaerobic digestion of cattle wastes with the effluent used to grow algae by the high-rate method has been criticized by Jones (1978). The Kollman–Kollman reports are not very detailed, and there are unanswered questions about the algal unit, the effect of traces of toxic materials in the feed stream, and the analytical procedures. Jones feels that requirements of mass transfer of carbon dioxide may be difficult to meet in shallow algal ponds without expensive, intense mixing. Certainly, more research and more information about high-rate cultivation of algae are needed.

Algal ponds are being evaluated in Israel for producing single-cell protein for animal feed (Moraine, Shelef, et al., 1979). Wastewaters from cities or industries are used, and the benefits are waste disposal, feed protein, and renovated water that may be used for irrigation. The most promising ponds use a high-rate design such as that shown in Fig. 2.11. A paddle-wheel mixer circulates the medium and maintains suspension, although some sewage sludge settles to the bottom and degrades anaerobically. The CO_2 and noxious gases from the anaerobic sludge pass into a photosynthetic zone where CO_2 is fixed and other gases are oxidized to harmless compounds. Supplementary oxygen can be added at night to keep the liquor aerobic and might be needed during cold or heavily overcast days, which occasionally occur during a 2-month period in Israel.

Light intensity and temperature have marked effects on growth of algae; thus seasonal variations are quite important. The populations that are established contain large numbers of bacteria and sometimes have significant amounts of fungi and protozoa. Detention time affects relative numbers because algae reproduce relatively slowly. Algae predominate at long detention times, but their relative abundance and population stability are lower with short detention times. Ponds built to handle wastewaters are of fixed size, and input flow varies. In Israel, wastewater is produced in greater amounts in the winter months, thus shortening pond detention when algal activity is least. Building larger ponds to optimize detention time would be prohibitively expensive; thus ponds are operated with trepidation during unfavorable months.

Composition of the algal population in an open pond is unpredictable, but almost always a few genera or just one species predominate greatly. A popula-

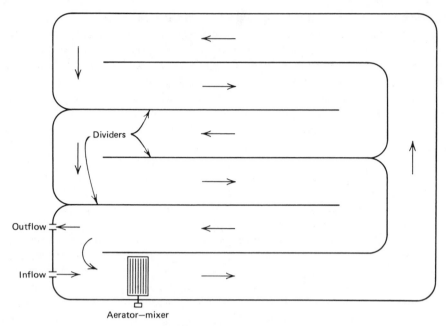

Figure 2.11 High-rate Israeli pond.

tion may be stable for months until climatic changes favor new species. *Micractinium, Scenedesums, Oocystis, Franceia,* and *Euglena* were important, but *Chlorella* was not encountered in high-rate ponds, which is surprising because it is very common in conventional oxidation ponds.

The residence time to maintain sufficient suspended solids for high productivity is 2 days in the summer and 7 days in the Israeli winter. Annual productivity is over 45 tons/acre (100 tons/ha). Collection of the algae by centrifugation or coagulation with chemicals and flotation is costly, and the flotation solids must be dewatered to make drying economical. Drying is essential to making animal feed but would not be necessary for a fuels-from-biomass operation.

One very important lesson can be learned from Israeli algal ponds that have near-perfect insolation and temperature conditions. With protein, which is much more valuable than a biomass energy feedstock, plus credits for waste treatment and renovation of water for irrigation, algal biomass is marginally attractive economically. As an energy feedstock that has minor credits for waste treatment because the required scale far exceeds the available sewage, algal biomass is much too expensive.

The problem of temperature rise of algal ponds during the day might be overcome by employing thermophilic species. Furthermore, it may be possible to develop dual-purpose systems for collection of solar heat and for culturing algae. In other words, some types of solar collector circulate water directly over the blackened surface, and it seems sensible to make further use of the aqueous phase by growing algae or operating a photosynthetic sheme for hydrogen generation.

Oswald's group at the University of California at Berkley has many years of research experience with algal ponds and has a respected leadership position in this field. Nevertheless, their reports on operating problems with ponds present a discouraging picture (Benemann, Weissman, et al., 1978b). Predominance of algal species shifted during the year so that large algae that are easy to harvest were replaced by tiny algae that escaped capture. Heavy infestations by rotifers that prey on algae reduced populations to unacceptable levels and required draining, cleaning, and restarting some of the ponds. In summer months there were numerous pond failures due to overgrazing by protozoa or crustaceans on the algae. It appears that biological problems exist in addition to the major engineering hurdles to be overcome before microalgae can become a cheap energy feedstock.

Harvesting of Microalgae

Collection of algal biomass could be so costly that it is impractical. Ordinary filtration does not work because the organisms blind or smear the filter surface. Addition of filter aid is expensive, and the added bulk, although biologically inert, interferes with subsequent processing. Coagulants or flocculants work, but alum must be added in equal weight to the algae or lime in twice the weight. Centrifugation works well for algal cells but consumes much energy and requires a large capital investment. Oswald has been using microstraining. A microscreen device is shown in Fig. 2.12. Typical openings in the screen are about 10 to 30

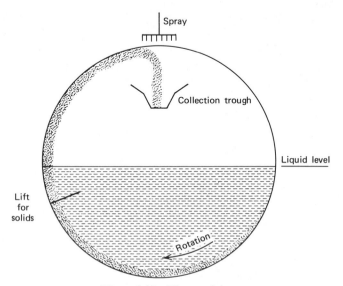

Figure 2.12 Microstraining.

microns. When algae grow in dilute suspension, concentration to about 2 percent is required, and above 4 percent is desirable. This means that large equipment and high throughput rates must be used.

Small microalgae are not captured by screening, and best yields with larger algae seldom exceed 90 percent because nonuniform screens have some larger openings. Filamentous algae or those with spines or extensions are more easily collected. Long, slender cells may bridge the openings in the screen or may be oriented for easy passage.

Benemann, Koupman, et al. (1977a) have attempted to control the species in algal ponds in order to maintain high collection efficiencies by microstraining. With no control, natural population shifts would lead to periods where small algae predominated and collection was very poor. Their control strategy was to recycle part of the concentrate from the microstrainer to the pond. Although this was effective in increasing pond density and in enriching collectible organisms, overall collection efficiency was not particularly good and sometimes dropped to only 40 percent. High recycle: harvest ratio were attempted, but still perform ance was mediocre. Weissman and Benemann (1979) have analyzed the mathematics of biomass recycling and species competition in continuous cultures and have tested their concepts with mixtures of algae. Selective recycle has a profound effect on relative numbers, but system productivity may be lowered. Although this recycle concept is clever, the field results have not been encouraging. With high proportions of recycle, collection costs would increase. Unless some dramatic breakthrough is achieved, microstaining will not be a satisfactory collection method because it is efficient only with certain species of algae that cannot be maintained with certainty.

Gregor and Cardenas (1976, 1977) have tested several species of algae, large and small, grown in the laboratory, and mixed species taken from sewage lagoons. Ultrafiltration units of a nonfouling polar polymer invented by Gregor were found to concentrate algae with high flux rates of water through the membrane. No chlorophyll was detected in the permeate; thus there had been no leakage. The upper limit to concentration was roughly 10 percent algae by weight, at which point the suspension was too thick for sufficient velocity past the membrane surface to keep it free.

Costs of a membrane process are most dependent on flux rate, initial cost, and membrane life. Gregor has used some membranes continuously for several years, and service might extend to 10 years. However, based on 2-yr replacement of membranes, a flux of 100 gal/day·ft$_2$ of membrane (4100 l/m^2·d) and the present commercial cost of about \$2/ft^2 (\$22/m^2) of membrane, collection of algae would cost about 20¢/1000 gal (5¢/m^3) of input stream. Since the algal suspensions are at mild pH, are not corrosive, and are processed at ambient conditions, it is very likely that inexpensive materials can be used to fabricate the membranes and the units. There is a potential for bringing the processing cost to only a few cents per 1000 gal. However, if the algal suspension were about 0.1 percent, there would be roughly 8 lb/1000 gal, and the collection cost would add about 1¢/lb, which is not encouraging when considering such goals as converting biomass to sugar which should be under 5¢/lb for economical fermentation to ethanol. This

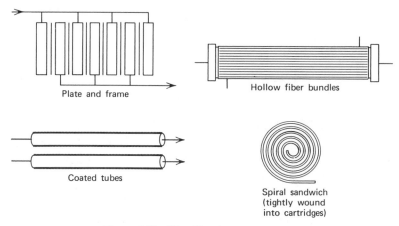

Figure 2.13 Ultrafiltration arrangements.

means that process development is needed to get more algae in the growth step and to get cheaper collection.

Some present configurations for ultrafiltration are shown in Fig. 2.13. Hollow fibers pack a great amount of surface in a compact volume but must be used with quite clean liquids because the passages are small. Plate and spiral wound are not as compact for a given surface but can be cleaned or replaced. Construction costs for the ponds and expensive harvesting methods are the main hurdles to commercialization. The high potential growth rates of algae offer great promise for collection of solar energy, but engineering breakthroughs are desperately needed.

Purification of products from algae would be facilitated by easy destruction of the cell wall. Redhead and Wright (1978) have isolated 70 microbial cultures that lyse blue–green algae. Almost all the isolates were fungi from the genera *Acremonium, Emericellopsis,* and *Verticillum,* and the other few were bacteria and streptomycetes. Whereas these studies were motivated by the possibility of microbial control of algal blooms, lysis could lead to easy fractionation of algal components for food and energy products.

Other pathogens for algae are known. The species shifts observed in nature may result from attacks on the predominant species so that another is favored. These shifts occur more rapidly than would be expected for simple interspecies competition in response to seasonal changes.

Additional Reading

For additional reading see Anderson and Tillman (1977); Christian and Zuckerman (1978); Ernest, Hamilton, et al. (1979); Goldman, Ryther, et al., (1977); Jones and Radding (1978); Report of the Alcohol Fuels Policy Review (1979); Russell-Hunter (1970); Saterson, Luppold, et al. (1979); Sayigh (1977); Tillman (1978); and Tillman, Sarkanen, et al. (1977).

3

Potential Products

Overview

When the goal is to obtain the most energy at the lowest cost, burning of biomass is probably best. Efficiency can be raised by using stack gases to dry the feedstock. However, such a use emphasizes the worst features of biomass—bulk and inconvenience. Other fuels; atomic reactors; and other technologies such as photovoltaic devices, direct solar heating, or windmills may be more attractive for power generation. In regions where other fuels are scarce, burning of biomass can be important, but the real opportunity for biomass is conversion to liquid fuels for transportation or to chemicals that can replace those now derived from petroleum.

The need for compatibility of fuels from biomass with existing distribution systems has already been mentioned. There have been tremendous investments in electrical transmission lines; transformers; pipelines; railroad tank cars; barges; tank trucks; petroleum depots; gas storage tanks; and the highways, streets, port facilities, and railroads that integrate with distribution. This investment has been gradual over many years, and it is highly doubtful that capital could be found to build a new distribution system quickly. This may not be a serious restriction for most of the proposed biomass fuels except for gasohol, a blend of alcohol and gasoline that requires special care in handling because of a great propensity to take up water.

One proposed fuel that can be produced either biologically or chemically is hydrogen. Although some hydrogen can be added to gas with satisfactory transport in pipelines, high concentrations of hydrogen under pressure are not compatible with existing pipelines and storage tanks. Hydrogen is highly diffusible and can penetrate metals; thus thick walls are necessary to prevent excessive losses. The capital costs would be staggering for modifying existing pipelines, storage tanks, and the lines to homes and factories, or building new systems.

Although proponents of new fuels usually claim that direct substitution for existing fuels will be possible, hydrogen and alcohols are not the only candidates with problems. Thermochemical processes can yield oils that are corrosive and reactive to threaten stability of the fuel and integrity of the distribution system.

Biomass fuels will have short-range, intermediate-range, and long-range markets. An immediate impact can be made on U.S. energy needs by burning biomass, either alone or in combination with coal. Thermochemical conversion and bioconversion have some options that merit larger-scale trials now and that could make significant contributions in just a few years. However, very recent breakthroughs in the laboratory may translate to much better processes that would be ready just 1 or 2 yr later than the current "best" processes. There is a danger that too rapid scale-up of the current processes could demonstrate unattractive economics and create hostile attitudes prejudicial to trials of the next generation of processes. Long after the current engineering research reaches fruition, there may be some highly sophisticated new processes such as biological fuel cells or direct biological production of hydrogen.

Chemicals from biomass must compete with petrochemicals. As oil prices rise, biomass can be expected to capture a significant portion of the feedstock market. However, there are also processes for converting coal to chemicals. Some of the coal conversion technology is well established, and its advocates also expect to capture markets as oil becomes increasingly expensive. O'Hara, Becker, et al. (1977) have reviewed petrochemical feedstocks from coal with regard to forecasts of supplies and demands and have summarized yields of various chemicals from the major processes for coal conversion. The products include light and heavy liquid fuels, gas, organic chemicals, sulfur, and ammonia. The aromatic fractions include benzene, toluene, and xylene; biomass, in contrast, will provide small amounts of aromatic compounds unless means are devised for economical conversion of lignin.

With coal as one very likely future source of chemicals now obtained from petroleum, research and development of biomass conversion processes must have carefully defined goals. Thermochemical conversion of biomass can use processes very much the same as those for coal and produce a similar product mix. This mix could have better proportions of chemicals in terms of value and market size. Nevertheless, more fresh biomass than coal will have to be processed to produce a given level of products; thus biomass plants must be larger. Both increased capital investment for the factory and high transportation costs for the biomass prejudice biomass as being superior to coal if the conversion technologies are analogous.

Whereas thermochemical conversion is a brute force approach, bioconversion allows subtle control of the pathways, and good yields of selected products can be achieved. Fractionation of the biomass permits optimization of separate processes for converting each fraction to the most suitable products. It seems much more sensible to seek bioconversion processes that result in finished, more valuable products instead of producing feedstocks that are used to make these finished products. As an example, why make ethylene from biomass and polymerize the ethylene for plastic products when a biopolymer might substitute for or supplement the product?

Goldstein (1975) suggests that wood be converted to plastics as substitutes for those made from petroleum. He points out that wastes, cull trees, and "green junk" such as woody shrubs could provide the biomass for plastics with minimal impact on the resources needed for lumber and for fiber. The polymeric constituents of wood could be utilized directly in plastics without degradation to monomers. Alternatively, wood can be decomposed to smaller molecules thermochemically or hydrolyzed to monomers that can be polymerized to plastics.

The tangled interdependencies of trade and government regulations are evident in the U.S. sales of chemicals on the world market. There is a complicated price control system that distinguishes between foreign and domestic petroleum, so-called new oil and old oil. With U.S. petrochemical companies using both, their average feedstock cost is lower than competing companies in nations that have no cheaper price-regulated old oil. Anderson (1977) estimated that U.S. chemical companies paid $10.38 per barrel when the world price was $13.93. From 1973 to 1976, U.S. petrochemical exports grew 73 percent. This was the reverse from previous years, when U.S. petrochemical sales were lagging compared to foreign competition. The feedstock advantage may disappear, however, because spokesmen for U.S. companies claim that a certainty of supply at the world price is better than uncertain supplies at a cheap price. Price deregulation of domestic U.S. oil and gas may encourage production but will raise prices so that fuels from biomass will have great incentives.

Shapiro (1978) cautions that return to older products may result in greater expenditures of energy. As examples he cites glass-reinforced nylon housings that consume 35 percent less energy than the zinc castings that were replaced, and cotton shirts that require 88 percent more energy than those from synthetic fibers. Since chemical feedstocks constitute a relatively low percentage of total oil and gas consumption, there may be an adequate supply for many years. Alternate feedstocks that ease demand for petroleum are desirable, but the research goals of any corporation must be return on investment and fairly rapid payoff. In other words, finding new feedstocks may not be the best way to spend corporate research dollars.

The top 50 chemicals in the United States in order of amount produced is shown in Table 3.1. Those with implications to biomass feedstocks are ammonia, ethylene, propylene, methanol, ethylene oxide, ethylene glycol, butadiene, acetic acid, phenol, adipic acid, and ethanol. Although ammonia contains no carbon, hydrogen for ammonia can be made by reacting a carbonaceous material such as biomass with water. Several other compounds are unlikely to be made from biomass, but the end products from these intermediates might be made by fermentation of biomass. A typical organic chemical at 5 million tons per year with a heat of combustion of 10,000 BTU/lb would represent 10^{14} BTU/yr, or only 0.1 Quad. This shows that the energy output of the chemical business is small but the dollar amount is significant because of the higher value and the larger number of the products.

Some rather important short-range aspects of the use of biomass are beyond the scope of this book. For example, retrofitting of boilers or special designs of grates for combustion of biomass are not covered. However, the form of the biomass and its treatment to make it suitable for combustion are of interest.

Direct Burning

Burning of wood in fireplaces, energy from lumbering and mill residues, and industrial burning contribute 1 to 2 Quads in the United States annually. Vermont has little coal and no oil, so burning of wood provides 40 percent or so of total

Table 3.1 Top 50 U.S. Chemicals

Rank	Chemicals	1978 Production, Thousands of Tons
1	Sulfuric acid	42.0
2	Lime	19.4
3	Ammonia	18.1
4	Oxygen	17.7
5	Nitrogen	15.0
6	Ethylene	14.6
7	Sodium hydroxide	12.4
8	Chlorine	12.1
9	Phosphoric acid	10.1
10	Nitric acid	8.6
11	Sodium carbonate	8.3
12	Ammonium nitrate	7.8
13	Propylene	7.2
14	Urea	6.8
15	Benzene	6.4
16	Toluene	5.9
17	Ethylene dichloride	5.9
18	Ethylbenzene	4.3
19	Vinyl chloride	3.8
20	Styrene	3.7
21	Methanol	3.7
22	Terephthalic acid	3.6
23	Carbon dioxide	3.5
24	Xylene	3.4
25	Formaldehyde	3.2
26	Hydrochloric acid	3.0
27	Ethylene oxide	2.6
28	Ethylene glycol	2.3
29	*p*-Xylene	2.1
30	Cumene	2.0
31	Ammonium sulfate	2.0
32	Butadiene (1,3–)	1.8
33	Acetic acid	1.7
34	Carbon black	1.7
35	Phenol	1.5
36	Acetone	1.3
37	Aluminum sulfate	1.2
38	Cyclohexane	1.2
39	Sodium sulfate	1.2
40	Propylene oxide	1.1
41	Calcium chloride	1.0
42	Acrylonitrile	1.0
43	Vinyl acetate	1.0
44	Isopropyl alcohol	1.0
45	Adipic acid	0.9
46	Sodium silicate	0.8
47	Sodium tripolyphosphate	0.8
48	Acetic anhydride	0.8
49	Titanium dioxide	0.7
50	Ethanol	0.7
	Total of above organics	95.3

Source: Chemical and Engineering News, May 5, 1980, p. 35.

energy needs. Pulp, paper, and lumbering generate scraps, sawdust, and wastes that commonly are burned to power the plants.

There is no great novelty to the equipment for burning wood. In a few cases there is provision for using combustion gases to dry the wood to achieve greater efficiency.

The other biomass most commonly burned is sugarcane bagasse. As fossil fuel costs have escalated, more and more cane manufacturers have been burning bagasse to generate steam for the factory. On a national basis, this energy is very small compared to that from burning wood.

Although there is potential for improving combustion equipment for biomass, more cost-effective research and development is possible by improvement of biomass itself. Wood wastes in large chunks or mixed with dirt or stones is unpromising, but other biomass may be well suited for processing to a more convenient, efficient fuel.

Wood-powered energy facilities have been compared to conventional technologies by Bliss and Blake (1977), who project that the cost of electricity from combustion of wood will decline slightly whereas that from coal will continue to rise. Nevertheless, by the year 2000, electricity from wood will still be 30 to 100 percent more expensive than from coal. Projections for gas from wood show a wide cost differential in favor of natural gas now but little difference by the year 2000. If natural gas prices continue to be regulated by law, cost projections are subject to much error.

Taylor (1979) reported on experiences of his company, Louisiana Pacific Corporation, with an electric power plant fueled with wood wastes. For 1978, the output was in megawatt-hours (MWh):

Power used by company	196,000
Power sold to utility	35,000
Power bought from utility	666
Total	232,000

During periods of maintenance, it was necessary to purchase electricity. The cost of producing electricity was $34/MWh, but the utility paid only $20. Correcting for the high purchase price of small amounts of electricity by the company, the net return from the utility was only $16/MWh. Generation of electricity from wood wastes has been successful because a waste disposal problem has been solved while creating an asset. However, the low price paid by the utility should serve as a warning to those contemplating conversion of biomass to pipeline gas or to electric power.

Densification

Reed and Bryant (1978) have developed a case for using densified biomass as a combustion fuel to replace coal, oil, or gas. Whereas coal has 5 to 20 percent ash and 1 to 5 percent sulfur which cause disposal and pollution problems, biomass has less than 1 percent ash and so little sulfur (0.1 percent) that pollution control may not be required. The problem is the low density of biomass, which means higher costs for transportation and storage. Compacting of biomass to a high bulk density could lead to much more economical handling.

Since the first U.S. patent on densification in 1880, many schemes have been advanced. Until recently, the usual goal was animal feed in a convenient form. Now there is considerable interest in forming sawdust and other residues into easily handled fuels. The types of processes are pelleting, cubing, briquetting, extrusion, and roller–compressing. The first three methods employ somewhat the same principle of forcing biomass into dies or cavities by mechanical force applied with rollers. Extrusion uses a screw to squeeze biomass through a constriction; cutters form appropriate sizes. The roller–compressor makes use of the tendency of forage crops to wrap around a rotating shaft. The finished rolls are cylinders several inches in diameter, which constitutes a more efficient form for cattle feeding when produced from hay. All the processes can be improved with binding agents such as pitch or paraffin, which add strength and increase fuel value. Figures 3.1 and 3.2 show operations for densification.

Fresh biomass has about $\frac{1}{3}$ the energy of coal per unit weight and about $\frac{1}{4}$ the energy of coal per unit volume. Densification can change these relationships to $\frac{2}{3}$ and almost $\frac{3}{4}$, respectively. Water plays an important role because the required pressure for densification is greatly increased if the biomass is either too wet or too dry. Optimal moisture content is between 10 and 25 percent. Although there is little change in cellulose below 250°C, lignin is fairly soft at 100°C. When the feedstock is preheated to 50 to 100°C, mechanical energy in densification results in a product temperature of about 150°C. Water cooling is sometimes needed to prevent the dies from becoming too hot. At these temperatures, moisture is driven out, lignin is pliable, and lipids in the biomass are mobile. Hot pellets are fragile and must be handled carefully until cool.

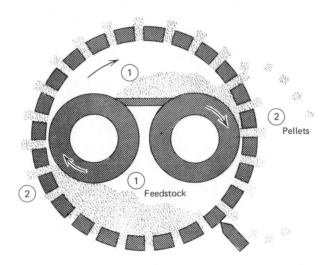

Figure 3.1 Pelleting for densification. Courtesy NTIS.

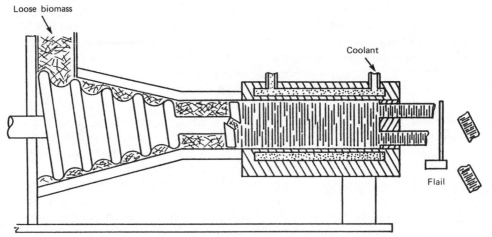

Figure 3.2 Extrusion for densification. Courtesy NTIS.

There are twelve or more manufacturers of densification equipment, mostly for preparation of animal feed. A common economic problem is excessive die wear caused by abrasive foreign matter. This could be very serious with biomass for energy because harvesting of the whole plant is anticipated, and there will be opportunities to include adhering stones and dirt.

The Woodex process operated by the Bio-Solar Company at a 420-ton/day plant in Brownsville, Oregon uses hammer milling, drying, and pelletizing. At least two other plants that use the Woodex process are operating and others are under construction. Alternative processes are being developed, and some produce commercial densified biomass fuel. Solid waste as the feedstock solves two problems, disposal and energy. Baltimore County and the Maryland Environmental Service operate a plant that fractionates municipal solid waste, shreds the combustibles, and compacts them in a pellet mill for sale as fuel.

Densified biomass usually has a water-repellent surface and can be safely stored unless soaked with water. The pellets have low moisture content; thus degradation by microorganisms is arrested. Most equipment for burning coal can accept densified biomass as a sole fuel or mixed with coal. Pollution control and ash disposal are much simpler with biomass.

There is an anomaly in the material balance for biomass densification in that roughly $\frac{1}{3}$ of the biomass goes for energy to run the process. However, the product is drier and sized for very efficient combustion so that the energy from the product is about 90 percent of that when feedstock is burned directly. A 300-ton/day Woodex plant costs about $1.25 million and sells its product (May 1978) at $22/ton ($1.35/million BTU). This competes quite favorably with coal in some locations. Obviously, densified biomass fuel is important now on a regional basis and should do much better as processes are improved.

Home heating with biomass fuel does not seem to present any major engineering hurdles. The customer used to the convenience of oil and natural gas will demand a totally automatic system. This means that a hopper will replace the oil tank, and mechanical conveying and metering will substitute for the oil pump. Conversion of existing equipment seems expensive, but new installations could cost little more than the oil furnace. Ash handling and disposal may trouble the user; thus compensation through lower fuel bills would be desirable.

Alcohol fuels from biomass or production of methane or chemicals may have more glamor, but densified biomass seems certain to figure in the U.S. energy picture and perhaps will play an important role. Biomass may have many facets: combustion fuels in both solid and liquid form as well as energy-related or energy-saving chemicals and fabricated products.

Sawmill residues that have long been a disposal problem can be dewatered mechanically and burned. Nyberg (1976) has discussed extrusion of wood residues whereby about 60 percent of the water is squeezed out. The product can be in the form of pellets or small logs, and the process is energy efficient.

Electricity

Electricity is considered the ultimate in energy convenience because it is a clean source of light, heat, and power. Cost is high relative to fuels, but electricity is ready for direct use without worries about conversion efficiencies as in combustion of a fuel. Generation of electricity by steam turbines is an add-on to combustion of a fuel to produce steam; thus a discussion of burning also serves as a basis for assessing the merits of electricity from biomass.

Substitute Natural Gas

Gas is also clean and convenient. Demand rises year after year. However, the Federal Power Commission has designated some industrial uses as having lower priority than residential and commercial space heating and food preparation. In times of shortage such as demand outstripping supply during severe winter cold spells, gas to industrial users is curtailed resulting in plant closings. Many factories are converting from gas back to coal.

Medium BTU Gas

Mixtures of carbon monoxide and hydrogen are the main type of gas with about 300 BTU/ft³ (2700 kcal/m³). Another medium BTU gas from biomass is methane diluted with carbon dioxide. Pipeline companies can accept limited amounts of medium BTU gas without noticeable impairment of performance, but upgrading to higher BTU value would be required to capture a large market.

Fuel Oil

Petroleum oils have established the technical nature for fuel oils. Substitutes from biomass should have similar viscosity, heating value, and corrosion properties to be well accepted in existing equipment and distribution systems. Combustible liquids such as gasoline and kerosene are not generally considered as oils. Fuels from biomass that are viscous and poorly characterized can be used for heating oil; low-viscosity fluids could replace gasoline.

Charcoal

Charcoal is a solid residue left after heating biomass in the absence of air. Some volatile material retained the char could be driven off by further heating; however, these volatiles contribute to easy ignition and even burning. Charcoal is very low in sulfur compared to coal and is desirable for the production of certain grades of steel. Good charcoal commands a premium price, but for a small market. Char in an oil slurry has promise as fuel for steam boilers; this would create a much larger demand.

Ammonia

There are many possible products based on synthesis gas, and ammonia requires only the hydrogen for reaction with nitrogen over a catalyst at high pressure. Ammonia from biomass is cost-dependent on the intermediate synthesis gas; thus the ammonia process itself does not need detailed discussion.

Generator Gas for Vehicles

Prior to World War II, several European countries had experimental projects for fueling engines with combustible gases from wood or charcoal. With restricted air, the burning of carbonaceous material provides gas containing enough carbon monoxide, hydrogen, and methane to power an internal combustion engine. Performance is poor in terms of power and vehicle acceleration.

When there was a general shortage of petroleum, gas generation for vehicles became very important. The experiences in Sweden are recounted in a book titled "GENGAS" available in translation (Reed and Jantzen, 1979). By late 1941 there were 70,000 generator cars registered, and further growth in numbers was limited by shortages of lubricating oil and rubber for tires. About 55 percent of the trucks and over 70 percent of the buses used generator gas. Farm equipment made good use of generators. After World War II most vehicles switched to gasoline because of greater convenience and performance.

Gas generators for vehicles are heavy and bulky. Automobiles have to sacrifice space or have to pull a unit mounted on a trailer. The units used in Sweden were massive and unsightly. Trucks and buses tended to have generators mounted on the frame, but trailers were also an option. Even boats were equipped with gas generators.

In addition to low power, vehicles with gas generators are inconvenient to start because it takes time to ignite the fire. Some vehicles could supplement with gasoline for starting and for better climbing of inclines. Despite these problems, gas generators can be used to convert existing vehicles to wood or charcoal fuel in an emergency.

Synthetic Fuels

Naturally occurring carbonaceous materials can be converted to fuels resembling petroleum or natural gas. The technology has been aimed mostly at coal, and a brief history of synthetic fuels provides a perspective for biomass processes. Hammond and Baron (1976) reviewed synthetic fuels and commented on their prospects. This section is derived from their article.

By 1815 the streets of London were lighted by gas manufactured from coal. The water–gas reaction to CO and H_2 was commercialized in France in the 1830s. The first coal-derived synthetic fuels were foul smelling and noxious because of H_2S and cyanogen, but scrubbing or contacting with lime and iron oxide were found to remove them.

For several years until the deposits were depleted, rich asphalts were heated to produce "coal oil," a convenient liquid fuel. Natural petroleum was an expensive rarity used medicinally until 1857, when Colonel Drake proved at Titusville, Pennsylvania that economical recovery by drilling and pumping was possible. Manufacture of gas for lighting continued until about 1900, when development of the electric light bulb sharply curtailed demand. However, gas for heating was a healthy industry that benefited greatly from strong, reliable pipelines and storage facilities. Manufactured gas declined to a small percentage of all gas sold.

A large synthetic fuel operation was vital to Germany's efforts in World War II. Their war machine survived on liquid fuels produced from lignite by the Bergius process, a high-pressure catalytic hydrogenation. This industry was phased out when access to foreign oil was restored. The Republic of South Africa, because of its racial policies, has uncertain relations with other nations and has been the target of embargoes. South Africa has little petroleum and can achieve independency for liquid fuels only by producing costly synthetic fuels.

Coal gasification must be run on a very large scale to have a chance of being economic. Some coals are much better feedstocks than others. There are problems with gas distribution and heat transfer to avoid overheating of catalysts. Finding uses for waste heat could provide valuable credits. The product gas is usually used for synthesis, and its use as fuel for heating offers little profit margin in view of federal controls.

There are three established approaches to coal liquefaction: pyrolysis, the Fischer–Tropsch process, and hydrogenation. At 1000 to 1400° C bituminous coal reacts to form coke, plus gaseous and liquid hydrocarbons. Lower temperatures give more liquid, but with undesirable levels of nitrogen and sulfur compounds that must be removed. Sales of gas and of char are essential to profitable operation, but the markets for high-value char are very limited.

The Fischer–Tropsch synthesis reacts CO and H_2 with organic liquids and water using an iron catalyst at about 225° C and high pressure. This process can be viewed as an extension of gasification. It is the main process used in South Africa.

Direct hydrogenation of coal requires that it be dispersed in a slurry of product oil. Conditions are roughly 500° C and 270 atmospheres, and metal oxide catalysts may be used. High-sulfur coal does not process well.

Liquefaction of coal also should be run on a great scale to be economic and thus has massive capital requirements. Preliminary economic estimates are not very attractive, but technological breakthroughs could change this.

Various processes are being studied for producing fuels from oil shale. Mining the shale so that it can be processed is very expensive, and reacting *in situ* presents tremendous engineering problems. It would seem that the vast shale resources of the United States will challenge the next generation of engineers as well as the present generation.

Hammond and Baron (1976) tracked energy costs since 1930 and showed the cost of electricity to be rising after a long period of price reduction. Natural gas is very cheap but rises in price as new legislation is passed. Oil took a big jump during the 1973 embargo and is in an upward spiral. A very interesting cost phenomenon for synthetic fuels was pointed out in that actual costs of a new and unproved process are about double the initial estimates. This is not because of errors in estimating, but rather a result of finding oversights or unrecognized problems in large-scale operation that require expensive solutions. The implications to biomass processes that are as yet unproven are obvious; an important saving factor for biomass is the much smaller scale anticipated for commercial processes; thus corrections and modifications should be relatively inexpensive.

Kleinpeter (1978) feels that methanol or Fischer–Tropsch gasoline can be produced from U.S. coal using existing technology, but the plant would cost $2 billion. The product would be methanol for 40¢/gal or gasoline at $1.33/gal. The announced Mobil Oil process for converting alcohols to hydrocarbons could upgrade the methanol to 97¢/gal gasoline. Selling price estimates are compared in Table 3.2.

The biomass people can really take heart from the predicaments of the coal people. Critics have been quick to attack the economics of biomass fuels, but clean fuels at prices competitive with those of synthetic fuels from coal seem definitely achievable. Whereas coal processes will require fantastic outlays of capital, biomass conversion is likely to be a distributed industry with many inexpensive factories near both feedstocks and markets.

Table 3.2 Selling Price Estimates

Product	Dollars per Barrel	Dollars per million BTU
Methanol from coal	17.35	6.20
Methanol to gasoline	39.05	7.10
Fischer–Tropsch	55.85	10.16
Oil from coal	30.00	5.00
Imported oil	14.00 (1977)	2.33
	22.00 (1979)	3.66

Methane

Anaerobic digestion of organic materials is an old, established method for producing gas composed of methane and carbon dioxide. The fuel value is low because heating of the CO_2 diverts useful energy. Removal of CO_2 gives methane-rich gas of high fuel value suitable for direct addition to the national pipeline system.

Methane has very few uses except for heating or as a cheap source of carbon for thermochemical reactions. It thus competes with natural gas or coal, which are still cheap materials. The main attractions of methane to a fuels-from-biomass program are the relatively inexpensive processes for its production, its acceptability for pipeline gas, and the great ease of product recovery because methane is insoluble in water. If other products from biomass were as easy to produce and to recover, interest in methane might fade.

Methanol

Methanol from biomass is not a new idea. Methanol is often called "wood alcohol" because it was produced by destructive distillation of wood, but today most methanol comes from synthesis gas. Nearly half of our methanol is converted to formaldehyde, some is used in solvents, and the remainder is used for a variety of syntheses.

The reaction from synthesis gas is

$$2H_2 + CO = CH_3OH$$

Usual conditions are 330° C and 100 to 200 atmospheres with a suitable catalyst. Almost any carbon source can be used to make synthesis gas, but natural gas has been preferred because of cost, convenience, and efficiency, which permits low capital cost for factories.

Methanol is a clear, colorless liquid that freezes at −73° C and boils at 65° C. Methanol is highly toxic on ingestion or by skin contact. It is an excellent antifreeze, as it is completely miscible with water. Its octane rating is high and it burns cleaner than gasoline in an internal combustion engine.

Methanol can be produced from biomass by first producing water gas. Hokanson and Katzen (1978) have performed an economic analysis of chemicals from wood. Gasifiers for wood, wood waste, or garbage operate at atmospheric pressure (coal gasifiers can go to 25 atmospheres) and produce a crude gas with H_2, CO, CO_2, traces of hydrocarbons, and some tar oil. Purification is required to bring this gas to the required synthesis gas specifications. If the partial oxidation is carried out with air, dilution by nitrogen gives a gas of very low fuel value. Use of oxygen avoids this dilution, and the product gas has medium value as fuel.

Stinson (1979) has an optimistic status report for methanol. At about $4 per million BTU, methanol might compete for electric power generation; this corresponds to 26¢/gal (7¢/l), which is much below the present price of 46¢/gal (12¢/l). Studies for the U.S. Department of Energy estimate that large-scale gasification of coal costing $25/ton could produce methanol at 24 to 30¢/gal, and use of cheap lignite might lower this to 15¢/gal (4¢/l). Two plants are planned at a 25,000-ton/day scale for coal gasification, one in Anchorage, Alaska and the other in North Dakota. A coal with only 1 percent sulfur would accumulate hundreds of tons per day of sulfur. Air separation to provide oxygen for one of the coal gasification factories would require a unit larger than any yet built.

Methanol can also be made from natural gas. Gas obtained with Alaskan crude oil is being reinjected into the ground because no gas pipeline is available. Sending methanol through the Alaskan pipeline seems preferable to building a separate pipeline.

Biomass might have some advantages over coal to compensate for the disadvantages of high collection costs. Coal slag at 1500°C operating temperatures is able to dissolve refractory ceramic materials very rapidly. Ash tends to escape removal systems and clog downstream units. Biomass slag and ash would be different from coal slag and less copious, and the lower amounts of sulfur would lead to savings for removal.

Ethanol

Ethanol is the only simple alcohol that is not poisonous, but ingestion of large amounts causes drunkeness and death. Its properties are:

Molecular weight	46.1
Specific gravity	0.789
Freezing point	−117.3°C
Boiling point	78.5°C

Except for beverage alcohol, almost all ethanol was previously made from petroleum by hydration of ethylene. Recently the world price of molasses was depressed, and alcohol could be produced more cheaply by fermentation. As oil prices have risen, the petrochemical companies have not been encouraged to produce ethanol because alternate products have been more profitable. Remis-

Table 3.3   Ethanol Production, Capacity, and Uses (1974).

	Ninety-five Percent Ethanol	
	Million Gallons per Year	Million Liters per Year
Synthetic ethanol capacity	361	1365
Production		
Synthetic ethanol, industrial	258	975
Fermentation ethanol, industrial	88.3	334
Fermentation ethanol, beverage	118	446
Total production	463.3	1755
Industrial uses (with percent of total)		
Chemical manufacture (26 percent)	89.8	536
Toiletries and cosmetics (20 percent)	69.1	261
Acetaldehyde (14 percent)	48.3	183
Industrial solvents (12 percent)	41.4	156
Detergents (10 percent)	34.5	130
Miscellaneous (18 percent)	345.3	1305
Synthetic ethanol plants		
Six petrochemical plants in sizes ranging 25 to 120 million gal/yr (95 to 454 million l/yr)		

sion of the 4¢/gal U.S. federal tax on gasoline is permitted if there is a blend containing 10 percent of alcohol from a renewable source. This has created a great demand for fermentation ethanol that allows it to command a price in the range of $1.70/gal (45¢/l), which is well above the price of $1.15/gal (30¢/l) of quite recent times. The production and uses of ethanol prior to the current expansion of fermentation facilities are shown in Table 3.3.

Hertzmark (1979) has assessed the effects of ethanol from grain on the U.S. agricultural sector. One interesting but minor consequence is eliminating a $20 to $30 million trade deficit in industrial ethanol because the output of just one new large ethanol plant matches our current imports. The short-range effect on grain prices could be small if presently idle lands are devoted to grain. Based on construction of five large alcohol plants per year, there might be a significant impact on traditional alcohol markets by 1985. However, the technologies for conversion of cellulose to ethanol should be ready by then. Hertzmark is optimistic about the benefits of building factories now for grain alcohol but feels that grain cannot be the alcohol feedstock of the future.

Alcohol Fuels

The Volkswagen Company has a sizable research and testing effort on alcohol fuels (Bernhardt, 1977b). A 45-vehicle test fleet has used conventional engines with only minor engineering modifications to run about 1.5 million total miles on methanol–gasoline fuel mixtures. In a series of studies using a single-cylinder engine, methanol produced lower exhaust concentrations of air pollutants, improved energy utilization, higher engine output and improved knock resistance. Methanol and methanol blends offer a true alternative to gasoline.

Methanol is a clean burning fuel that increases engine power by about 10 percent. Low volatility of pure methanol leads to poor starting when ambient temperature falls below 8°C, but a volatile additive such as butane or gasoline or a flame preheater can be placed in the intake manifold. After starting, cold-weather driving with methanol is excellent. Because of methanol's lower energy content, mileage is noticeably poorer than that for gasoline.

For pure ethanol, engines were modified by using hot exhaust gases to warm the inlet manifold system. The carburetor had to be adjusted for a higher ethanol flow rate. Preliminary tests have shown good emission behavior as with methanol and even better fuel economy.

About 1 billion gallons of ethanol were produced in Brazil in 1979, enough to reach a 17 percent blend in gasoline. With the government subsidy for ethanol, it can be used as a chemical feedstock. Major products from ethanol are acetaldehyde and acetic acid, which are competitive with that derived from petroleum; butadiene and ethylene are also produced but are cheaper when made petrochemically (*Chemical Engineering News,* April 23, 1979, p. 12). Indian companies also use fermentation alcohol as a feedstock; some products are acetic acid and low-density polyethylene (Davies, 1974).

Volkswagen engineers have experimented with diesel engines using methanol or ethanol. Improved efficiency, less soot formation, and improved odors are expected. Other devices such as a stratified-charge engine and a gas turbine have been tested with alcohol fuels with quite encouraging results. Innovative power units should do a better job with alcohols than with gasoline. A study by Exxon, sponsored by the U.S. Environmental Protection Agency, showed that the cost of the fuel was only 10 to 20 percent of the total annual operating cost of a vehicle; therefore, the motorist should realize that the painful increases in gasoline prices are not changing operating costs very much.

The use of methanol–gasoline blends is still controversial. If the engines are not adjusted, there may be problems with vapor lock and drivability. This is complicated by the distribution problem and the need to keep methanol dry. The advantages are improved thermal efficiency and lower emissions. More study is needed of the phase stability, fuel distribution, and attack of methanol on materials. Some present-day plastic and aluminum parts are corroded by methanol. Widespread use of methanol would require establishment of a distribution and storage system that excludes water because more than 4 percent contamination with water considerably impairs performance.

Gasohol

Gasohol, a blend of alcohol and gasoline, has had intense coverage in the news media and heated discussion by legislators and government agencies. Farmers and their representatives have seized on gasohol as a means to increase demand for grain and thus drive up corn prices. This is highly attractive to a politically powerful group, and the farmers have snatched at all facts that might support their case. However, claims for increased mileage are debatable, and some of the economics being bandied about are very misleading.

There is no doubt that gasohol is a reasonably good fuel. Customers with curiosity, feelings of patriotism, or other motives buy gasohol and come back for more. The debates should not haggle over gasohol because it is a great idea if the price is right. Much of this book advocates alcohol fuels but not alcohol from grain. Grain is but one of many possible sources of biomass that could be converted to alcohols, and grain is not at the top of the list in terms of great potential for cheap fuels.

Ethanol prices are not regulated by law, but gasoline is a regulated commodity. Gasohol is classified with gasoline. The Energy Tax Act of 1978 gives tax relief to encourage use of gasohol but the alcohol must come from renewable resources. Arco Petroleum already has 7 percent alcohols in their gasoline to help reach octane specifications, whereas other companies have up to 5 percent, probably not alcohol from renewable sources. For perspective, we should remember that outlawing tetraethyl lead in the fuel for newer automobiles has required about 6 percent more petroleum to create the blends to get antiknock properties. As the web becomes more tangled, confidence in the wisdom of government regulations weakens.

Heitland, Czaschke, et al. (1977), Yand and Trindade (1979), and Lindeman and Rocchiccioli (1979) have reported in detail on the alcohol–fuel situation in Brazil. There was a program in 1923 for use of ethanol in automobiles, and it sold at one-half the price of gasoline. In 1931, legislation mandated the use of 5 percent alcohol in gasoline with the intention of stabilizing the sugar industry. Similar desires in 1966 led to a law that permitted blending up to 25 percent alcohol in gasoline. Between 1974 and 1976 performance of various blends of gasohol was evaluated. As most Brazilian engines are operated with a rich mixture (high fuel: air ratio in the carburetor), little change in mileage or drivability was detected over the range of alcohol concentrations tested. Octane rating was improved, and performance was fully satisfactory with standard engines up to 20 percent alcohol. A fleet of Volkswagen vehicles has been operated since 1977 on straight alcohol. The compression ratio was 12:1, which is higher than the standard practice of 7:1. There are two ways of handling gasohol: blending tanks at distribution centers from which tank trucks deliver to filling stations and blending as the tank truck is filled.

The output of Brazilian alcohol factories almost doubled to 400 million gal (1.5 billion l) in 1977, partly because idle capacity of existing distilleries was quickly utilized. By the end of 1978, over 200 projects had been approved by the National Alcohol Committee of Brazil to increase capacity by 1 billion gal (3.9 billion l) per year. By 1985 fermentation alcohol will supply about 2 percent of Brazil's energy requirements.

Brazil has several years of experience with nationwide use of alcohol fuels, and the results are very positive. Nevertheless, some caution is advised in translating their results to the United States because warm-climate fuels may not work equally well in very cold weather.

A product called *Hydrofuel* is promoted by United International Research, Inc. of Hauppauge, N. Y. (*Biomass Digest*, June 1978). An additive to gasoline and alcohol in a 3:1 blend keeps water emulsified so that there is no phase separation in cold weather. There would be an economic advantage to emulsifying the water instead of removing it to get absolute ethanol for blending, and a more complete evaluation seems warranted for Hydrofuel.

Pratt (1978) has summarized briefly the old history of gasohol in the United States. Gasoline–alcohol mixtures were widely used in Europe after World War I, and by 1934 consumption was more than 700 million gal (2.6 billion l) annually. Alcohol-burning tractors and trucks were made in the United States for sale overseas. In the U.S. Congress, a bill was introduced in 1933 that would have required all internal combustion engine fuels to contain at least 10 percent fermentation alcohol.

A fermentation alcohol plant in Atchison, Kansas produced 2500 gal (9450 l) of anhydrous alcohol daily and during 1938/39, 20 million gal (76 million l) of gasoline/alcohol blends was sold through independent dealers and farm bureaus in 10 Western and Midwestern states. There were 250 dealers in Nebraska alone. Unreliable supplies of cheap corn, advancing technology for producing high-octane gasoline, and aggressive marketing by the petroleum industry were factors in the demise of this new industry. Nevertheless, gasohol worked well and was nearly a commercial success.

Anderson (1978) presented an excellent status report on gasohol with data on energy content, prices, agricultural impacts, and the like. Except for a few advocates with favorable costs and energy balances based on rather optimistic assumptions, most experts feel that grain alcohol is too costly and has a negative net energy balance because the feedstocks and processing represent more energy investment than the fuel value of the product. Biomass other than grain has a chance of changing this picture.

Alcohol fuels must not be discussed only in the context of today's needs because markets will have expanded before the alcohol plants are built. By 1990 the demand for gasoline may be 115 billion gal/yr (14.3 Quads of energy). For a 10 percent blend of alcohol in gasoline, this 11.5 billion gallons translates to very large plants. A fairly large factory for alcohol by fermentation of biomass would produce about 30 million gallons (110 million l) of alcohol per year; thus 350 factories would be needed if all the blended alcohol came from fermentation. If the broth is 10 percent ethanol, each plant would ferment 300 million gallons (1130 million l) per year. These sizes are fantastic in terms of present-day fermentation capacities. This much fermentation broth would cover a football field to a height of 1300 ft (400 m).

Diesel Fuels

Solutions or suspensions of lignin in alcohols or other solvents could be good fuels for diesel engines. Smith (1977) mentions work he performed in 1934 on suspensions of powdered petroleum coke in heavy fuel oil using rosin soap stabilizers. This was not original because the Germans developed similar fuels for their submarines in World War I because of petroleum shortages. General Motors has reported on recent successful testing of powdered coal suspended in heavy fuel oil. The first diesel engines could run on air suspensions of powdered coal. It would seem that biomass fuels of almost any nature could be used in diesel engines and the decisions to use them would be based on cost, stability of the suspension, and heating value.

Conversion of Alcohols to Hydrocarbons

The Fischer–Tropsch process dating from 1933 uses catalysts and reducing conditions to convert carbon monoxide to a mixture of hydrocarbons. In analagous reactions, alcohols can be reacted with themselves to form hydrocarbons. Pearson (1979) describes the reaction of methanol with phosphorus pentoxide at 185 to 300° C to give alkanes, cycloalkanes, alkenes, and aromatic hydrocarbons. There is no need to separate these compounds if the goal is a fuel similar to gasoline. With ethanol, the weight yield of ethylene is about 65 percent, which is close to the theoretical yield. Water slows the rate somewhat, but yields are still excellent with 80 percent ethanol. The catalyst is composed of phosphorus oxides; thus regeneration becomes more costly as more water must be removed. The Mobil Oil Company with support from the U.S. Department of Energy is developing a process for making high-octane gasoline from alcohols with selective zeolite catalysts (Voltz and Wise, 1976). With methanol, 36 percent yield of hydrocarbons was obtained, while the yield with absolute ethanol was 60 percent. Water is a by-product; thus it is not necessary to use a dry feedstock. The reactions are exothermic; therefore, maintenance of high temperature will not be consumptive of outside fuels. The pore size of the catalyst keeps the hydrocarbons in the product from reaching too high a molecular weight because large molecules are retained and cracked to smaller size while smaller molecules can escape from the pores. Mobil estimated in 1976 that the conversion would cost about 5¢/gal of product. However, 2.37 volumes of methanol or 1.5 volume of ethanol are required to produce one volume of gasoline, so feedstock costs will keep prices high for this product.

Multiple Products

Ideally, biomass could feed the world and provide fuel. Fractionation to obtain edible portions is implicit in some of the current proposals such as corn for food with corn stover for energy and sugar from cane with bagasse for energy. Another attractive idea is to squeeze the protein-rich juices from fresh alfalfa and to send the rest for conversion to fuel (Tsao, 1978). Alfalfa juices coagulate very soon after cutting, so a harvesting machine followed by a roller unit for immediate squeezing could be a technological breakthrough. This legume has excellent ability to fix nitrogen and thus needs no fertilizer, and productivity is fair although inferior to corn or sugarcane. Alfalfa also appears attractive for rotation with corn to replenish soil nitrogen. Food fractions are much more valuable than energy, and credits for food could make biomass for energy financially viable.

Bioconversion processes will have excess microbial cells that might be used as a food. However, anaerobic digestion with elective mixed cultures results in a variable cell mixture that is poorly defined. It could prove very difficult to obtain approval to feed these cells to animals that become human food. In contrast, fermentation with essentially pure cultures whose properties are consistent and defined would provide excess cells for human or animal nutrition. For example, some yeasts are well-known human food.

It is sobering for a biomass-to-energy advocate to realize that a number of compounds being considered as fuels have already had several years of testing as fermentation substrates for single-cell protein. A unit weight of methane yields 0.85 units of dry cells plus the liberation of some metabolic heat, and the cells can be more than 80 percent protein. Ethanol and methanol have been used as fermentation substrates. A number of organisms can produce protein from cellulose as the main carbon source.

The current status of single-cell protein by fermentation of various organic compounds or of waste materials such as cheese whey is reviewed by Litchfield (1978). In the United States, soybeans are a low-priced protein source that provide tough competition for microbial protein. Nevertheless, the point is raised as to whether it is more economical to recover such materials as methane, methanol, or ethanol or to convert them directly to single-cell protein. Even more questionable is to produce fermentable sugars from cellulosic biomass and to go to ethanol when it is easy to convert sugars to single-cell protein. The answer is reasonably simple because the current demand for single-cell protein is small. If there were vast markets, it would make economic sense to deemphasize energy and to concentrate on food. However, the potential resources for cellulosic biomass are enormous and fairly well matched to our energy needs. Even with recycle and reuse of microbial cells, there will be very large amounts of spent cells. Furthermore, fermentation is uneconomic if prolonged to exhaust the substrate; thus waste streams with fairly high nutrient levels will be generated. Biological waste treatment will produce more microbial cells of potential value as feed for animals or even for recycle as fermentation nutrients. Adjustments of the processes to optimize profits as markets develop for single-cell protein are possible; thus some flexibility should be included in the design.

Tapping Plants for Fuel Products

Instead of harvesting whole plants, the nuts, the fruit exudates, or the clippings can be collected periodically. It is particularly attractive to consider exudates that are potential fuels because processing costs might be very low. An existing industry for oleoresins from pine trees illustrates the pitfalls for tapping plants because the large amount of labor required keeps prices high, and the industry has been in a prolonged period of decline.

Nobel laureate Melvin Calvin (1976, 1978) has been studying plants that produce hydrocarbonlike materials (Nielsen, Nishimura, et al., 1977). This is an outgrowth of studies during World War II when the United States had severe shortages of rubber because Japan controlled many rubber plantations and threatened the Pacific sea routes. It is doubtful that rubber plantations could be economical in the United States even if climate permitted because tapping and collection at our labor rates would be too costly. Nevertheless, in wartime emergency it appeared feasible to grow *Hevea brasiliensis,* which exudes a material similar to natural rubber and could be a satisfactory substitute. At the end of the war, the research was terminated. Calvin's group has grown several plants that produce hydrocarbonlike materials and has analyzed benzene and acetone extracts of air-dried, ground materials. Extractables constituted up to 40 percent of the weight of seeds and up to 25 percent of the weight of some leaves. Whole plants had a wide concentration range, but some had over 15 percent extractables. Of more interest is the composition of latex that can exude from the plants. In most of the plants tested, latex was predominantly sterols with 75 percent by weight of the latex sometimes noted. One plant, *Hevea brasiliensis,* was 87 percent rubber and only 1 percent sterols in the latex.

Assuming 10 tons of biomass per acre on a dry basis and 10 percent "oil" in the plant, one ton equaling seven barrels could be harvested. If cultivation costs were $150/acre, the unprocessed oil would cost a little over $20/barrel. These cost estimates by Calvin's group must be criticized if based on exudates because "oil" yields would be lower than the amount obtained in the laboratory by exhaustive extraction. Collection costs for exudates would probably be prohibitive anyway unless automated harvest proved practical. Whereas *H. brasiliensis* latex is highly viscous, other species have free-flowing latex more amenable to automated collection. Calvin's group with test plantings of *Euphorbia lathyris* and *Euphorbia tirucalli* has shown yields translating to over 8 barrels per acre (8 Bbl/acre), which they deem a favorable indication for "petroleum plantations." Total plant harvesting was mentioned as a way to get both "oil" and residual biomass to be processed for additional credits.

In perspective with other energy options, products from latex present a hazy picture. The cost of $20/Bbl is probably far too optimistic, but perhaps it is unwise to compare latex and its wealth of interesting biochemicals with petroleum, which is mostly aliphatics plus some aromatics and traces of exotic compounds. As chemical feedstock or as a starting material for rubber, special latex could be economical. Yields are not very attractive at only 1 ton/acre·yr compared to biomass, which, of course, is not valuable latex. Common tree species reach 10 tons/acre·yr, several plants can yield 20 tons, and algae have great potential if light and CO_2 limitations can be lifted. Even at high yields, enormous growing areas will be required; thus it is very unlikely that low-yield hydrocarbonlike materials could be obtained without overcommitment of good land and valuable water.

Loomis (1978) has severely criticized the proposals to grow hydrocarbon-producing plants on marginal lands. Calvin's report of 10 Bbl of extractable oil per acre is the basis for suggesting yields of 10 to 20 Bbl/acre in areas of the southwestern United States that have 5 to 20 cm of annual rainfall. Back calculating from these yield assumptions, there would need to be 19 tons/acre (42 tons/ha) of above-ground biomass that exceeds the high yields of similar plants with ideal rainfall. The nitrogen requirement is far above that available in desert soils and even that of many good organic soils.

Typical desert plants survive because of adaptation to highly restricted growth. Loomis feels that the marginal arid lands could support only very low annual crop yields, and there may be a calculation error in the high estimates that have had so much publicity. These lands could be irrigated and fertilized to obtain high yields, but then food crops of greater value than the energy crops could be grown.

A quite favorable cost estimate for producing *Euphorbia* showed the cultivation costs per unit of land area to be close to those of corn or alfalfa (Mendel, 1979). However, yields on a commercial scale would be about 6.8 dry tons/acre (15.2 tons/ha) instead of 17 dry tons/acre (38.1 tons/ha) experienced on carefully tended experimental plots. The preliminary production cost estimate of oil from *Euphorbia* was $33.76/Bbl for a factory handling 3000 tons of biomass per day.

Jones, Kohan, et al. (1979) have refined the preliminary cost estimates for oil from *Euphorbia* by designing a plant and developing the associated costs. With a feedstock cost of $20/dry ton, the oil would have to sell at almost $50/Bbl. This is not as far out of line as it would have seemed a few months ago because OPEC price increases have been so frequent. Nevertheless, other biomass options appear to be more economical. *Euphorbia* oil should be considered for specialty uses, and not as a high-priority approach to finding cheap fuel.

The Brazilian tree *langsdorffia* from the genus *Compaifera* grows wild. Mature trees yield about 12 gal (45 l) per year of hydrocarbon oil that could substitute for diesel fuel. Calvin feels that selective breeding and improved methods for tapping the trees could give much better yields. These trees would grow only in the very warm areas of the United States. Its wood has some use in Brazil, and a few trees are tapped for their hydrocarbon exudate. There is an undocumented report of testing this material directly as a diesel fuel. After an hour or two, deposits of gum stopped the engine. Nevertheless, fractionated or refined oils from exotic plants could become very important, first as substitutes for petrochemicals and perhaps later as cheap fuels.

A plant exudate that may be used in simple automotive engines is the oil from eucalyptus. Takeda (1979) has compared gasoline, 70/30 oil to gasoline, and 100 percent eucalyptus oil and found them to have about the same power output and fuel consumption. The cost was estimated at $35/gal ($9.25/l) for the oil today, but development work on cultivation of the trees and extraction of the oil could lead to a competitive price.

Although fuels from oily plants are not likely to be competitive with fossil fuels in the foreseeable future, chemicals from plants are already achieving commercial importance. Guayule and jojoba, which were mentioned briefly in the previous chapter as feedstocks, are both being grown in large amounts. During World War II, when the United States was cut off from its suppliers of natural rubber, there was a crash effort to develop native sources. Many species were evaluated, and *Parthenium argentatum* (guayule) appeared best. The other species deserve reevaluation because circumstances are different now.

Guayule rubber can substitute for natural and synthetic rubber (*Chemical Engineering News*, August 28, 1978). Goodyear has made radial tires from guayule rubber that pass the Department of Transportation high-speed endurance tests. Mexican government officials have exhibited tires from guayule after 50,000 miles of wear; there were no visible defects. Campos-Lopez, Neavez-Camacho, *et al.* (1979) reviewed the prospects for guayule.

Benedict and Inman (1979) have reviewed current research on hydrocarbon-producing plants. The U.S. Department of Agriculture and the University of Arizona have developed *Simmondsia chinensis* (jojoba), a native plant of the Arizona desert, as a commercial crop. The seeds contain an oil that can substitute for sperm whale oil. Several thousands of acres in Arizona and California support a growing new agricultural industry. The oil is in great demand, especially in Japan.

Gum Naval Stores

Pine trees in wild forests are tapped for their exudate from which turpentine and rosin are produced. As the rosin was widely used to seal and preserve planking in wooden ships, the products were called "naval stores." This industry is in a sad decline. Turpentine laborers walk from the nearest roadway or trail into the forest and approach each tree, take its few ounces of yield, and spray the wound with acid to stimulate flow. When their buckets are full, they walk back to dump them into the gathering vehicles, and another collection trip is started. There are other hand operations such as slashing the tree and hanging a trough. Attempts at automation have had little success. Added to high labor costs are investment costs. Pine trees reach maturity in about 25 yr and produce gum for about 8 yr. Pine gum competes with gum from petroleum and has lost ground steadily.

The gum naval stores industry would have perished long ago were it not for the other uses of the trees in lumber and pulpwood. If trees become the biomass for energy, turpentine and gum could be available in highly significant quantities.

In the search for better, safer sprays than sulfuric acid for increasing flow from the tree wounds, many organic chemicals were tested. The pesticide, paraquat, was ineffective on flow, and the experiment was deemed a failure. Later when the trees were cut down and sectioned, copious deposits of oleoresins were found near the wounds. Further testing of paraquat showed that it greatly stimulates oleoresin production within the tree. Up to 10 times the amount of oleoresin of an untreated tree is found after paraquat treatment.

The entire tree above ground is considered for energy. As the tree is harvested, the oleoresins are also collected. Concentrations of oleoresins and volatile organic chemicals are high in the pine needles and small branches, and the main additional cost in their stimulation is the application of paraquat several months before the trees are felled. Stimulation of oleoresins acts powerfully to attract insect pests. This is bad for lumber and may discolor pulpwood but is of little consequence to biomass for energy.

Distillation of pine gum gives gum spirits of turpentine or simply turpentine, an excellent solvent and a thinner for paints and varnishes. Oil-based paints have shifted to petroleum-derived solvents of lower cost and less objectional odor. Oil-based paints have also lost much ground to latex paints.

Old stumps can be a source of gum naval stores. Collected stumps are water washed, chipped, and extracted with an organic solvent. Distillation recovers the solvent and provides a turpentine fraction. Other pine chemicals are obtained as by-products.

Sulfate turpentine is a by-product of sulfate pulping for paper. Pulpwood trees are generally smaller and younger than those tapped for turpentine. Cooking with alkali and sulfides breaks the structure of wood, leaving fibers high in cellulose. Vapors from pulping are condensed, and an organic layer must be treated to remove odiferous sulfides to yield turpentine.

Gum rosin is the residue from distillation of the exudate of pine trees. It is used to size paper and fabrics. Rosin is a mixture of organic acids, predominantly abietic acid, which has three rings. The other acids are less stable than abietic and tend to rearrange to form abietic acid. As the diversity of acid composition decreases, rosin is more easily crystallized, and this is undesirable for its applications. Rosin floats to the surface in steps subsequent to sulfate pulping of wood, but cooking results in acid rearrangement to give an inferior grade.

Pulping operations are not oriented toward gum naval stores, and major losses are tolerated. Much is left in the forest as about 40 percent of the tree is trimmed. Nearly half of the more volatile components escape while the logs are stored awaiting pulping. More losses during debarking and chipping and poor yields of naval stores during pulping result in only 20 percent or so of theoretical yields. Crude gum naval stores contain components of relatively high value. Some structures are shown in Fig. 3.3.

β-Pinene is more reactive than α-pinene; thus it commands a higher price. If ample, reliable amounts were available at reasonable prices, their useful structures as chemical intermediates would attract wide markets for pine chemicals.

Trees as an energy crop have potential for by-products of considerable value. Other species than pine have volatile constituents, but there is a good technological base for gum naval stores. By breeding productive species and stimulating production with compounds such as paraquat, 10 percent or more of the biomass could be oleoresins. Liquid fractions are established commercial solvents, and their use as fuels might be possible. However, the chemical structures of pine chemicals indicate far better uses as intermediates for synthesis than as cheap fuels. By-product credits from large amounts of oleoresins would mean attractive prices for the remainder of the tree as a feedstock for conversion to fuels.

It should be noted that there is a commonality between production of hydrocarbons directly by the plants proposed by Calvin and the production of turpentine by pine trees. Turpentine and other gum naval stores are probably too valuable for use as motor fuel, but the yield from paraquat-treated forest could be over 5 tons/acre · yr. This is a by-product to the tree biomass but is greater than the total plant yield targeted by Calvin. If direct production of a hydrocarbon fuel by plants is to become practical, gum naval stores appear most promising.

The potential of obtaining both oleoresins and fermentable sugars from pine trees was investigated by Morris (1978). The Purdue process (see Chapter 7) for

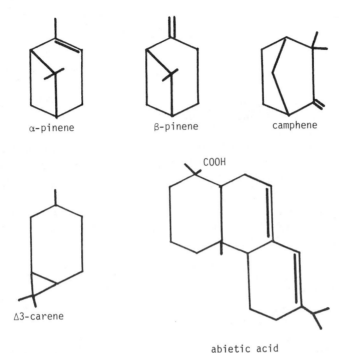

α-pinene   β-pinene   camphene

Δ3-carene

abietic acid

Figure 3.3   Pine chemicals.

pretreatment of cellulosic biomass was run on loblolly pine chips that were high in oleoresins because of treatment of the trees with paraquat. Morris found that none of the cellulose solvents were particularly helpful in extracting oleoresins; thus solvent recovery of terpenes, lipids, and rosin prior to the Purdue process was recommended. Based on a 1000-ton/day input of pine chips, recovery of oleoresins would add about $12,000/day of salable products with an additional operating cost of only about $2000/day. The value of the other products, fermentable sugars and lignin, would be roughly $89,600/day. Obviously, valuable by-products such as oleoresins can contribute significantly to profitability, and yields higher than those used by Morris are quite possible.

O'Neil (1978) has reported on economic feasibility of a new industry based on fuels and chemicals from paraquat-treated coniferous trees. Basic features are short-rotation forestry, effective treatment with paraquat, utilization of the complete tree, extraction of oleoresins, and conversion of remaining biomass to products. Complete-tree utilization leads to much greater production of chemicals than obtained as by-products of pulping. Mechanized harvesting followed by immediate chipping with either rapid processing or enclosed storage avoids the great losses of volatiles that results from the customary felling of trees and long-term storage of logs or chips. Pools of oleoresins just beneath the bark are lost in the debarking step of lumber or pulp manufacture but are retained in the Georgia Tech (Georgia Institute of Technology, Atlanta) process.

O'Neil provided the following projection:

Current production in the United States   18,500 tons of gum rosin
                                                 92,000 tons of wood rosin
                                                 172,000 tons of tall oil rosin

Assume (1) 30 percent of the trees can come into a paraquat program and (2) 10-fold increase in rosin in treated trees; then the total increase is 775,000 tons/yr for an increase in income of $383 million for the industry for rosin. In addition, total turpentine would increase from 22 million gal (83 million l) to 82 million gal (309 million l), with an increment of $45 million. While the prices may fall somewhat with such dramatic increases in supply, rosin and turpentine are both minor factors in a large market including petrochemicals that have an upward price trend. Certainly, the first few plantations for energy could expect very sizable by-product credits if paraquat-treated pine trees were the crop.

Uses of Lignin

Just in terms of its fuel content, lignin is worth about 2.5¢/lb. Its ratios of C:O and H:O are higher than those of carbohydrate fractions of biomass; thus lignin has more potential for oxidation. However, lignin is composed of aromatic chemicals that sell at attractive prices. Drew (1978) has reviewed the status and potential of chemical uses for lignin.

Chemicals now made from lignin include vanillin (the basis for vanilla flavor), dimethyl sulfide, methyl mercaptan, and dimethyl sulfoxide. None of these has potential markets anywhere near approaching the massive amounts of lignin that will be available from biomass processed for energy. High-temperature–high-pressure reactions have been patented for hydrocracking of lignin to phenol, methyl phenols, and higher phenols. Yields need improvement, but such processes deserve attention. Hydrogenation of lignin gives the following approximate yields:

Gases	25	percent
Neutrals	14	percent
Phenols	37.5	percent
Catechols	9	percent
Overheads	2	percent

Lignin isolated from cellulosic biomass or recovered from the residue after bioconversion might be used in commercial thermosetting resins, of either amino or phenolic type. Lignin already appears competitive in cost with these resins and thus has excellent prospects as materials made from petroleum soar in price. However, the world market for amino resins was 2.5 million tons and for phenolic resins was 1.4 million tons in 1975. Even with dramatic growth, these markets are miniscule compared with the amount of lignin that may become available.

Pulp and paper companies have been trying for many years to develop markets for lignin, but with little success. Their lignin is damaged by pulping and has lost much of its reactivity. Lignin from mild processing as in the bioconversion of plant matter should be much more reactive and thus may capture much of the existing markets. These are:

1. *Dispersants.* When adsorbed on fine particles, the charge of lignin chemicals causes electrostatic repulsion to hinder coagulation.

2. *Binders.* Lignin forms chemical and physical links to other molecules.

3. *Sequesterants.* Hydroxyl, sulfonic, and carboxyl groups of lignin chemicals complex metal ions.

4. *Passivators.* The sequestering properties and the coating of surfaces by lignin or its metal complexes provide resistance to corrosion.

5. *Emulsification.* Electrostatic repulsion and the coating action on small droplets stabilize emulsions.

6. *Humectants.* Lignin holds water and keeps formulations moist.

7. *Cement additives.* Lignin prevents agglomeration during grinding and retards setting of Portland cement.

8. *Drilling muds.* Dispersion and thickness are improved.

9. *Fillers.* Unreactive fillers such as wood flour add inexpensive bulk to plastic resins. Undamaged, reactive lignin would have much more value.

A proposed use for sulfonated lignin is in tertiary oil recovery, where surfactant properties aid the penetration of water to displace oil from porous strata. Reactive lignin might be converted to detergents better than sulfonated lignin from pulping. The market would be roughly 1 million tons/yr.

Forintek Canada Corporation in Ottawa has developed plywood and chipwood boards using lignin as a substitute for phenolic resin binders. The conditions of the pressing step have to be modified, but the product meets all specifications for this type of material. Lignin from sulfite pulping is damaged so severely that it is not suitable as a binder. Lignin from kraft pulping is satisfactory, and lignin from the mild treatments proposed for making biomass fuels should be superior. The author visited the laboratories to observe pressing of boards with lignin binder and was told that phenolic binder costs 60¢/lb. This price is expected to rise very soon. Roughly a dozen fuel–alcohol plants each producing 20 million gal/yr could satisfy the market for lignin as a wood binder, and the by-product credit would be more than the value of the ethanol. Whereas the amount of lignin from a national program for fuel alcohol would overwhelm the plywood market, this acceptance would stimulate the search for other applications. With lignin revenues defraying part of the plant costs, fuel-grade ethanol could be sold well below the price of gasoline. There are many patents for producing organic chemicals by hydrogenation of lignin, but production costs are high and separation of the various products is expensive. Improvement of technology makes production of chemicals from lignin more likely.

Conversion of lignin to useful chemicals by biological means is difficult. Anaerobic microbial attack that might allow accumulation of valuable intermediates has not been demonstrated. Aerobic metabolism of lignin has slow rates, and oxidation of the aromatic rings greatly predominates over splitting of the ether linkages between the rings. The small organic compounds that are produced do not accumulate because their utilization by microorganisms is very rapid.

Enzymes attack lignin and modify side chains with no depolymerization of the ring structure. Discovery of an enzyme or a microbial process to produce smaller, aromatic compounds from lignin would be a great advance.

Reddy has isolated 22 bacterial strains that degrade lignin. (*Chemical Engineering News*, November 21, 1977, p. 8). One goal of the work is predigested feed for cattle and sheep that would make use of lignin while also allowing more efficient use of cellulose by weakening its protection by lignin.

Complete microbial metabolism of lignin occurs constantly in nature to continue the cycle of carbon. However, rotting of trees and stumps can take years. About the only commerical possibility of fermenting lignin by known technology would be to grow microbial cells for single-cell protein. However, the slow rates, use of a solid substrate, and problems in product purification from unreacted substrate are bad features. In summary, using lignin in its polymerized form seems wiser than trying to break it down.

Fuel Cells

Fuel cells are catalytic devices for carrying out chemical reactions so that electrons can be captured or released at electrodes. The terms "biological fuel cell" or "biofuel cell" describe a system where electrochemically active substances are converted by cells or enzymes into molecules that power a fuel cell. Electric potentials in microbial systems were observed early in this century, and proposals for biological fuel cells have been around for several years (Brake, Townsend, et al., 1965). However, current densities are low and electrodes are easily fouled. The types of electrodes used in the laboratory contain rare metals such as platinum and palladium; thus the cost would be prohibitive for large units unless effective, cheap substitute electrodes are devised. The biological fuel cell with electrodes in contact with cells may be of interest in the distant future, but biochemical fuel cells operating with clean feeds seem more practical now.

Silverman (1976) has outlined current research on biochemical fuel cells. The biological conversion rate is controlled by cell or enzyme activity, pH, temperature, reagent concentrations, and concentrations of inhibitors.

A method for getting electricity from glucose is to use *Clostridium butyricum* to form hydrogen and reacting it with air or oxygen in a fuel cell. The reactions are

$$
\begin{array}{ll}
\text{Anode} & H_2 \rightarrow 2H^+ + 2e \\
\text{Cathode} & O_2 + 2H_2O + 4e \rightarrow 4\ OH^- \\
\text{Net} & 2H_2 + O_2 \rightarrow 2H_2O
\end{array}
$$

The maximum current density is about $1\ mA/cm^2$, which means that this type of fuel cell must have large electrode areas to provide significant power.

Another method for utilizing glucose is enzymatic conversion to gluconic acid in the presence of an electron acceptor such as methylene blue. Reduced methylene blue is reoxidized electrochemically. An enzyme cofactor, in this case FAD, is required, as is typical of biochemical oxidations and reductions. However, the cofactor is regenerated so that only small concentrations are needed.

An interesting configuration has the fermentation chamber separated from the fuel cell by a semipermeable membrane. Oxidizable small molecules pass through the membrane while cells and large molecules are retained.

Suzuki, Karube, et al. (1978) have studied biochemical fuel cells with *Clostridium butyricum* immobilized in a cell with electrodes of platinum and of lead oxide. Wastewater from an alcohol factory was the fuel. As current was generated, the biochemical oxygen demand (BOD) of wastewater circulated through the cell decreased.

Fuel cells should see increasing use because of their high efficiency in converting from chemical to electrical energy. Their role in fuels from biomass could be important, but probably not in direct conjunction with a biological reaction. More likely is the production of a suitable fuel from biomass and a separate fuel cell operating at temperatures and other conditions that are optimum.

Waxes

Reese (1976) has reported on proposals to obtain wax esters from zooplankton. It is estimated that at least 3×10^8 tons of wax esters are produced in the oceans of the world each year. In Arctic and Anarctica waters, up to 75 percent of the dry weight of zooplankton may be wax esters. Oil slicks at sea may result from zooplankton, although others attribute them to phytoplankton because certain marine algae are known to produce waxy hydrocarbons when their medium is deficient in nitrogen. There is no practical technology at present for recovering wax esters from nature or for producing them in culture. However, these compounds are possible targets for the future.

Furfural and Its Derivatives

In the presence of acid and heat, sugars condense to heterocyclic compounds. Pentose sugars yield furfural, and hexose sugars yield hydroxymethyl furfural.

The structures are

Pentose $C_5H_{10}O_5$ ⟶ Furfural

Hexose $C_6H_{12}O_6$ ⟶ $HOCH_2$ Hydroxymethyl furfural

Numerous plant residues such as oat hulls, corn cobs, bagasse, cottonseed hulls, rice hulls, olive pits, and scrap wood have sufficient pentosans (hemicellulose) to be used as commercial feedstocks for manufacturing furfural. Hydroxymethyl furfural is not an important commodity because it is priced too high and because there are better uses for hexose sugars. Lipinsky, Birkett, et al. (1978a) have collected information about furfural manufacturing and found that yields are from 50 to 70 percent of theoretical based on pentosan content. Ready availability of large amounts of feedstock at low cost is more important than pentosan content.

The total market for furfural approaches 100,000 tons/yr, an amount far too low to fit into the energy picture. However, the present cost of over 50¢/lb puts furfural out of contention for many potential uses. When a massive alcohol fuels program is implemented, there will be tremendous amounts of pentosans and pentose sugars available so that improved technology could provide furfural at a much lower cost. This would make furfural very attractive for old uses in markets that have been taken over by petrochemicals and for many new uses.

Furfural is used as a solvent for purifying lubricating oil and rosin, for extractive distillation of butadiene and isoprene, and in polymers. Reduction gives furfuryl alcohol, which is used principally for making resins used in preparing molds in foundaries. Hydrogenation of the ring and removal of the aldehyde group gives tetrahydrofuran, an excellent solvent. Almost any long dicarboxylic acid can serve in the manufacture of nylon, and for many years tetrahydrofuran was a prime reactant for the DuPont Company. Low-cost butadiene displaced tetrahydrofuran commercially. The reactions are

$+ O_2 \longrightarrow HOOC—CH_2—CH_2—CH_2—COOH$

Tetrahydrofuran   Adipic acid

$CH_2{=}CH—CH{=}CH_2 + O_2 \longrightarrow HOOC—CH_2—CH_2—COOH$

Butadiene   Glutaric acid

The production of furfural is described by Paturau (1969). The major producer is the Quaker Oats Company. Plant residues are loaded into digesters with steam and sulfuric acid at 153° C and about 4 atmospheres of pressure. Vapors from the reactor carry out furfural over a 6- to 8-hr period. By-products include small amounts of methanol and methyl acetate. Another by-product, acetic acid, is present in larger amounts than is furfural. If the feedstock were relatively pure pentosan from an alcohol fuels factory, little acetic acid would be expected because most of it comes from lignin. Digester residue is squeezed to remove water before being used as boiler fuel. The squeezings have potential value because sugars are present.

Research is needed on deriving furfural from pentose solutions. If the extracts of pentose sugars from fractionation of biomass could be kept in concentrated form, there would be a good chance of economical production of furfural. Expenditure of energy to concentrate pentose solutions penalizes the process.

Xylitol

Hydrogenation of xylose gives xylitol, a low-calorie sweetener. Factories for alcohol fuels could derive some profit from sales of xylitol, but the market is small. Proponents of xylitol point out that its appearance in nature and in some foods should lead to easy approval by the FDA. However, recent findings that xylitol causes stomach cancer in rodents seems to have squelched interest.

Carbon Dioxide

The gas evolved from alcohol fermentation is carbon dioxide. It is ideal for manufacturing dry ice, but the market is tiny compared to the amounts of CO_2 that will become available from a fuels program. Another relatively small market could be in greenhouses or on hydroponic farms where the productivity of plants is increased manyfold by growing in high concentrations of CO_2. If an inexpensive, transparent enclosure could be devised, all the CO_2 from bioconversion could be recycled to the growing areas.

Acrylic Acid

Acrylic acid is an industrial chemical important in the manufacture of polymers. The reaction is

$$CH_2{=}CH{-}COOH \longrightarrow \left(\begin{array}{c} CH_2CH{-} \\ | \\ COOH \end{array}\right)_n$$

      Acrylic acid   Polyacrylic acid

Often copolymerized with other compounds, it provides the carboxyl side chain. Acrylic acid from a fermentation process could have large markets as a substitute for petroleum-derived acrylic acid.

There are several possible fermentations routes. A group at the Massachusetts Institute of Technology, Cambridge (MIT) (Wang, Cooney, *et al.* 1978a) have a program for exploring the feasibility of commercial fermentations for acrylic acid. It occurs naturally as an intermediate, usually as the coenzyme-A derivative, an analogy with most biochemical pathways involving organic acids.

Clostridium propionicum has been used by the MIT group to ferment β-alanine; small amounts of acrylic acid were detected in the broth. The only other substrate to give similar results was α-alanine, and cheaper substrates are needed for a practical process. Another organism, *Megasphaera elsdenii*, consumes lactic acid and forms traces of acrylate. However, acetate and propionate are the main products, and there is little hope of diverting very much of either to acrylic acid.

Other Chemicals

A number of fermentation biochemicals that have not found commercial uses should be reconsidered for possible fuels or feedstocks. Diacetyl, acetylmethyl carbinol (acetoin), and 2,3-butanediol are

$$
\begin{array}{ccc}
\underset{\text{Diacetyl}}{CH_3-\overset{\overset{O}{\|}}{C}-\overset{\overset{O}{\|}}{C}-CH_3} &
\underset{\text{Acetyl methyl carbinol}}{CH_3-\overset{\overset{O}{\|}}{C}-\overset{\overset{H}{\underset{|}{O}}}{C}H_2-CH_3} &
\underset{\text{2,3-Butanediol}}{CH_3-\overset{\overset{H}{\underset{|}{O}}}{C}H_2-\overset{\overset{H}{\underset{|}{O}}}{C}H_2-CH_3}
\end{array}
$$

These compounds are liquids that might be used as motor fuels. However, dehydration could give unsaturated hydrocarbons that would be good fuels or excellent feedstocks for polymers and synthetic rubber. Laboratory-scale production of acetoin plus diacetyl was reported by Gupta, Yadav, et al. (1978). The yield was 35 g of acetoin from 50 g of sucrose.

It may be possible to extract some of these compounds from fermentation broths with immiscible solvents. This would be much more energy efficient than distillation and could lead to better economics than those for ethanol.

Organic acids have also been proposed as energy products from biomass. The Kolbe electrolytic process converts acids to hydrocarbons. Organic acids are higher boiling than is water; thus recovery by distillation must first volatilize all the water. Not only is distillation expensive, but destruction of some acids is possible. For example, substituted acids such as lactic acid or unsaturated acids as fumaric acid can react with themselves to form tars. Membrane processes have been proposed for recovering organic acids and seem to have merit, but no commercial applications have yet appeared.

The calcium salts of some organic acids are sparingly soluble and could be precipitated from a fermentation broth. Addition of sulfuric acid would regenerate the organic acid while precipitating calcium sulfate, but this is expensive because of raw materials cost and disposal costs for the messy sludges. Weiss (1976) has suggested the use of anaerobic digestion to produce methane and organic acids with recovery of the calcium salts of the acids. Pyrolysis of calcium acetate gives acetone and calcium carbonate that can be recycled to the acid-recovery step. Other organic acids yield analogous ketones. A typical reaction is

$$Ca(-O\overset{\overset{\displaystyle O}{\|}}{C}-CH_3)_2 \xrightarrow{\text{heat}} CH_3\overset{\overset{\displaystyle O}{\|}}{C} CH_3 + CaCO_3$$

Calcium acetate   Acetone   Calcium carbonate

If algal biomass is used, CO_2 and organic materials from digestion and product recovery can be recycled to the algal ponds. The heating value of acetone is inferior on a per-carbon-atom basis to that of ethanol; thus acetone yields less of the calorific value of the feedstock. Acetone and other ketones have more value as solvents and chemical intermediates than as fuels.

Glycerol from Algae

Biological production of glycerol would spare some petroleum and represent a small energy saving. Halophilic green algae of the genus *Dunaliella* accumulate glycerol intracellularly to adjust osmotic pressure. The Israelis are attempting to develop a commercial process, and their work is proprietary. Williams, Foo, et al. (1978) have reported some features of a *Dunaliella* process on a laboratory scale. The organism is 10 to 12 microns long and 8 microns wide with no cell wall. It thrives in 2.5-M sodium chloride (5 times seawater concentration), but this does not guarantee protection against contamination because a fungus was observed growing on glycerol leaking from the algae. Brine filtrate from harvested cells is recycled several times after supplementation with nutrients containing nitrogen and phosphorus.

Although a membrane process has been proposed, the existing recovery process is based on concentration of glycerol by evaporation of water. Recovery from dilute solution is too costly to be practical. Brine solutions would pose serious purification problems, so it seems prudent to ignore the glycerol leaked from the cells and to isolate that contained within them. Some strains are more leaky than others, but some strains release very litle glycerol to the medium. Titers are about 1 g/l, bound glycerol.

Dunaliella have a specific gravity of 1.08 to 1.1 and settle at a rate of 1.6 cm/hr in water but float in brine. Laboratory collection is fairly easy, and the cells can be washed quickly to remove adhering salt. Lysis of the cells in a small amount of water gives a concentrated glycerol solution, and the spent cells have value as animal feed. With a current price of glycerol of about 50¢/lb, there is some margin for using expensive purification steps. However, the best chance seems for low technology–low investment processes.

Plastics

The main types of plastics are thermoplastics, which soften with heat, and the thermoset plastics, which react to become tough and durable. The thermoplastics have large markets and are relatively cheap, 30 to 35¢/lb. The thermoset plastics range from $0.40 to $1.00/lb and had sales of about 1.4 million tons in 1979. Using fractions from biomass to substitute for plastics would mean very attractive credits for a conversion factory. However, the total plastics market is small compared to the fuels market, so only a relatively minor amount of biomass would saturate the demand.

4

Thermochemical Processes

The simplest way to obtain energy from biomass is to burn it. At $2 to $3 per million BTU, the cost of burning wood is competitive with that for fossil fuels. However, biomass is bulky, relatively hard to handle, difficult to burn efficiently, and produces ash. As alternatives to burning, biomass in several different processes at elevated temperatures yields clean convenient fuels. It takes energy and investments to change one fuel to another, better fuel; thus the costs on a BTU basis must be significantly higher for fuels from thermochemical conversion of biomass.

Biomass Pretreatment

Biomass as collected contains moisture and is contaminated with dirt and various inorganic materials. The possible pretreatment processes prior to thermochemical processes are cleaning to remove inert matter, drying, and subdivision to small sizes. Aquatic biomass is very wet, and most other plant materials have 40 to 70 percent water. Thermochemical processes will utilize temperatures far above the boiling point of water, which means that the heat of vaporization of the water must be supplied. A problem is physical size of the biomass because reactions take place at the surface or with gases driven from the material. For a given weight, large chunks have relatively little surface, and heat penetration and ease of escape of gases suffer from having longer paths.

Cleaning is costly because of losses of biomass, operating expenses, and disposal. For example, water sprays would have some costs associated with the makeup water, pumping costs, wastewater recycle costs, and handling costs for the collected dirt and junk. It is unlikely that cleanup can be justified because it is cheaper to pay for increased maintenance in subsequent processing steps and heat lost in the hot, inert residue. In contrast, biological conversion processes use water anyway, so cleaning is a very minor additional expense.

Drying seems crucial to thermochemical processing. Inexpensive methods of water removal such as filtration, sedimentation, or squeezing would be mandatory for very wet material. Use of conventional dryers powered by fossil fuels would do great damage to the hopes of use of biomass to alleviate the energy crisis, but small amounts of fossil fuels might be used during startup. The main energy burden for drying must be placed on the biomass itself, either by diverting some of it for direct burning, by using hot stack gases, or by using a hot process stream. Several of the thermochemical processes incorporate drying close to the conversion step. With countercurrent operation, hot gases pass through the incoming biomass and drive out water.

One type of process avoids drying. The reaction $C + H_2O = CO + H_2$ with water as a reactant needs a proper ratio of biomass and water. Instead of drying, there may be blending of wet batches and drier batches of feedstocks.

Subdivision can be expensive. Grinding or milling biomass to a powder requires as much energy as is contained in the biomass; thus a process that requires fine powder is not worthy of consideration. Chipping or shredding biomass is relatively cheap, adding perhaps 5 percent to the purchase price, and some thermochemical conversion works well with coarse chips or small chunks.

Direct Burning

Burning of wood began in prehistory and was carried out with few improvements until the time of Benjamin Franklin. Even today, huts can be found in Asia, Africa, Central and South America, and elsewhere with open fires and a hole in the roof to let the smoke escape. Franklin devised an improved grate and a flue such that combustion was better and the hot gases could heat the room instead of going immediately up the stack. Further refinements with burning of wood and of coal have led to quite good equipment designs.

Most small installations use human power to toss logs or to shovel in the fuel and to shovel out the ashes. Factories use automatic conveying. For home heating, ash removal is annoying and time consuming. Coal-fired furnaces usually have dampers that can restrict the air supply to slow the rate of combustion and thus adjust the temperature. Restricted air means inefficient combustion and greater percentage of carbon monoxide and smoke going out the chimney.

Many U.S. homes in the 1930s had automatic stokers. These did not reduce the amount of shoveling because coal had to be put into the hopper, but a screw feed added coal to the furnace over a period of time and freed the shoveler from a rigorous schedule. Similar hoppers could handle biomass, but householders accustomed to the convenience of oil or natural gas would complain bitterly about return to a 50-yr-old technology.

Steam Generation

In a conventional boiler furnace, heat radiates to the receiving surfaces, and then conduction and convective heat transfer boil the fluid. Newer, less common systems convey heated gases from wood combustion to banks of tubular heat absorbing surfaces with transfer primarily by convection. High-pressure steam is produced for heating and to provide power for generating electricity. Except for some use in Vermont and Oregon, generation of electricity from wood biomass in central stations is practically nonexistent in the United States. On the other hand, industrial steam generation is widely practiced in the wood industry, which uses waste wood as fuel.

Wood fuel is normally hogged (hammer-milled) to chips for feeding boilers. Boiler efficiencies are less than those for coal-fired installations, and the boiler sizes for the same power ratings tend to be larger for wood. Process control is less reliable; thus automation is not as effective, and higher labor costs are encountered for burning wood. When moisture content is high, the heating value of biomass suffers, as do boiler ratings and efficiency. High-moisture wood can be dried with flue gases but is not easy because large volumes of gas must be circulated through the chips. The flue gas can also serve to exchange heat with the inlet air, so a separate drum drier may be a better choice for drying green chips. There are often fines from chipping that could fuel a drying step.

Municipal solid waste is burned to produce energy by several towns and cities. One method uses incinerators with tubes in the walls for boiling water. The temperature must not become too high because glass and metals would form a slag. Oil burners are used to provide supplementary heat for temperature control because moisture changes in the wastes affect flame temperature.

Conversion from oil to coal is taking place at central power stations. When the coal contains excessive sulfur, some method of removal is needed to comply with the current emission standard for sulfur oxides in stack gases. The technical acceptability and cost-effectiveness of sulfur emission control has been a controversial matter.

A modern coal–fuel plant could have 880-MW generating capacity, which translates to 3200 tons/hr of coal. The method of burning is in suspension within the boiler furnace as pulverized coal. The ash from coal is removed, depending on furnace temperatures, as a liquid slag (wet bottom) or as a powder (dry bottom).

Wood biomass as a prepared pulverized fuel analogous to pulverized coal could avoid premature equipment obsolescence at central stations. Such a use of wood biomass, however, poses formidable technical problems. The only common materials that approach the size of pulverized coal are sanding dust and sawdust. Preparation of both appears to be costly and requires dry wood. Wood flour, prepared only on a relatively small scale, may give insights into wood pulverization. Wood flour is presently used for linoleum manufacture, for explosives formulations, and in plastics.

Oxygen in wood serves to lower the furnace temperatures and redistribute the heat transferred by radiation. The overall effect could be a derating of boiler capacity. On the other hand, nitrogen oxides formation may be reduced.

Small-capacity boilers for wood firing tend to be shop-fabricated units and of a water-tube design. The vertical type can adapt to different stoker designs, horizontal and inclined, and has capacities of 38 tons/hr of steam at 45 atmospheres and 400°C final steam temperature. The horizontal design employs stationary grates and can be built up to 20 tons/hr at 32 atmospheres and 345°C final steam temperature.

The most common wood-fired boilers range from 8 to 50 tons/hr of steam, but boilers that burn fossil fuel plus wood have been installed in sizes greater than 250 tons/hr of steam. Nearly half of the wood-burning boilers recently sold list coal, oil, or natural gas as an alternate fuel. The most common alternatives are oil and natural gas, with coal accounting for only 2 percent.

Fluidized-bed combustion could provide countercurrent operation in which spent hot gases could dry incoming biomass quite efficiently. Fluid beds are stable when the solid particles are carefully sized and gas velocities are controlled. To provide such conditions for feeding irregularly and erratically shaped wood waste and inert materials of unpredictable character, the fluid bed may be established and maintained with a prepared material such as sand. The waste fuel is fed in at a slow, uniform rate. Combustion air is the fluidizing medium. Excess air and moisture are regulated so that the bed temperatures are considerably below the melting or fusion point of the bed material and generally range between 630°C and 980°C.

Hamrick (1978) has studied the problems of very large power-generating stations with wood fuel. Few of the present furnace and boiler systems with wood exceed a rating of 25 MW, but central power stations with ratings of over 500 MW are needed for economical operation. Coal fractures and is easily pulverized, but wood shreds. The time required for combustion of a particle varies inversely with the surface area : particle weight ratio; powdered coal burns about 13 times as rapidly as wood particles that will pass a half-inch screen. However, the combustion rate is profoundly affected by temperatures; thus preheating the combustion air can compensate for larger particle size. The calculated air temperature for wood is 565°C to obtain the same combustion rate as for powdered coal with the air at 345°C.

Retrofitting of existing boilers that use oil, gas, or coal could be difficult. Most of these systems have air preheaters that would have to be modified to reach the higher temperature needed for wood. Converting gas-fired boilers to coal is faced with small combustion space in which the coal is only partly reacted so that particles impinge on the tubes for a buildup of slag. Wood probably will perform better than coal in this equipment because wood has little ash and any buildup on the tubes could be flushed periodically with steam blowers. If the combustion space has to be lengthened for wood, rebuilding the equipment could be prohibitively expensive.

Pollution from a wood-burning power plant is shown in Table 4.1. The values are easily converted to an input-energy basis by multiplying by the assumed thermal efficiency of 0.213 for the plant. An electric power plant takes almost four times the land as a steam-generating plant of equal power rating.

A highly detailed analysis has been made of plantations for energy where trees are harvested for burning and side streams are treated and recycled (Intertechnology/Solar, 1978). Several potential plantation sites were selected, and a conceptual design was used for cost estimation. One novel proposal was integration

Table 4.1   Some Characteristics of
Wood-Burning Power Plants[a]

Parameter	Amount of Million BTU of Output	
	Electric Power	Steam
Sulfur as SO_2	1 lb	0.27 lb
Nitrogen (NO_x)	5 lb	1.4 lb
Particulates	Trace	Trace
Solid waste	9.8 lb	2.7 lb
Aqueous waste	390 gal	27 gal
Freshwater intake	770 gal	100 gal

Source: Kohan, Barkhordar, *et al.* (1979).
[a] Basis: 1000 tons/day of wood.

of algal ponds with the combustion process to utilize CO_2 and minerals from stack gases and ash. The algae would be digested to methane, and water with digester sludge would help irrigate and fertilize the forests. Net cost of energy was estimated at $1.30 per million BTU, which is very attractive. There did not appear to be a very good match between the combustion operation and the algae ponds; thus an initial venture into plantations might well leave the ponds for a later refinement. Even though the estimates were performed by people enthusiastic to the concept, their optimistic costs are so favorable that further work deserves high priority.

The economics of an energy plant associated with a paper fiber operation have been presented by Eimers (1978). The bases were (1) 2100 tons/day of wood for burning, (2) 76,000 acres (31,000 ha) of short-rotation forest, (3) 450,000 acres (182,000 ha) of forest for fiber supply, and (4) $20 million total investment for feedstock and power plant. A significant savings for wood burning is elimination of pollution-control equipment because wood is very low in sulfur compared to most coal and cheap oil. In the example, this translated to a 30 percent lower cost for the combustion and boiler units. Unfortunately, however, control of SO_2 emissions could be a serious problem. Government regulations usually demand the maximum removal of pollutants that is possible with existing commercial equipment. Although this may be greater removal than is necessary to maintain excellent air quality, the reasoning is that legislating high removal ensures good operation of the pollution-control equipment. Wood has a low sulfur content, but enough to just exceed the stack gas specifications for some locations in the United States. The existing methods for SO_2 removal are not efficient for low SO_2 concentrations, and some experts think that sulfur would have to be added to wood to obtain satisfactory operation. Either more reasonable laws must be passed, or different pollution-control equipment will have to be developed.

Working backward from a value of $27/Bbl for fuel oil, moisture-free wood for burning would be worth $39.40/ton in a new factory and $59.20 in an existing power plant. When compared to the cost of wood at $15 to $30/ton, there is some margin for profit.

A sensitivity analysis for energy from short rotation crops shows that productivity is critical. Higher productivity comes from higher yield or shorter rotation time. In Eimers' analysis, the yield must be more than 7.5 tons/acre (17 tons/ha) per year at a 4-yr cycle. Genetics and species selection could provide very fast growing trees. Improved harvesting costs could help nearly as much as increasing the yield, and automation of harvesting young trees looks very promising in light of new machines that will cut, chip, and load into trailers. An excellent point is made with respect to the form of energy produced because electricity is more valuable and more easily marketed than the steam used for its production.

Other approaches to cost improvements such as better combustion or heat-transfer efficiency may work for coal, oil, or wood. A fluidized bed of wood with steam tubes projecting into the bed would provide high heat-transfer coefficients, but the materials would have to resist abrasion and corrosion in a severe environment.

Eimers concluded that fuel plantations are currently uneconomic for generating steam because the probable return on investment was only about 5 percent. The rapid increase in the cost of energy since this analysis may have improved the prospects for plantations, but equipment costs have also risen. The best opportunity seems to be to grow a dual-purpose crop with some components leading to high-value products whereas residues are used for power generation. The example of short-rotation forestry integrated with fiber production may not accurately reflect what could be accomplished with very large plantations solely for energy, but the indications are that wood to steam is not the most attractive alternative for using biomass.

Not much R&D expenditure seems advisable for direct burning. Although it is unwise to state categorically that no new principles will be uncovered, there seems to be little chance of a startling breakthrough for direct burning. The equipment manufacturers should be able to refine their designs without support from government and the taxpayers.

Pyrolysis

In the 1930s the Ford Motor Company closed a plant that pyrolyzed wood to produce chemicals. Cheap chemicals from petroleum had captured their market. The process, also known as "destructive distillation," places wood in large retorts that are heated. Pyrolytic destruction of the organic constituents leads to volatile compounds, tars, and charcoal, all of which have value. The volatiles are composed of acetic acid, methanol, acetone, esters, and traces of various other compounds. Water is present in large amounts to complicate the separation and purification steps. The charcoal is very good for manufacturing steel. Figure 4.1 is flow sheet for pyrolysis.

Above 300° C pyrolysis of wood proceeds rapidly. Catalysts such as zinc chloride permit reaction at lower temperatures. The tars contain unreacted or partially reacted sugars mixed with polymerized material. Prolonged reaction gives lower quality tar; thus a fluidized-bed reactor is a good choice for minimizing

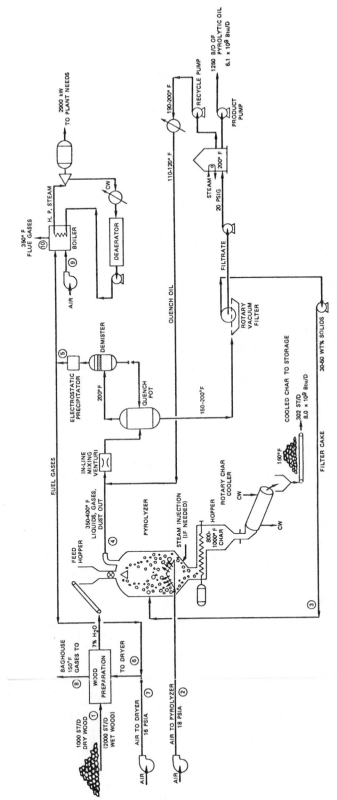

Figure 4.1 Flow sheet for pyrolysis. Courtesy NTIS.

exposure to high temperature. Vacuum pyrolysis to draw off volatiles gives yields of up to 75 percent desirable tar.

In Argentina, coal and iron are found at great distances from each other. It is economical to grow Eucalyptus trees near the iron mines and to pyrolyze them to charcoal for steel manufacturing. This is unusual to go for charcoal as the main product; in most other countries charcoal has limited markets because its density, and thus its energy content per unit volume is low. Research in the United States is aimed at increasing other pyrolysis products at the expense of char.

Bliss and Blake (1977) have analyzed conversion technologies. For pyrolysis, there are several processes that might be commercially successful with biomass. The Nichols–Herreshoff process (Nichols Engineering and Research Corporation, Belle Mead, N. J.) uses a multiple-hearth furnace that has long been used for roasting minerals. Solid material falls through the center of one stage, moves to the periphery to fall to the next stage, then moves again to the center, and so on. There are rotating arms in each stage to move the solids. Air moves countercurrently and combusts material in the lower stages to form hot gases that pyrolyze solids in the upper stages. Hot gases are also used in the drier for feed. A layout is shown in Fig. 4.2. The product gas has a heating value of 200 BTU/ft^3 (1800 kcal/m^3) by further combustion and additional energy as its sensible heat at 425 to 650°C. This means that this gas must be used on site and would be of little interest if it had to be cooled for use as a supplemental pipeline gas. This process is good for making charcoal, but charcoal is unlikely to be important for large-scale energy purposes.

The Engineering Experiment Station of Georgia Tech developed a process to be sold by the Tech-Air Corporation of Atlanta, Georgia. Waste materials are dried with hot, spent gases before conversion to char and volatile compounds by combustion using restricted supply of air. The volatiles are partially condensed to produce a pyrolytic oil, and the remaining vapors are burned for heat. Product heating values as percentage of input energy are represented as follows: char 35 percent, oil 35 percent, combustible gases 22 percent, and lost energy 8 percent. Again, char is a major product that is not particularly desirable in a fuels-from-biomass program.

The occidental flash pyrolysis process (Occidental Research Corporation) was developed with municipal waste in mind but should work well using wood. Finely shredded waste has a short residence time in a flash-pyrolysis unit. The products are char, oil, and medium BTU gas. A layout is shown in Fig. 4.3. Char, oil, and gas leave the pyrolysis reactor at 510°C and pass through a cyclone separator to remove the char. Vapor passes to a gas–oil separation column where it is water quenched to 82°C.

A pyrolysis system with fine biomass passed through a long, heated tube has been studied by Diebold and Smith (1979). Yield of hydrocarbons other than methane reached 24 percent of the feedstock weight and had 53 percent of the feedstock energy. This fraction could substitute for gasoline. Char, tars, and combustible gases can be by-products or can fuel the pyrolysis.

Electromagnetic plasma has pyrolyzed biomass (*Chemical and Engineering*

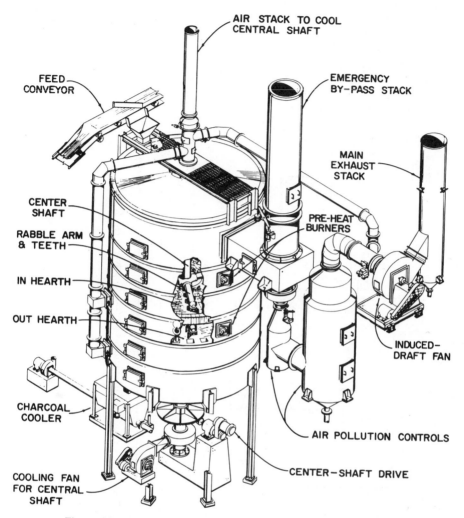

Figure 4.2 Nichols–Herreshoff process. Derived from trade literature.

News, April 16, 1979, p. 37). Pellets of lignin were placed near the focus of the microwave guide, and inert carrier gases flushed volatile products. This arrangement provides high energy intensity with short exposure times. In contrast to conventional thermal pyrolysis, which favors methane and saturated organic compounds, pyrolysis with plasma produces unsaturated hydrocarbons such as acetylene. This technique has merit for research on the mechanisms of pyrolysis, but it may be a long time before commercial pyrolysis with plasma can be developed.

One type of reactor for lignin was developed by Hydrocarbon Research, Inc. several years ago and used with dried lignin from the kraft process (Hellwig, Alpert, et al., 1969). A fluidized bed of catalyst and lignin particles held at about $240°\text{C}$ in reducing gases produces mostly phenols and benzene. Lignin from

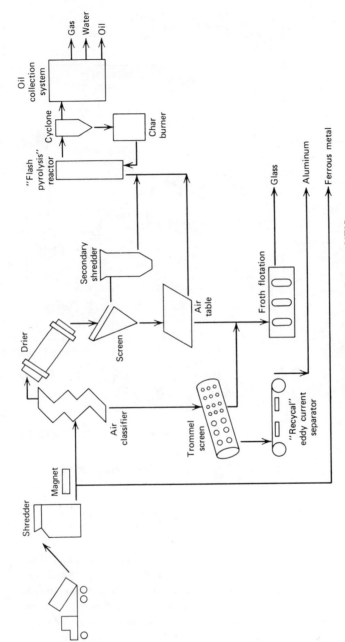

Figure 4.3 Occidental flash pyrolysis process. Courtesy NTIS.

mild processing of biomass should be more reactive than lignin from wood pulping. Hydrocarbon Research, Inc. has reevaluated their process and concluded that a 20 percent return on investment is possible (*Biomass Digest,* February 1980). The price assumptions were:

Phenol	30¢/lb
Benzene	$1.50/gal
Fuels	$3/million BTU
Lignin	5¢/lb

Process yields from lignin were (weight percentages): 20.2 phenol, 14.4 benzene, 13.1 fuel oil, and 29.1 fuel gas. These are high yields compared to other processes; thus this constitutes an attractive path from lignin to chemicals.

The GERE Process

The proprietary Garrett Energy Research and Engineering Company (GERE) biomass process is designed to convert a variety of biomass materials into an industrial fuel gas of medium calorific value. Claimed advantages are that it (1) may be economical even on a fairly small scale, (2) has high overall thermal efficiency, and (3) uses standard, compact, simple equipment.

The process is based on a type of multiple-hearth furnace that has been successfully employed for continuous high-temperature processing in the chemical and metallurgical industries for over 100 yr. The reactor consists of several vertically stacked compartments with a common central shaft. Rabble teeth are mounted on arms attached to a central shaft, and slow rotation moves the solid material. The downcomers through which the solid drops onto the hearth below are located alternately near the inner and outer peripheries. The hearths are sufficiently isolated to enable the conditions at each hearth to be optimized for its particular function. Since the solid is spread in a thin layer with some constant raking and tumbling action, the residence time is adjustable, and efficient heat and mass transfer are achieved.

High thermal efficiency is obtained by double-effect drying, in which most of the latent heat of evaporation of the moisture of the raw material is usefully recovered in a second, vacuum drying stage. The heat for pyrolysis is supplied by the combustion of char in a separate compartment. Thus the gaseous products of pyrolysis are not diluted by combustion flue gas (unlike many partial oxidation systems), and the BTU content is considerably higher.

The process is sketched in Fig. 4.4. The raw material, first chopped and mechanically dewatered if necessary, is loaded into a hopper and conveyed to the top of the multiple-hearth converter by means of a jacketed screw conveyor. The material is partially dried under a modest vacuum by flue gas flowing countercurrently through the conveyor jacket and through a hollow screw, if more heat transfer surface is needed.

One or more of the uppermost hearths are devoted to direct-contact drying

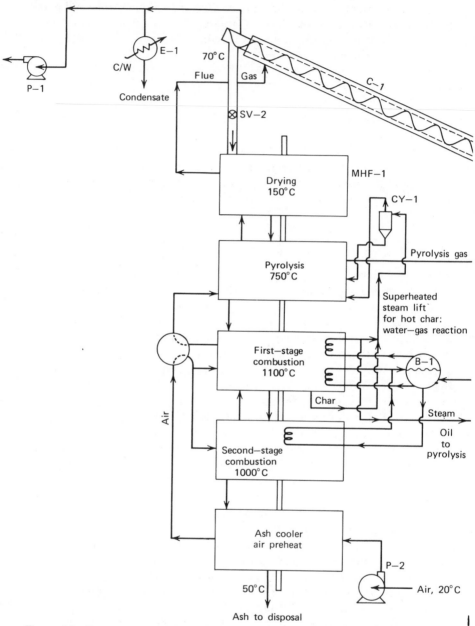

Figure 4.4 Garrett energy research and engineering process: B-1, boiler; C-1, conveyor–vacuum drier; CY-1, cyclone; C/W, cooling water; E-1, barometric or surface condenser; E-3, quench tower cooler; E-4, amine–glycol cooler; E-5, reboiler; E-6, gas cooler; FEL, front-end loader; H-1, feed hopper; LB-1, lump breaker; MHF, multiple-hearth pyrolysis reactor; P-1, vacuum pump; P-2, combustion air blower; P-3, pump; P-4, reciprocating compressor; R-1, ramp; S-1, settling drum; SV-1, star valve; SV-2, star valve; T-1, quench tower; T-2, CO₂, H₂S absorber; T-3, stripper–regenerator. Courtesy NTIS.

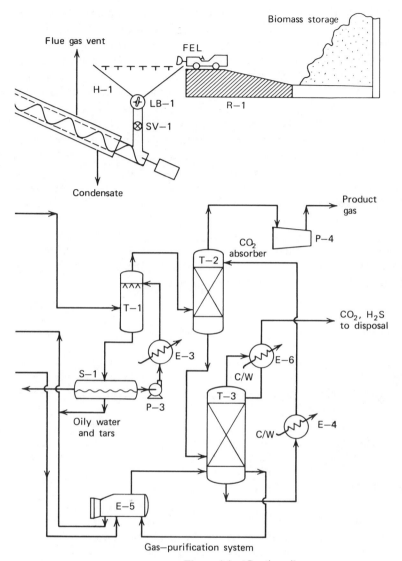

Figure 4.4 (Continued)

by hot flue gas from the combustion zone. The nearly dry material falls to the pyrolysis hearth to be heated mainly by hot char from the combustion zone and by hot gases accompanying the char. The char is introduced through an external steam lift in which some synthesis gas is formed as the char reacts with superheated steam. Synthesis gas with gaseous products of pyrolysis is cooled and condensed. Water-insoluble tar in the condensate may be recycled to the pyrolysis zone.

Char from the pyrolysis zone drops onto the combustion hearth, where it is burned to generate steam. The hot flue gas preheats the incoming air in an external heat exchanger and then goes to the drying section at the top. Most of the hot char is directed back to the pyrolysis zone, but some drops to the cleanup burner, where it is completely burned. The ash drops to the lowest hearth, where it is cooled by preheating the incoming air. The ash is then discharged from the reactor for disposal.

Gas leaving the condensate collector may be utilized directly or sent to a scrubber in which H_2S, CO_2, and moisture are removed. This product gas may be added to a natural gas pipeline. The only waste product is ash, which could have application as a fertilizer or source of chemicals.

Gasification

Reaction of carbonaceous materials with steam and oxygen to produce a mixture of carbon monoxide, hydrogen, carbon dioxide, methane, unreacted steam, some tar, and char is termed *hydrogasification* or simply *gasification*. The principal reactions are

$$C + \tfrac{1}{2}O_2 = CO_2 \qquad \text{Heat is evolved}$$
$$C + H_2O = CO + H_2 \qquad \text{Heat is required}$$
$$CO + H_2O = CO_2 + H_2 \qquad \text{Heat is evolved}$$
$$C + 2H_2 = CH_4 \qquad \text{Heat is evolved}$$

Wet biomass supplies the water needed, thus avoiding expensive drying that penalizes the use of biomass in other thermochemical processes.

Bliss and Blake (1977) have tabulated various processes for coal gasification (Table 4.2). Several of these processes operate near atmospheric pressure. The Purox process developed by Union Carbide Corporation uses oxygen instead of air and thus avoids dilution with nitrogen. Used with municipal solid waste, the Purox process discharges a molten slag at 1650°C. The crude gas is quite high in moisture and is at 95°C which means that little cooling is needed.

Synthesis gas is desired at 20 to 28 atm because pressure has a beneficial shift on equilibrium concentrations of subsequent reactions. Except for ammonia synthesis where nitrogen is a reactant, it is better to minimize concentrations of inert gases to save on compressing costs. The goal with coal gasification is to gasify at high pressure because this seems cheaper than compressing the product gas, but analagous processes for biomass have not been reported.

A reactor for coal gasification is shown in Fig. 4.5. After crushing and drying, finely divided coal is mixed with steam and oxygen to produce a mixture of CO, CO_2, CH_4, and H_2. Reaction with oxygen supplies the needed heat. Quenching of the hot gases recovers heat and separates some heavy oils and tars. The H_2:CO ratio is too low for good production of methane, so a "shift conversion" is carried out by reacting with added water for $CO + H_2O = CO_2 + H_2$. Carbon dioxide and H_2S are removed by various means. Although little H_2S is present, it is a deadly poison for the methanation catalyst. Methanation takes place at about 400° C and 70 atmospheres, usually with a nickel or iron catalyst. After another CO_2 removal step and drying, the gas is suitable for the pipeline system.

Two of the best catalysts found thus far for gasification are calcium oxide and wood ash. Dry catalyst can be just as effective when blended with dry wood as is an aqueous slurry added at the high-temperature reactor stage. Wood ash can be mixed with water and sprayed on the feedstock (Feldman, 1978).

Coffman (1978) has described a gasification process that can be termed "catalyzed steam pressure cooking" at 21 atmospheres and 590° C. It is accomplished in a long, slender rotary kiln pressurized by evolution of gases. The chips tumble and mix as the kiln rotates with passage through a gradient of temperatures. The gas has the approximate composition:

Methane	18 percent
Other hydrocarbons	8 percent
Hydrogen	28 percent
Carbon monoxide	1 percent
Carbon dioxide	45 percent

Tar and char are by-products. Projected gas costs are claimed to be low, and commercial feasibility could come quickly as fossil fuel costs rise. Hooverman (1979) feels that the principal problems relate to scaling up to economic sizes, although there are some technical difficulties with rotating seals, heat transfer between process streams, and methods for startup and shutdown.

Feldman, Chauhan, et al. (1979) have developed a fluidized-bed gasifier in which particles of sand assist in fluidization and are separated from the exit gas and mixed with the feed stream to transfer heat. The process should attain 75 percent thermal efficiency. Product gas has 400 BTU/SCF (3560 kcal/m^3), and there is 6900 BTU of gas produced per pound (3800 cal/g) of feedstock after subtracting the energy to power the process. This technology is very attractive for converting bulky biomass to a convenient gas, but the transportation cost for low- or medium-BTU gas is relatively high. If the user is more than 50 mi (83 km) from the gasification factory, upgrading the gas should be

Table 4.2 Reactors for Coal Gasification

Reactor Type	Reaction Conditions		Properties of Product(s)						
	°F	PSI	% CO	% CO$_2$	% H$_2$	% CH$_4$	% C$_2$H$_6$	% N$_2$	Btu/SCF (gas)
Moving Bed									
Lurgi	1150–1600	450	18.7	29.9	40.9	9.6	1.0	—	294
Wellman-Galusha	1000–1200	Atm.	—	—	—	—	—	—	268
Wilputte	—	Atm.	39.6	16.5	40.6	—	—	—	~300
Bureau of Mines (stirred reactor)	1000	100	21.7	6.8	15.8	3.0	—	52.7	164
Woodall-Duckham	—	Atm.	27.8	6.3	16.2	1.7	—	68.0	160
Fluidized Bed									
Winkler	2000–2200	Atm.	35.5	16.8	43.5	3.2	—	1.0	294
Exxon	1300–1800	25–45	—	—	—	—	—	—	—
BCR	2100	250	25.7	5.2	23.4	—	—	45.5	153
Acceptor	1500	150–300	8.3	7.7	65.7	12.6	—	5.7	374
Char-Oil-Energy-Demo	650–1600	Atm.	(Products—Btu basis: Gas—70%, Oil—16%, Char—4%)						—
Synthane	1100–4500	600	16.7	28.9	27.8	24.4	0.8	0.8	405
Battelle-Carbide	—	—	0.0–38.6	3.3–28.2	47.9–66.2	1.1–6.9	—	—	Variable
Westinghouse	2100	300–450	19.2	9.3	14.4	2.7	—	54.3	136
U-Gas	1900	300	19.5	10.1	13.2	4.7	—	52.5	153
U-Gas	1900	1000	13.7	14.9	12.7	7.8	—	51.0	104
Ignifluid	2000	300	—	—	—	—	—	—	—

	Temp. (°F)	Pressure (psi)						
Entrained Flow								
Koppers-Totzek	2700	5–7	55.4	7.0	34.6	—	1.0	290
Garrett	—	—	(Products: Char—56.7%, Tar—35%, Gas—6.6%)					700
HRI	1700	450	28.0	41.8	25.9	4.3	—	—
Babcock & Wilcox	2700	0–400	52.8	9.9	33.7	1.1	2.5	286
BiGas	2700	500–1500	44.0	14.0	24.4	15.6	0.6	378
Heating by Recirculated Solids								
Toscoal	800–1000	Atm.	(Products—Per ton coal: Gas—.438 MSCF, Oil—0.25 Bbl.. Char—1100 lb)					600
Lurgi Ruhrgas	1700–1800	Atm.	56.0	1.0	25.0	2.0	1.0	512
Molten Media Heat Carrier								
Patgas	2600	1000	63.5	—	36.0	—	—	315
Kellogg	1750	1200	33.6	13.3	45.0	7.5	0.4	329
Otto-Rummel	2500	350	53.3	11.0	33.9	—	1.1	284
Hydrogenation								
Hydrane	1400–1650	1000	22.9	—	73.2	—	—	826
Hygas	1000–1500	1000	30.2	24.5	18.7	—	—	495

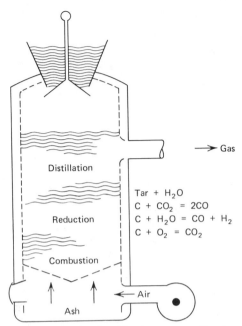

Figure 4.5   Schematic of gasification reactor. Courtesy NTIS.

considered. When the user is very close to the supplier, the hot gas can be used directly to supply extra heat and the net efficiency is somewhat higher.

A group at Texas Tech University (Beck, 1978) incorporated an original top-feed concept into a fluidized-bed system devised by Burton and Bailie (1974) and have treated cattle manure. Feeding into the hot rising gases results in a very short residence time for volatile materials from the manure so that the product mix is markedly different from that seen in other thermochemical processes. A sketch of the unit is given in Fig. 4.6. Less than 15 percent of the solid matter of the manure is not gasified. Huffman, Halligan, et al. (1977) have presented results with a countercurrent fluidized-bed reactor with feed rates of 17 to 52 lb/hr (7.7 to 23.6 kg/hr) of feedlot manure that had about 10 percent moisture and 25 percent ash. The most notable product feature is relatively high concentrations of ethylene, which is very desirable as a starting compound for organic synthesis or polymerization.

Typical results are presented in Table 4.3. There are significant shifts in product composition with temperature and other operating conditions. Total gas increased with temperature, as did the yields of hydrogen and ethylene. Tar is another product from this system; it can be burned or pyrolyzed to gases and char.

The Texas Tech process shows some promise for fuel gas but has its best chances for gases that can be fractionated for chemical synthesis. The hydrogen and nitrogen can be used for ammonia synthesis, but additional hydrogen

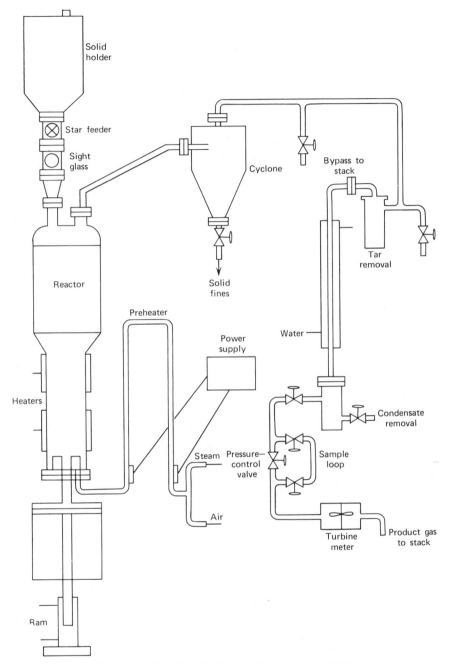

Figure 4.6 Top-feed fluidized gasifier. Courtesy NTIS.

Table 4.3 Top-Feed Fluidized-Bed Data

Operating Conditions	Run Number						
	1	6a	6b	7	8	9	10
Manure feed rate, kg daf/hr[a]	5.22	7.21	16.15	12.97	12.34	12.70	8.26
Manure feed rate, kg ar/hr	7.76	10.60	23.61	18.96	18.05	18.01	11.75
Air feed rate, kl/hr	1.149	1.700	1.487	1.904	1.402	4.249	0.765
Steam feed rate, kg/hr	5.44	4.54	4.54	3.63	3.08	2.72	3.72
Average temperature, °C	711	695	641	617	629	668	628
Product Gas Data[b]							
Total Dry Gas, l/g daf[c]	1.19	(0.667)	0.580	0.406	0.455	(0.718)	0.318
Heat value (HHV), cal/l	2855	2918	3790	3380	3523	2624	3345
Gas composition, vol percent							
H_2	25.2	22.2	20.0	28.2	17.4	15.1	20.9
N_2	14.6	27.8	15.1	23.2	26.7	36.8	24.2
CH_4	12.8	7.7	12.6	9.2	14.1	8.9	11.7
CO	11.6	15.3	21.3	16.4	21.2	20.3	22.4
CO_2	30.8	20.7	22.1	15.4	14.1	14.2	14.8
C_2H_4	4.7	6.4	8.5	4.9	5.8	4.2	5.5
C_2H_6	0.3	0.5	0.4	2.7	0.7	0.5	0.5
(Reformed H_2 without C_2's): N_2 ratio	6.0	2.5	6.1	3.5	3.6	1.9	3.7

[a] daf = dry, ash-free; ar = as received.

[b] All data are average values from at least two samples. Individual gas samples were analyzed on the gas chromatograph using at least two injections.

[c] Values in parentheses are back-calculated values using a nitrogen balance.

from reforming the CO_2 and CO would be required to achieve the proper proportions.

Costs of manure pyrolysis have been analyzed by Garrett (1977), who feels that the economics of large scale operation are negated by hauling costs. In other words, there is not enough energy in manure to cover the energy expenditures for transporting it very far. Hauling could amount to 25 percent of annual expenses. Feedlot size varies for regions of the United States, with some Western states having a large average size. In a state such as California, where 50 percent of the lots have more than 66 tons/day of manure, a fairly large gasification plant (250,000 tons of manure per year) could serve several lots and not incur excessive hauling costs.

High-temperature baths of molten salts or molten metals have been tested for gasification of coal. Rapid rates and high yields are possible, but ash and sulfur in the feed accumulate in the bath. Economical recovery of the medium is likely to be difficult. A pilot process for coal using molten sodium carbonate is under construction, and biomass feedstocks could be tested. This technology is at an early stage of development, and biomass can follow the leads for coal if economic prospects become attractive.

Using the Purox gasification as a prototype because it is nearer to commercialization than most other processes, the capital and operating costs using wood as a feedstock have been estimated by Desrosiers (1979). The plant sizes and capital costs were

400 oven-dried tons/day	$9 million 1978 dollars
800 oven-dried tons/day	15 million 1978 dollars
1600 oven-dried tons/day	24 million 1978 dollars

With a reasonable cost for the wood, the product gas would have a break-even cost of $4 to $5 per million BTU, which is high compared to the expected cost of natural gas for the next few years.

The construction of coal gasification plants in South Africa is probably the largest industrial project ever undertaken. Heylin (1979) has reported on the project and its problems and implications for the United States. The estimated cost is $7 billion, and the construction force will reach 25,000 to achieve completion by 1983. It would take the United States 6 or 7 yr to build such plants. If petroleum prices escalate at the present rate, oil from coal could be highly profitable. However, there is sufficient uncertainty about petroleum prices to cause great reluctance for a U.S. company to undertake a similar project. In a sense, coal gasification overshadows any biomass process. However, biomass can be ready much sooner and at a smaller scale with far less financial risk.

If the gas is to be used in pipelines instead of being converted to liquid fuels, there is a different perspective. Reaction of solid fuel with water and limited oxygen was a widespread practice in the United States until World War II. There were about 1200 factories producing gases with heating values ranging from 300 to 500 BTU/ft^3 STP. Development of large pipelines for natural gas caused unfavorable economics for synthetic gas, and almost all the factories shut down. With the increased cost of natural gas, gas production plants are beginning to appear again with coal as a feedstock. In some locations biomass could be an attractive alternative feedstock.

An excellent review of gasification (Reed, 1979a, b) provides some history and a status report on various options. The greater reactivity of biomass compared to coal appears to be a very significant advantage.

It is interesting that many people, including officials at the U.S. Department of Energy, consider gasification to be the top priority for biomass conversion. This is based on simple substitution of biomass for coal in one of the many known processes for gasification. However, it is not likely that biomass can be more economical than coal. Collection costs are high for biomass, and the conversion factories will be larger and much more expensive than those of the same productivity using coal. There are indications that biomass is more reactive than coal; thus the throughput rates may be higher, and the gasification productivity per unit mass may be better than that for coal. Bulky biomass would probably need larger reactors than those for coal unless the biomass reaction rates are relatively much higher. The prime constraint on very large-scale gasification is the exorbitant capital expenditures that would be required; thus the increase in capital and collection costs for biomass practically dictates choice of coal except under very specialized, regional circumstances.

Liquefaction

Hydrogenation of coal to produce fuel oils is well known. Conversion of biomass is not analogous in that the main reactions are not additions of hydrogen but are removals of oxygen. The reducing agent for this removal can be carbon monoxide obtained from the biomass itself. Bliss and Blake (1977) have proposed the term "carboxylolysis" for the reductive conversion of cellulosic materials to oil, but the more common name is "liquefaction." Alkaline materials are catalysts that allow the reactions to have acceptable rates at reasonable temperatures. Wood ash contains alkaline compounds that are good catalysts; sodium carbonate is slightly better although more expensive.

At the temperatures for liquefaction, steam and carbon monoxide undergo the shift reaction to hydrogen and carbon dioxide; thus hydrogen could be the principal reductant. However, carbon monoxide is thought to be of most importance. The reactions are probably:

$$\text{Cellulose (empirical formula } C_6H_{10}O_5) + CO = C_6H_{10}O_4 + CO_2$$
$$\text{Cellulose} = C_6H_{10} + CO_2$$

The former is a reduction, and the latter is a disproportionation. Reaction products are a mixture containing a considerable amount of partially oxidized hydrocarbons.

Research by the U.S. Bureau of Mines and the Pittsburgh Energy Research Center led to a process that can react biomass with carbon monoxide to produce an oil. In the early 1970s, technical and economic study deemed the processes feasible, and a design of a 3-ton/day pilot plant was prepared. Bechtel Corporation developed detailed recommendations in 1975 for a design by Rust Engineering Company, and the Energy Research and Development Administration (now the U.S. Department of Energy) awarded a contract for construction of a pilot plant at Albany, Oregon. On completion in 1977, the first product oil was obtained just a few days before the first oil came out of the Alaska pipeline. Unfortunately, liquefaction of biomass is presently too expensive.

The liquefaction process uses a portion of the product oil to prepare a slurry of biomass. The slurry plus catalyst is pumped into a reactor in which the pressure is 350 to 450° C and the pressure is 30 to 40 atmospheres. Hydrogen or carbon monoxide liberated from the biomass provides reducing power. A flow sheet is given in Fig. 4.7.

Checkout and startup of the Albany pilot plant are reported by Bechtel National, Inc. (1979). Many small problems were encountered such as the need to find special gasket materials to resist the severe conditions and the oil itself. The most serious problem was failure of the oil, water, and solids to separate in the centrifuge step, and true continuous operation is impossible unless part of the product oil can be cleaned and recycled. Production costs without provision for profit are given in Table 4.4 Note that amortization of the capital

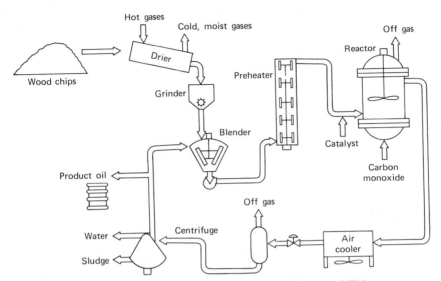

Figure 4.7 Process for liquefaction of wood. Courtesy NTIS.

investment is the major cost, in sharp contrast to other conversion processes where the biomass feedstock is usually the dominant cost.

Elliot and Walkup (1977) have reported bench-scale experiments that support the Albany, Oregon facility. Figure 4.8 shows data obtained with a small autoclave charged with wood, aqueous sodium carbonate catalyst, and anthracene oil as a carrier. The temperature was 350° C, and carbon monoxide was used to establish the desired pressure. The definition of total conversion was

$$\frac{\text{wood - insoluble residue}}{\text{wood}}$$

Oil yield was
$$\frac{\text{benzene solubles}}{\text{wood}}$$

Table 4.4 Liquefaction Cost Estimates Based on Albany Plant

Item	Cost per Barrel of Oil
Amortization	$13.8
Wood	7.4
Electricity	4.8
Taxes and insurance	3.7
Labor	3.2
Consumables	2.0
Total	$34.9

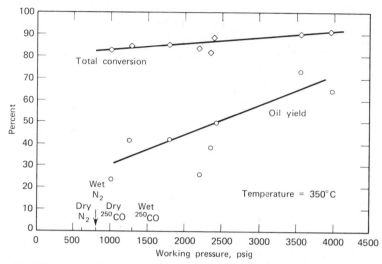

Figure 4.8 Effect of working pressure on conversion of wood flour to oil. Courtesy NTIS.

Note that increase in the pressure increases yield, but that extremely high pressures are required for good yields. Anthracene oil has a significant vapor pressure, which makes it more difficult to achieve the desired partial pressure of CO. Other experiments are summarized in Table 4.5. These data indicate that H_2 is not desirable and that the feed materials generate reducing activity such that some oil is obtained with no added CO.

Although there is a pilot plant for liquefaction, practical, reliable technology cannot be claimed to be established. There is a problem with the biomass–oil slurry feed system. A fraction of product oil is recycled back as carrier oil for the biomass. A slurry of 30 percent solids by weight can be prepared with powdered biomass, but it is a thick paste. Pumping this to the reactor has been difficult with plunger pumps having check valves. Fouling of check valves is likely during extended operation.

An alternate feed method is a lock hopper system. Pulverized biomass would be added to vessels that are then closed and brought to reaction pressure. Rotary vanes would dispense feed to the reactor. The proposed liquefaction process operates above the pressure of usual types of lock hoppers; thus this aspect of process development will be a significant advance in chemical engineering. There must be at least two lock hoppers so that one can be filled while the other is in use.

Sodium carbonate solution is added continuously by means of a high-pressure plunger pump. Carbon monoxide process gas is taken from trailers and compressed to operating pressure in a nonlubricated, noncontaminating diaphragm-type compressor. The compressed process gas is preheated in a gas–gas interchanger and sparged into the reactor. In a commercial plant synthesis gases of various $H_2:CO$ molar ratios will be used as the process gas.

Table 4.5 Effect of Gas Mixtures on Liquefaction

Gas Composition	Total Conversion, Percent	Oil Yield, Percent
100 percent CO	78.0	52.8
50 percent CO, 50 percent H_2	77.0	58.5
50 percent CO, 50 percent N_2	76.0	61.8
100 percent N_2	68.0	41.0
50 percent N_2, 50 percent H_2	68.4	39.6
100 percent H_2	65.6	34.6
33 percent CO, 33 percent H_2, 33 percent N_2	75.5	56.6

Liquid from the reactor is cooled to 200°C. As the pressure is reduced, some liquid will flash, and the remainder is collected in a bottoms tank and centrifuged to separate oil and water. The oil is filtered to remove any solids and transferred to a hold tank. Residual gases contain carbon monoxide, carbon dioxide, and hydrogen.

There has been continuing progress and discussion of biomass liquefaction (Berry, 1979; Elliot and Giacoletto, 1979; Ergun, 1979; Figueroa and Ergun, 1979; Seth and Ergun, 1979). There are many variables thus the research plan is complicated. Numerous biomass feedstocks such as different woods, crop residues, and solid wastes have been tested successfully. An interesting development is pretreatment for liquefaction (Schaleger, Yaghoubzadeh, et al., 1979). The purpose is to change the rheological properties of the biomass slurry in oil so that higher concentrations can be pumped. The original design of the Albany pilot plant called for 30 percent slurries, but pumping them is almost impossible. Furthermore, to even approach this concentration, it has been necessary to pulverize the feedstock, and the power cost has been prohibitively high. Pretreatment of wood chips in a water suspension taken to 260°C and 47 atmospheres for 60 min weakened them for easy pulverization, and the biochemical structure was more amenable to catalytic reduction. Further experimentation showed that 0.05 percent sufuric acid at 180°C for 45 min was even more effective. The sugars and oligosaccharides formed by hydrolysis should be much more reactive than the parent carbohydrate polymers. It must be noted that these pretreatments are much the same as those for producing fermentable sugars (see Chapter 7). As sugars are more valuable than fuel oil, the wisdom of liquefaction instead of fermentation must be questioned.

A new advance in coal liquefaction announced by Dow Chemical Corporation may also have applications to biomass (*Chemical Engineering News*, September 25, 1979). The process is based on emulsion technology by which tiny, highly active catalyst particles are contacted with a slurry of coal in product oil. At a temperature of 350°C the coal softens prior to reaction and is more easily hydrogenated. Hydrocyclones provide separation for recycle of the catalyst, and liquid–liquid extraction provides a solids-free oil.

Liquefaction produces a clean, convenient fuel equivalent to common heating oil, but the capital cost of the equipment and the energy requirements are high. Compared to other conversion technologies, liquefaction seems to have very tough problems to overcome to achieve significant cost reductions. Unless petroleum prices jump by a factor of 3 or so, there would be little market for the product.

5

Biological Concepts

Photosynthetic biomass comes from green plants, and some knowledge of biology is required to appreciate the problems involved in growing large amounts for energy purposes. Interactions with parasites, predators, symbionts, and disease organisms are important factors. In addition, understanding conversion of these materials to fuels through biological means (bioconversion) requires certain foundations in biology and biochemistry. One must be acquainted with those biochemicals that are present in large amounts; pathways of synthesis, metabolism, and fermentation; and the regulation of cellular activities.

The cell is the structural and functional unit of life. Cells are often highly specialized in multicellular organisms and function by cooperation with other specialized cells, but many organisms are free-living single cells. Although cells may differ in size, shape, and specialization, there is a basic common structure. Every cell contains *cytoplasm,* a colloidal system of large organic molecules and a complex solution of smaller organic molecules and inorganic salts. The cytoplasm is bounded by a semielastic, selectively permeable *cell membrane* that controls the movement of molecules into and out of the cell. Threadlike *chromosomes* suspended in the cytoplasm bear a linear arrangment of *genes.* Information carried on the genes controls every cellular activity, and, as the units of heredity, genes determine the characteristics of cells passed from one generation to the next.

In most cells the chromosomes are surrounded by a membrane to form a conspicuous *nucleus.* Other organized membrane-surrounded intracellular structures serve as specialized sites for cellular activities. For example, *chloroplasts* in the cells of green plants play an essential role in photosynthesis. Chlorophyll, the principal photosynthetic pigment, is contained within the layered membranous structure of the chloroplast. Other intracellular structures, called *mitochondria,* are the site of chemical reactions that make energy available to the cell. Cells that possess organized nuclei are described as *eukaryotic.*

In bacteria and blue–green algae the chromosomes are not surrounded by a membrane, and there is little apparent subcellular organization. The chlorophyll of blue–green algae is associated with loosely arranged membranes within the cytoplasm; bacterial chlorophyll, when present, is located in vesicular *chromatophores.* Because they lack a discrete nucleus, these organisms are said to be *prokaryotic.*

Many cells are surrounded by an outer covering, external to the cell membrane. Plant cells, bacteria, and blue–green algae are protected by rigid cell walls. Certain algae and protozoa are sheathed by gelatinous layers of inorganic materials such as silica.

The distinctive and sometimes elaborate shape exhibited by many unicellular organisms is an inherited characteristic. However, evidence gathered in the culture of isolated cells suggests that in multicellular organisms, cell shape is environmentally determined.

The smallest known cell, a pleuropneumonialike organism (PPLO), is approximately 0.1 micron in diameter, and the largest, the ostrich egg, about 150 mm in diameter. Most cells, however, have diameters of 0.5 to 40 microns. Because all the substances required by the cell must enter through the surface membrane, one of the most important limitations to cell size is the surface:volume ratio. The ease with which a given substance passes through the membrane and its rate of diffusion through the cytoplasm and utilization rate by the cell are related to cell size. Another important dimensional factor is the proximity of the most distant parts of the cell to the genes, which continuously monitor cellular activity.

Microbiology

Three groups of microorganisms are of special concern to bioprocesses to obtain energy: bacteria, algae, and fungi. A fourth group, the protozoa, are of less significance in bioprocesses. Certain viruses, called *phages,* are also important in that they can infect microorganisms and may destroy a culture.

Bacteria

Bacteria are tiny single-cell organisms ranging from 0.5 to 20 microns in size, although some may be smaller and a few exceed 100 microns in length. The typical bacterial cell is surrounded by a rigid protective membrane made up of compounds called *mucoproteins* that are peculiar to bacteria. It is this wall that imparts characteristic shape to the cell. Bacteria may be round or ovoid, rod shaped, or spiral. Some bacteria can vary in shape, depending on culture conditions; this is termed *pleomorphism.* Certain species are further characterized by the arrangement of cells in clusters, chains, or discrete packets.

Bacteria are prokaryotic (possessing no organized muclei or organelles). They do, of course, contain genetic material, both DNA and RNA, and the cytoplasm may contain numerous granules composed of carbohydrates, fats, and other nutrients. When chlorophyll is present, it is of a type unique to bacteria. Many bacteria exhibit motility by means of one or more hairlike appendages called *flagella.* Bacteria reproduce by dividing into equal parts, a process termed *binary fission.*

Under unfavorable conditions, certain bacteria can transform to spores that germinate on return to a favorable environment. Many species of bacteria may, under appropriate circumstances, become surrounded by a gelatinous material. If a number of cells share the same gelatinous mass, it is called a *slime;* if the cells are separately surrounded, each is said to have a capsule. The slime or capsule affords the cell a means of attachment and provides a measure of protection against drying and predators.

Bacteria are classified on the basis of pathogenicity, morphology, and physiological characteristics. They cover the entire spectrum of nutritional requirements from photosynthetic autotrophs to the most fastidious of heterotrophs. Many possess exocellular enzymes that allow them to break down a variety of complex substrates to molecules that can enter the cell to be further metabolized. Each species of bacteria grows best within certain ranges of temperature and pH, commonly between 25° C and 40° C and not too far from neutral pH. There are species, however, that thrive at temperature or pH extremes. Many bacteria synthesize pigments that impart distinctive colors to their colonies.

With the use of special staining techniques, individual cells of most species are visible with the light microscope; some, however, are best observed by electron microscopy.

Algae

Algae contain photosynthetic pigments that enable them to synthesize structural materials and storage compounds from carbon dioxide and water. The distinctive colors imparted by these pigments are one of the criteria by which algae are classified. Some species are unicellular and microscopic; others are filamentous, branched, or colonial. Some have life cycles in which both unicellular and multicellular forms arise, but the most common mode of reproduction is simple cell division. On the basis of pigmentation, storage compounds, cell organization, and morphology, biologists divide the algae into as many as nine groups.

Cyanophyta (blue–green algae) may be unicellular, colonial, branched, or filamentous. The photosynthetic pigments are not organized into discrete structures (chloroplasts or chromophores) but are dispersed throughout the cytoplasm. The cell wall is usually very thin and may be enclosed within a gelatinous sheath. Cyanophyta contain glycogenlike storage compounds instead of starch. The blue–green algae, like the bacteria, possess no distinct nucleus and are, therefore, prokaryotic. Also, like some bacteria, cyanophyta have the ability to convert atmospheric nitrogen to ammonia (nitrogen fixation), which can be used in the synthesis of organic compounds or excreted into the medium.

Chlorophyta (green algae) may be free swimming or attached. The cells are eukaryotic; that is, each has a distinct nucleus and chlorphyl is contained in chloroplasts. Starch is the predominant storage compound. Individuals may be branched, filamentous, colonial, or single cells; often they are microscopic, but may become so numerous as to be visible as an algal "bloom" or scum on the surface of standing water.

Chrysophyta (yellow–green or yellow–brown algae) are unicellular or colonial. All species are motile and surrounded by a thick cell wall; in some forms (the diatoms) the wall is impregnated with silicon. Starch is not present and food is stored as lipids, which often gives members of chrysophyta a metallic luster.

Fungi

As a group, fungi are characterized by simple vegetative bodies from which reproductive structures are elaborated. All fungal cells possess distinct nuclei and, at some stage in their life cycles, reproduce by spores formed in specialized fruiting bodies. The fungi contain no chlorophyll and thus require sources of complex organic molecules; many species grow on dead organic material, and others live as parasites. Many can live on carbohydrate, inorganic nitrogen, and salts. Food is stored as glycogen or oil.

Fungi are classified as slime molds, true fungi, or yeasts, based on vegetative and reproductive structure.

Yeasts are unicellular organisms surrounded by a cell wall and possessing a distinct nucleus. Yeasts reproduce by a process known as *budding;* a small new cell is pinched off the parent cell, but under certain conditions an individual yeast cell may become a fruiting body, producing four spores. The spores are more resistant than vegetative cells to extremes of temperature and prolonged periods of drying, enabling yeasts to survive unfavorable environmental conditions.

Protozoa

The protozoa are a widely diverse group of organisms of 15,000 to 20,000 species. Most are microscopic, although some attain a length as great as 5 mm. A cell membrane encloses the cytoplasm, and within the cytoplasm are found a number of cellular inclusions, or organelles, which are the sites of specialized cell functions. For this reason, protozoa are often referred to as *acellular* rather than unicellular.

Viruses

Viruses are particles of a size below the resolution of the light microscope. They are not cellular in structure and are composed mainly of nucleic acid surrounded by a protein sheath. Lacking metabolic machinery, viruses exist only as intracellular parasites. They are highly host specific, infecting only a single species or closely related species. Plant and animal viruses are generally named for the diseases that they cause, such as tobacco mosaic virus or influenza virus.

Not all types of microorganism appear to be susceptible, but bacteria and certain molds are subject to invasion by virus particles. Those that attack bacteria are called *bacteriophages* and may be either *virulent* or *temperate.* Virulent bacteriophages divert the cellular resources to the manufacture of phage particles; as new phage particles are released to the medium, the host cell dies and disintegrates. Temperate bacteriophages have no immediate effect on the host cell; they become attached to the bacterial chromosome and may be carried through many generations before being triggered to virulence by some physical or chemical event.

Biochemistry

All organisms require sources of carbon, nitrogen, sulfur, phosphorus, water, and certain trace elements. Some have specific vitamin requirements as well. Water, in addition to its role as a reactant and as the principal component of cytoplasm, serves as the medium through which molecules are transferred into and out of the cell. All cells, even those of terrestrial organisms, require constant moisture to remain active.

Green plants need only carbon dioxide, nitrate or ammonium ions, dissolved minerals, and water to manufacture all their cellular components. Photosynthetic bacteria require an additional specific source of hydrogen ions, and the chemosynthetic bacteria must have a specific oxidizable substrate. Some microorganisms have the ability to "fix" atmospheric nitrogen by reducing it to ammonia. Organisms that use only simple inorganic compounds as nutrients are said to be *autotrophic* (self-nourishing).

Organisms that require compounds that have been manufactured by other organisms are called *heterotrophs* (other-nourishing). Because organic molecules frequently are too large or insoluble to pass through the cell membrane, many heterotrophs produce enzymes that act outside the cell. These *exoenzymes* hydrolyze large molecules to smaller units that can readily enter the cell.

Once inside the cell, nutrient molecules are used by the cellular machinery to synthesize new compounds that meet that cell's specific requirements. Compounds present in high concentration in biomass are usually polymers, either in the form of structural elements, such as cell walls, or in storage materials such as starch.

Carbohydrates

Carbohydrates are polyhydroxy compounds, usually with carbon, hydrogen, and oxygen in a ratio of 1:2:1 ($C_nH_{2n}O_n$). Simple sugars are termed *monosaccharides*. They are the monomers from which polysaccharides are formed; complete hydrolysis of complex carbohydrate polymers produces monosaccharides. Some simple sugars of importance are shown in Fig. 5.1. Except for fructose, which has a ketone group, these sugars each have an aldehyde group. However, both aldehyde and ketone groups are active carbonyl functions that react reversibly with hydroxyl groups. Thus sugars cyclize readily, and five- or six-membered rings are favored because little strain is placed on the carbon to carbon bonds. Figure 5.1 also shows D-glucose depicted several ways. In solution, an equilibrium is set up with ring forms predominating over the open chain. Formation of polysaccharides locks the sugars into rings. A carbon atom linked to four different atoms or groups can have them in two different arrangements called *stereoisomers*. If considered as a tetrahedron with the groups in the corners and the carbon at the center, the corners are A, B, C, and D. Grasping A and looking at the base of the tetrahedron, the arrangement of the other groups may be B,C,D or its mirror image B,D,C. Organic compounds with different arrangements have different properties. The differences are pronounced in carbohydrates because the hydroxyl groups interact by hydrogen bonding and form rings with carbonyl groups with spacing critical. Sugars are numbered starting with the aldehyde group or with the end nearest to the ketone group. The first or last carbon atoms do not have stereoisomers because they are not four different groups, for example.

$$R-\overset{\overset{\displaystyle O}{\|}}{C}-H \quad \text{or} \quad R-\overset{\overset{\displaystyle H}{|}}{\underset{\underset{\displaystyle H}{|}}{C}}-OH.$$

The central carbon atoms, if not ketone groups, do exhibit stereoisomerism. The number of possible isomers is:

3—carbon sugar (triose)	2
4—carbon sugar (tetrose)	4
5—carbon sugar (pentose)	8
6—carbon sugar (hexose)	16
7—carbon sugar (heptose)	32

Sugar nomenclature is simplified by observing the hydroxyl group on the next-to-last carbon atom: it points to the right for dextrorotatory (D) sugars and to the left for levorotatory (L) sugars. In nature, L sugars are uncommon except for some of the pentoses found in hemicellulose, plant gums, and mucilages.

When sugars twist to form a ring, the new hydroxyl from reaction with the carbonyl group can point either way. In the Haworth formula for glucose this is the-OH on carbon 1. It is designated α if it points down and β if it is up. This may seem to be a minor change, but the properties of polysaccharides are vastly different if linked α instead of β.

Two simple sugars may react by the elimination of water to form a disaccharide. Some disaccharides that are relevant to energy from biomass are shown in Fig. 5.1. Maltose, a dimer of glucose, is the repeating unit in starch and occurs in starch hydrolysate. Cellobiose, another glucose dimer, is the repeating unit in cellulose and is found during its breakdown. Sucrose is a component in sugarcane, sugar beets, and sweet sorghum; it is acceptable as the main nutrient for most fermentations.

Hemicellulose is found associated with cellulose. Whereas complete hydrolysis of cellulose yields exclusively glucose, hemicellulose is composed of roughly

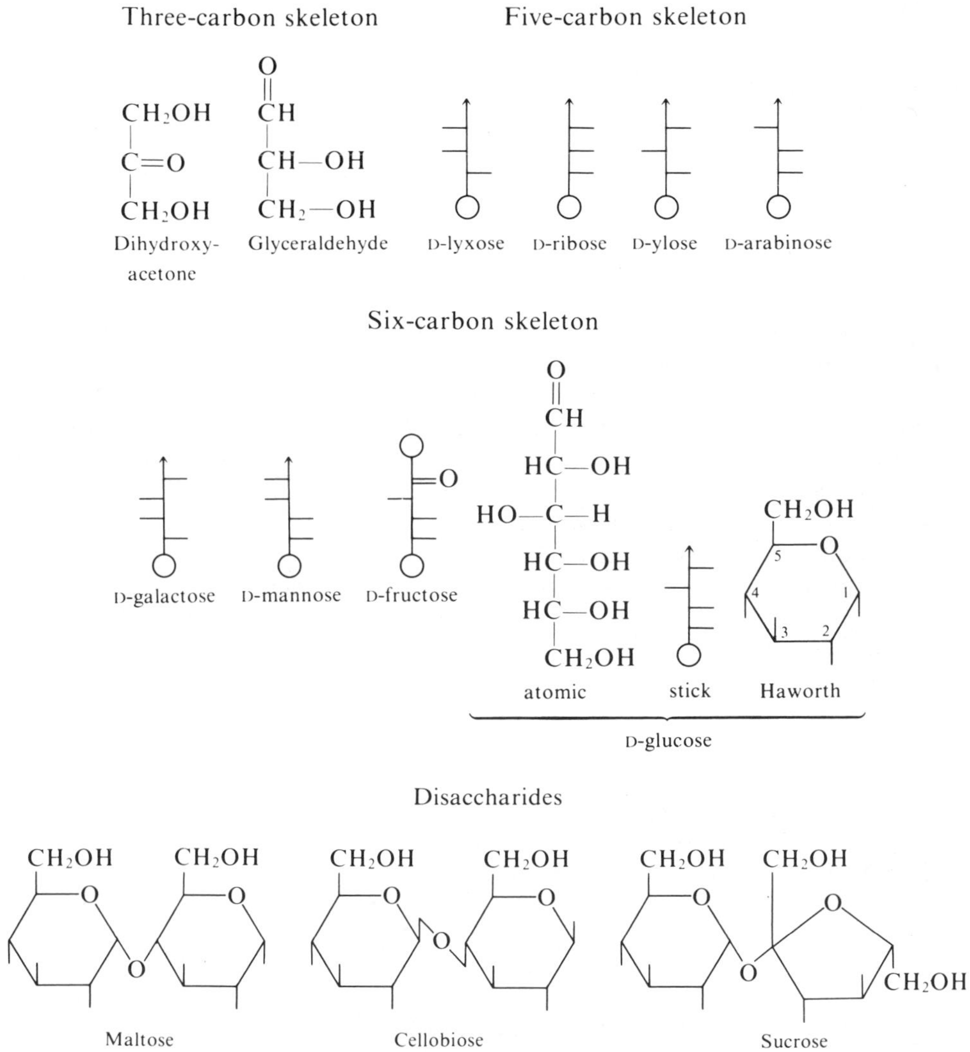

Three-carbon skeleton   Five-carbon skeleton

Dihydroxy-   Glyceraldehyde   D-lyxose   D-ribose   D-ylose   D-arabinose
acetone

Six-carbon skeleton

D-galactose   D-mannose   D-fructose

atomic   stick   Haworth

D-glucose

Disaccharides

Maltose   Cellobiose   Sucrose

Figure 5.1   Important sugars.

80 percent pentoses plus small amounts of hexoses and uronic acids. Uronic acids are derived from sugars; the terminal alcohol group is oxidized to a carboxylic acid. Glucuronic acid is depicted in Fig. 5.2. Xylose and arabinose are the most common pentoses in hemicellulose, and corn cobs are a very rich source of xylose.

Lignin is a relative of the carbohydrates, although the structural similarities are not readily evident. Aromatic rings with one or more OH groups and with various side chains are linked in three dimensions to form lignin. There is further discussion of lignin in Chapter 7 and a structure shown in Fig. 7.3.

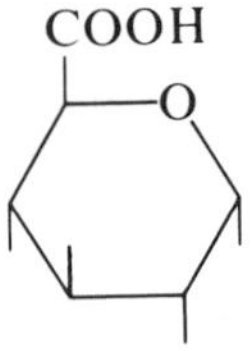

COOH

Figure 5.2 Glucuronic acid.

The aromatic rings in lignin come from dehydration and cyclization of sugars. The links between the rings are ethers formed by elimination of water from two OH groups. Ether links are very resistant to cleavage, and the entire lignin structure is resistant to microbial attack. Some microorganisms break down lignin slowly in aerobic processes by first oxidizing the aromatic ring.

Lignin is a much smaller molecule than cellulose. There can be 10,000 or more glucose units in one molecule of cellulose, but there are only about 25 aromatic rings in a lignin molecule. The softening temperature of lignin carefully extracted from wood is about 180° C; treated lignins soften at 120 to 175° C depending on the nature of the substituents and the fragmentation of the polymer. Lignin can be fractionated in various solvents to give materials of different softening points and degrees of color from brown to black.

Lipids

Lipids are soluble in nonpolar solvents and relatively insoluble in water. Fats are common lipids composed of gylcerol esterified with fatty acids that are characterized by a carbon chain and a carboxyl group:

$$-\overset{O}{\underset{}{C}}-OH.$$

Figure 5.3 shows a triglyceride fat; R, R', and R″ are the carbon chains of different fatty acids. The chain of a fatty acid may be from 2 to over 20 carbon atoms in length, but 16- and 18-carbon chains are most common, 14 is not uncommon, and odd-number lengths are very rare. Double bonds may occur in the chain, usually at the 9th, 12th, and 15th carbon atom. This unsaturation affects physical properties, tendency to become rancid, and dietary proclivity for cardiovascular disease. Fats are mixtures of various triglycerides and thus do not exhibit sharp melting points. They function as storage compounds and as constituents of cell membranes where their repulsion of water affects permeability. If recoverable fats are in biomass produced for energy, they are too valuable as food to be considered for combustion fuels.

Fatty alcohols are hydroxy analogues of fatty acids. Waxes are esters of a fatty acid with a fatty alcohol. Waxes protect the skins of animals and surfaces of

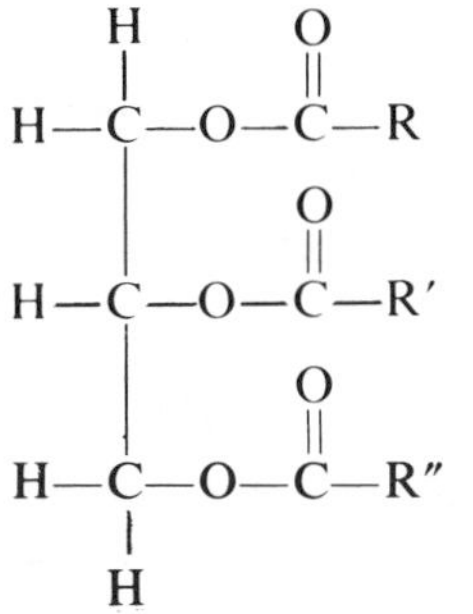

$$
\begin{array}{c}
\overset{\displaystyle H}{|} \qquad \overset{\displaystyle O}{\|} \\[2pt]
H-C-O-C-R \\[2pt]
| \qquad\qquad \overset{\displaystyle O}{\|} \\[2pt]
H-C-O-C-R' \\[2pt]
| \qquad\qquad \overset{\displaystyle O}{\|} \\[2pt]
H-C-O-C-R'' \\[2pt]
| \\
H
\end{array}
$$

Figure 5.3 Triglycerides structure.

plants, particularly the leaves. Waxes are likely to be present in biomass for energy, but not in major amounts. They would be more valuable as by-products for industry than as combustion fuels.

Other lipids such as sterols and phospholipids play vital roles in regulation, nerve tissue, and membranes. However, their concentrations are insignificant to fuels from biomass.

Proteins

Proteins are composed of units called *amino acids*. There are about 20 amino acids that are found regularly in naturally occurring proteins. Amino acids are characterized by a carboxyl group and an amino group, $-NH_2$. Some representative amino acids are shown in Fig. 5.4.

With the removal of a molecule of water, two amino acids may be joined by a *peptide bond* (also in Fig. 5.4) to form a *dipeptide*. Several amino acids bonded in this manner form a *polypeptide*. A naturally occurring polypeptide of many amino acids is called a *protein*. Because of the great length of protein chains and

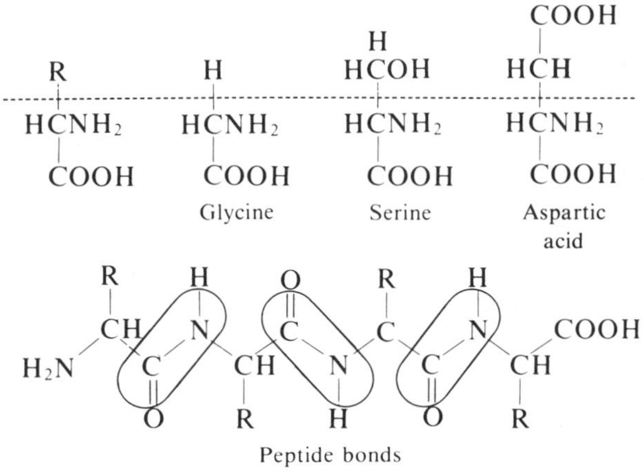

Peptide bonds

Figure 5.4 Typical amino acids.

the various sequences of amino acids, the number of possible proteins is astronomical. In addition to peptide bonds, other bonds may be formed, giving the molecule a complex and distinctive configuration. In the presence of certain chemical reagents, excessive heat, radiation, unfavorable pH, and so on, the protein structure may become disorganized. Proteins are important components of cell membranes and of muscle. The antibodies that protect organisms against invasion by foreign proteins are themselves proteins.

A special class of proteins, the *enzymes,* plays a vital role in all cellular activity. To initiate any chemical reaction, a certain amount of energy is required. Heat could provide the necessary *activation energy,* but the amount of heat that would be needed to initiate many biological reactions would destroy the cell itself. Enzymes are biological catalysts that expedite reactions by lowering the amount of activation required.

Virtually every cellular reaction requires the presence of an enzyme. As the reactant molecules come into contact with the enzyme surface, an enzyme–substrate complex is formed. When the reaction is complete, the complex dissociates, freeing the enzyme for further reaction so that only small amounts are needed.

The surfaces of enzymes are shaped and have patterns of chemical groups and charges that account for their specificity; most enzymes will catalyze only a single reaction or a few closely related reactions. The optimum pH for most enzymes is not far from neutral; most lose activity quickly at temperatures above 60°C.

Enzymes function in conjunction with another special class of compounds known as *coenzymes.* Coenzymes are not proteins; many of the known coenzymes include vitamins, such as niacin and riboflavin, as part of their molecular structure. It is the coenzymes that carry reactant groups or electrons between substrate molecules in the course of a reaction. As coenzymes serve merely as carriers and are constantly recycled, only small amounts are needed to produce large amounts of biochemical product.

Enzyme Kinetics

Enzymes provide focusing points for biochemical reactions. The place on the enzyme where reaction takes place is termed the "active site." When concentrations of reactants are lower than those of enzymes, it is easy to find an available active site. As reactant concentrations increase, many sites are in use and the reaction rate becomes limited by how rapidly molecules can diffuse to an available active site and how rapidly products can diffuse out of the way of the incoming reactants. The effect of reactant concentration on rate in a system with a fixed amount of enzyme is shown in Fig. 5.5. The rate rises steeply and nearly linearly with concentration and then levels off to approach a maximum rate asymptotically. From mechanistic considerations, the equation

$$k = \hat{k}\,\frac{S}{K_M + S}$$

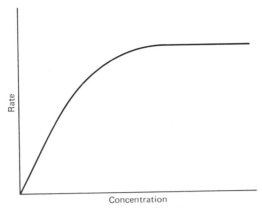

Figure 5.5 Effect of concentration on rate.

has been derived. It is termed the *Michaelis–Menten* equation, where k is the reaction rate coefficient, $\hat{k}$ is the maximum rate, S is reactant or substrate concentration, and K_M is the Michaelis constant. The maximum rate is approached when the active sites are saturated, which denotes that molecules are in heavy competition to find a place to react. It can be seen from the equation that at $k = \frac{1}{2}\hat{k}$, $K_M = S$. Since this corresponds to half of the maximum or saturation rate, K_M is referred to as the *half-saturation constant*. The values of $\hat{k}$ and K_M are characteristic of an enzyme and are thus useful for comparisons.

Enzyme kinetics often depart markedly from the Michaelis–Menten equation because of various types of inhibition. A number of equations have been proposed for modeling inhibition.

Nucleic Acids

Two kinds of nucleic acid are found in living organisms: deoxyribonucleic acid (DNA) and ribonucleic acid (RNA). Nucleic acids are chains of *nucleotides*. Each nucleotide consists of a nitrogen-containing organic base, a sugar, and a phosphate group. The sequence in which the nucleotides are arranged is actually the code that determines the amino acids to be assembled, and in what order, to form proteins.

The ultimate control of all cellular activity rests with the nucleic acids. Enzymes are required for each cellular reaction and thus have immediate control. But it is nucleic acids that dictate the synthesis of enzymes. Moreover, it is nucleic acids that are responsible for the maintenance of genetic continuity. When any organism reproduces, equivalent DNA molecules are transferred to each offspring. Even a slight alteration in the nucleotide sequence of a DNA chain may result in some permanent change, or *mutation*, which will persist through succeeding generations.

Growth Kinetics

The relationship between nutrient concentration and growth rate of a micro-bial culture often closely resembles that shown in Fig. 5.5. By analogy with the Michaelis-Menten equation, Monod formulated a very similar expression:

$$\mu = \hat{\mu}\ \frac{S}{K_S + S}$$

where μ = specific growth rate coefficient
$\hat{\mu}$ = maximum specific growth rate
S = concentration of growth-limiting nutrient
K_S = a constant

Occasionally growth rate data fit poorly to the Monod equation, but it is extremely useful in providing an understanding of growth rate dependency. The Monod equation is not applicable during transient states when microor-ganisms are responding to sudden upsets because it is a steady-state, time-independent relationship.

Metabolism

Each activity implicit in the word "life" requires energy. In living things energy is stored and transferred as chemical bond energy. The multitude of reactions that take place within a living system are collectively termed *metabolism*.

Certain metabolic reactions, once activated, proceed spontaneously with a net release of energy. Hydrolysis and molecular rearrangements are examples of spontaneous reactions. The hydrolytic splitting of starch to glucose, for in-stance, results in a net release of energy.

A great many biochemical reactions are not spontaneous and thus require an energy input. In living systems this requirement is met by coupling an energy-requiring reaction with an energy-releasing reaction. The synthesis and breakdown of biochemical compounds is achieved through pathways involv-ing the formation of energy-rich intermediate compounds. In this way energy can be transferred in a stepwise manner.

If a sufficient amount of energy is produced by a metabolic reaction, it may be used to synthesize a high-energy compound. Adenosine triphosphate (ATP) is such a compound; it is a nucleotide consisting of the nitrogen-con-taining compound, adenine, the sugar ribose, and three phosphate groups (Fig. 5.6). Although ATP has adequate stability for the short term, it hydro-lyzes spontaneously in water. The symbol ($\sim$) is used to represent a high-energy bond. When the terminal phosphate linkage is broken, adenosine di-

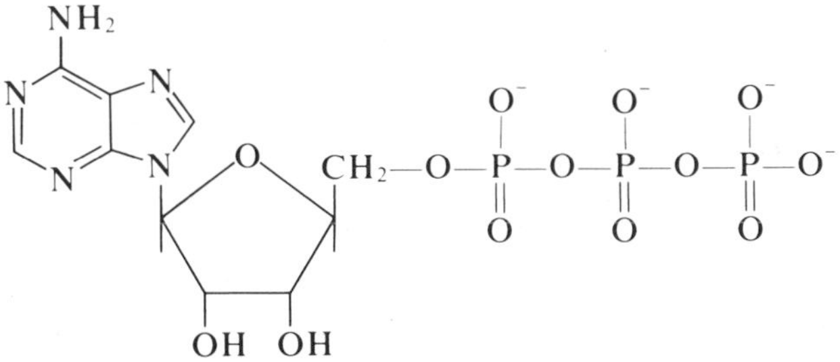

Figure 5.6 Structure of ATP.

phosphate and inorganic phosphate are formed, and energy is provided. When sufficient energy becomes available, ATP is reformed.

Oxidation and reduction are very common steps in metabolism. Reduction reactions store energy in the reduced compound, whereas oxidation liberates energy. In biological systems the most frequent mechanism of oxidation is the removal of hydrogen, and, conversely, the addition of hydrogen is the most frequent method of reduction. When this takes place within a cell, hydrogen is transported between donor and acceptor molecules by coenzymes. Nicotin-amide–adenine dinucleotide (NAD) and nicotinamide–adenine dinucleotide phosphate (NADP) are two coenzymes that function in this manner.

Photosynthesis

Every living thing can synthesize ATP, but only green plants and a few microorganisms have the capacity to make it from energy-poor materials. Through the process of *photosynthesis*, these organisms can convert light energy to chemical bond energy and reduce carbon dioxide to carbohydrate.

The photosynthetic apparatus of higher plants is located in membrane-surrounded organelles called *chloroplasts*. Chloroplasts are generally either globular or plasmoconvex in shape and range in diameter from 1 to 10 microns. Contained within the membrane of the chloroplast is a proteinacious matrix known as the *stroma;* throughout the stroma a continuous membrane is arranged in *lamellae*. In certain areas, called *grana*, the lamellae appear tightly packed and layered. Electron microscopy has revealed that the grana are made up of widened vesicular portions of lamellae known as *thylakoids*. Portions of lamellae that are not part of a granum are referred to as *intergranal lamellae*. Figure 5.7 depicts the structure of a chloroplast. The number of chloroplasts per cell varies with species and environmental conditions. In higher plants numerous chloroplasts are found in the cells of the mesophyll tissue of leaves, whereas algae of the genus *Chlorella* contain a single chloroplast.

Prokaryotic organisms—bacteria and blue-green algae—do not possess chloroplasts. Instead, their photosynthetic systems are associated with the cell

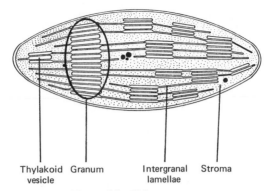

Thylakoid Granum   Intergranal   Stroma
vesicle   lamellae

Figure 5.7   Chloroplast structure.

membrane or with lamellar structures located in organelles known as *chrom-atophores.* Chromatophores, unlike chloroplasts, are not surrounded by a membrane.

All photosynthetic organisms contain one or more of the group of green pigments called *chlorophylls.* In addition, many contain accessory pigments that impart characteristic colors to the cells. Photosynthetic bacteria contain *bacteriochlorophyll,* which differs slightly in structure from the chlorophylls of other plants, and which occurs only in bacteria. Photosynthetic pigments and the enzymes needed for photosynthesis are located in the membranes of the thylakoids and intergranal lamellae. Some appreciation of the complexity of a lamella may be derived from Table 5.1.

When light strikes a photosynthetic organism, energy is absorbed by the chlorophyll. Accessory pigments broaden the useful wavelength band, delocalize energy absorbtion, and protect against oxidation. Each pigment has a characteristic relationship of energy absorption to wavelength of incident light. Although the rate of photosynthesis peaks at about 657 mμ and drops off rapidly at wavelengths greater than 700 mμ, the efficiency is greater for mixed wavelengths.

High light intensity saturates the photosynthetic apparatus and gives no further increase in the rate of photosynthesis. In fact, the pigments are bleached by bright light so that photosynthesis is impaired and recovery is slow on return to dim light. Energy is used to convert ADP to ATP and to reduce NADP by the addition of hydrogen ions donated by water. The reactions are

$$2ADP + 2P_i + 2NADP^+ + 4H_2O + \text{light energy} \rightarrow$$
$$2ATP + 2NADPH + O_2 + 2H_2O$$

Because light is essential to the production of ATP and the reduction of NADP, these reactions are known as the *light reactions* of photosynthesis. The light reactions in photosynthetic bacteria differ somewhat from those in the green plants. Bacteria do not use water as a source of hydrogen ions, and

Table 5.1   Composition of a Single Lamella from Spinach Chloroplast

Constituent	Number	Constituent	Number
Chlorophyll	230 molecules	Galoctosyl glycerides	500 molecules
Carotenoids	48 molecules	Sulfolipids	48 molecules
Quinones	46 molecules	Cytochromes	2 molecules
Phospholipids	116 molecules	Nonheme iron	10 atoms
		Other metals	4 atoms

oxygen is not formed. Some use organic molecules; others use hydrogen sulfide (H_2S) and give off sulfur.

The remaining photosynthetic reactions can take place whenever ATP, NADPH, and CO_2 are present; they are, therefore, called the *dark reactions*. The biochemical pathway was first elucidated by M. Calvin and his associates and is frequently referred to as the Calvin cycle. A CO_2 molecule combines with a five-carbon sugar that immediately splits to form two molecules of the three-carbon compound phosphoglycerate (PGA). Five-sixths of the PGA is used to regenerate the five-carbon sugar, ribulose diphosphate, through a complicated series of reactions so it can again combine with CO_2. The overall equation for the photosynthetic formation of hexose by this pathway can be written

$$6CO_2 + 18ATP + 12NADPH + 12H^+ + 12H_2O \rightarrow$$
$$hexose + 18P_i + 18ADP + 12NADP^+$$

The dark reactions of photosynthesis are outlined in Fig. 5.8.

The reactions of the Calvin cycle occur in all known photosynthetic plants.

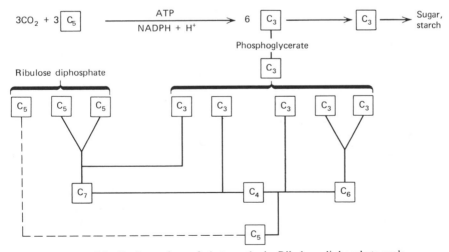

Figure 5.8 Dark reactions of photosynthesis. Ribulose–diphosphate cycle.

In certain plants, however, the three-carbon compound is not the first inter-mediate into which CO_2 is incorporated. In such plants CO_2 is first fixed in the four-carbon compound oxaloacetate, ultimately released, and again fixed via the Calvin cycle. This group, known as the "C_4 plants," includes many tropical species, as well as corn, sugarcane, and sorghum. Plants in which PGA is the first intermediate are termed "C_3 plants."

The overall equation for CO_2 fixation in the C_4 plants is

$$6CO_2 + 30ATP + 12NADPH + 12H^+ + 24H_2O \rightarrow$$
$$\text{hexose} + 30ADP + 30P_i + 12NADP^+$$

The net ATP requirement for CO_2 fixation is greater in C_4 plants; neverthe-less, they are, in reality, far more productive than the C_3 group. The reason for this difference is the occurrence of photorespiration in C_3 plants.

Photosynthesis binds energy into compounds such as glucose and com-pounds derived from glucose. All plants utilize the energy bound by photosyn-thesis through the oxidative breakdown of these molecules during respiration, which is discussed in a later section. The C_3 plants, however, also exhibit a second type of respiration, *photorespiration*, which occurs in the light at the same time as photosynthesis. During photorespiration molecular oxygen is reduced at the expense of products formed in photosynthesis and without the production of energy-rich compounds. The rate of photorespiration decreases with increasing partial pressure of CO_2; when the CO_2 concentration of the gas phase is elevated, C_3 plants can about double their rate of growth (Zelitch, 1979).

Bassham (1977) has discussed possible ways of increasing photosynthetic efficiency by reducing photorespiration in C_3 plants. The legume alfalfa, with its nitrogen fixing properties, might be a suitable crop for such experimenta-tion. Plants cannot independently fix nitrogen; only a few kinds of organism (certain bacteria and fungi and the blue–green algae) have the capacity to convert nitrogen from the air to compounds such as ammonia. In leguminous plants, however, nitrogen-fixing bacteria live in a symbiotic relationship within root nodules of the plant. The bacteria derive the energy required for nitrogen fixation from nutrients supplied by the plant, and the plant is repaid with fixed nitrogen. Genetic engineers have suggested the addition of genes for nitrogen fixation to C_4 plants. Biological nitrogen fixation is highly energy consumptive, however, and might overtax the plant's energy supply.

Chemosynthesis

The ultimate source of energy for most living things is the sun, but certain groups of bacteria require neither light nor organic energy sources. These or-ganisms derive energy from the oxidation of inorganic substances and are called the *chemosynthetic* bacteria.

For example, one species of nitrifying bacteria oxidizes ammonia to nitrite, and another species oxidizes nitrite to nitrate:

$$NH_3 + 3O_2 \rightarrow HNO_2 + H_2O + energy$$
$$HNO_2 + \tfrac{1}{2}O_2 \rightarrow HNO_3 + energy$$

Certain sulfur bacteria oxidize elemental sulfur to sulfate:

$$3S + 3O_2 + 2H_2O \rightarrow 2H_2SO_4 + energy$$

One species of hydrogen bacteria oxidizes hydrogen gas, reducing carbon dioxide to methane:

$$4H_2 + CO_2 \rightarrow CH_4 + 2H_2O + energy$$

Respiration

The oxidative breakdown of organic molecules is called *respiration*. It is through this process that the cell recovers energy stored in organic substances. Respiration is really a controlled series of dehydrogenations in which small amounts of energy are released at several stages. The released energy is incorporated into ATP, where it is readily available for other reactions. As in all metabolic pathways, a specific enzyme is required at each step. In higher plants and animals the mitochondria are the sites of respiration. In prokaryotic organisms the respiratory enzymes are associated with the cytoplasmic and intracytoplasmic membranes.

There are two main, alternate schemes for breakdown of sugars. In one, glucose is oxidized to form two molecules of the three-carbon compound pyruvate. The series of eight reactions, termed *glycolysis*, is outlined in Fig. 5.9.

In many organisms, including humans, respiration can proceed only in the presence of molecular oxygen (*aerobic* respiration). There are organisms, however, that can carry on respiration in the absence of oxygen (*anaerobic* respiration). Anaerobic respiration occurs in many microorganisms and, under certain conditions, in the muscle cells of animals. Most bacteria are *facultative anaerobes*, growing in the presence or absence of oxygen. Some microorganisms, the *obligate anaerobes,* require the absence of oxygen. *Obligate aerobes,* on the other hand, must have molecular oxygen.

The aerobic oxidation of pyruvate is outlined in Fig. 5.10. Carbon dioxide is removed from pyruvate, leaving a two-carbon acetate group. Acetate is carried by a coenzyme (coenzyme A, or CoA) into the *citric acid cycle*. No oxygen is taken up in the citric acid cycle, but a series of oxidations takes place in which hydrogen is transferred to coenzymes, and further removal of carbon dioxide occurs. The hydrogen is passed from the coenzymes through a series of carrier molecules called "the respiratory chain," or cytochrome system, and finally to oxygen. Energy produced during the reduction of the cytochrome molecules is used to convert ADP to ATP. Oxygen is required only as the final hydrogen acceptor.

A second set of major reactions for hexose catabolism diverges from gly-

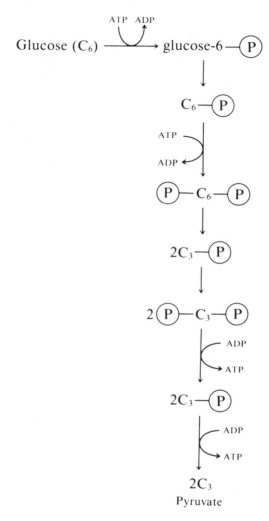

Figure 5.9 Glycolysis.

colysis at the level of glucose-6-phosphate and thus is termed the hexose-monophosphate "shunt" or phosphogluconate pathway. The same coenzymes appear as in glycolysis, and the result is again ATP generation. However, the different intermediates, especially the pentoses, can lead to products not possible with glycolysis.

Molecules other than oxygen must act as hydrogen acceptors for anaerobic respiration. Usually pyruvic acid itself acts as a hydrogen acceptor, and lactic acid is formed. Three-carbon acids may be converted to ethanol and carbon dioxide. A culture fluid rich in organic compounds may result from an anaerobic process, in contrast to aerobic respiration, which carries the reactions to

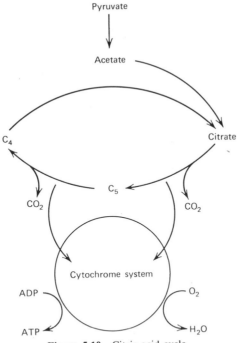

Figure 5.10   Citric acid cycle.

near completion, resulting in much carbon dioxide and only traces of organic compounds.

Glycolysis produces a net gain of two ATP molecules. Further anaerobic oxidation of pyruvic acid adds none; thus anaerobic respiration of one molecule of glucose yields only two ATP molecules. On the other hand, aerobic oxidation of one glucose to carbon dioxide and water, via glycolysis and the citric acid cycle, produces a total of 38 new ATP molecules.

In addition to carbohydrates, cells regularly oxidize fats as a source of energy. Proteins and amino acids are also metabolized. In general, one or two biochemical reactions lead to a compound that can enter into the usual pathways for carbohydrates.

Fermentation Pathways

Products Formed from Pyruvate

Anaerobic organisms cannot employ the citric acid cycle or phosphogluconate shunt because the oxidation–reduction potential is too low for the coenzymes to be recharged. In lieu of complete oxidation of pyruvate to carbon dioxide and water, anaerobic pathways may oxidize some compounds while reducing others to gain small amounts of energy.

Keto acids such as pyruvic acid can undergo oxidative decarboxylation. Products such as acetic, propionic, and butyric acid are formed. Coenzyme A, which links to carboxyl compounds through its sulfyhydryl group, is widely employed in reactions of organic acids. The net reaction for pyruvate is

$$
\underset{\text{Pyruvate}}{CH_3\overset{O}{\overset{||}{C}}-\overset{O}{\overset{||}{C}}O^- + H^+} \longrightarrow \underset{\text{Acetaldehyde}}{CH_3\overset{O}{\overset{||}{C}}H} + CO_2
$$

The continued operation of the system depends on the ability of the enzymes of the respiratory chain being able to reoxidize the mole of NADH formed per mole of pyruvate. Pyruvate itself can act as a hydrogen acceptor in reoxidation of NADH, and lactate is formed. The reaction is

$$
\underset{\text{Pyruvate}}{CH_3\overset{O}{\overset{||}{C}}-\overset{O}{\overset{||}{C}}-O^-} + 2NADH \longrightarrow \underset{\text{Lactate}}{CH_3\overset{OH}{\overset{|}{C}}H-\overset{O}{\overset{||}{C}}-O^-} + 2NAD
$$

A commercial fermentation process for lactic acid has been known for many years.

Acetaldehyde is toxic to life, and its accumulation would inhibit the fermentation organisms. It is held in a reactive compound, acetyl coenzyme A, and quickly transformed. An enzyme, alcohol dehydrogenase, working in reverse readily converts acetaldehyde to ethanol while reoxidizing NADH, as shown:

$$
\underset{\text{Acetaldehyde}}{CH_3\overset{O}{\overset{||}{C}}H} + 2NADH \longrightarrow \underset{\text{Ethanol}}{CH_3CH_2OH} + 2NAD
$$

Other biological products from acetaldehyde are

Although 2,3-butanediol is difficult to separate from water, it is of interest as a possible fuel or source of diene that might be used in synthetic rubber.

Acetone and Butanol

Some anaerobic bacteria of the *Clostridium* family take acetyl coenzyme A to a key intermediate acetoacetyl coenzyme A

$$2 \, CH_3\overset{O}{\underset{\|}{C}}\sim S—CoA = CH_3\overset{O}{\underset{\|}{C}} \, CH_2\overset{O}{\underset{\|}{C}}\sim S \, CoA + HS—CoA$$

Acetyl CoA   Acetoacetyl CoA   Coenzyme A

Further reactions are

Acetoacetyl CoA $\longrightarrow$ $CH_3\overset{OH}{\underset{|}{C}}H—CH_2\overset{O}{\underset{\|}{C}}\sim S—CoA$ $\longrightarrow$ $CH_3CH=CH—\overset{O}{\underset{\|}{C}}\sim S \, CoA$

β-Hydroxybutyrl
CoA

Crotonyl CoA

$\longrightarrow$ $CH_3\overset{O}{\underset{\|}{C}} \, CH_3 + CO_2$

Acetone

$CH_3CH_2CH_2\overset{O}{\underset{\|}{C}}\sim S—CoA$

Butyrl CoA

$CH_3CH_2CH_2CH_2OH$

Butanol

The need for NADH regeneration leads to production of both acetone and butanol, the former as the oxidized molecule and the latter as the reduced partner.

Products Based on the Citric Acid Cycle

Organic acids related to the citric acid cycle that have been produced by commercial fermentations are citric acid, fumaric acid, and glutamic acid. None is suitable as a fuel, but each has potential as feedstock for chemical synthesis. Present costs preclude large volume uses to which petrochemicals are put, but cheaper sugars for fermentation substrates would have a large impact on costs. The pathways to these acids are steps from the citric acid cycle. Selected cultures accumulate the desired compounds, or mutants may be found that cannot carry out a subsequent step. Insoluble salts such as calcium citrate may be precipitated during the fermentation to shift an equilibrium or to prevent the product from reaching toxic levels.

Other Fermentations

Many organic compounds are produced commercially by fermentation. Antibiotics, vitamins, and others represent low conversion yields based on sugar and thus are not relevant to energy from biomass. Glycerol can be made by fermentation as an offshoot of glycolysis, but difficulty in separating it from water makes costs unattractive. Many thousands of compounds have been detected in trace amounts in microbial culture fluids, and some of these might be produced in commercial amounts by certain microbial strains. Acrylic acid has been detected, and its wide use in plastics justifies interest in developing a commercial fermentation.

A rather roundabout relationship of fermentation to energy is the production of xanthan gum. The bacteria *Xanthomonas campestris,* a pathogen for cabbages, is used to polymerize glucose commercially to a gum that gives solutions of very high viscosity. This gum is used to thicken foods and to aid oil well drilling muds in carrying cuttings up the shaft. It also can be pumped into old oil fields to force residual oil into pools that can be collected. Very large amounts of gum solution are needed, but high yields of oil are expected. Cheap sugars from biomass would give lower costs for xanthan gum and make oil recovery more practical.

Nitrogen Fixation

Before World War II, the primary method of providing nitrogen to agricultural lands was cultivation of legumes. Interest in legumes declined when inexpensive nitrogenous fertilizers became available and the value of land and food crops rose. The use of fertilizers has had a particularly dramatic impact as the main factor in a 3 percent annual increase in world cereal grain production that has persisted for a quarter of a century. Less developed countries hope to increase from the present average of less than 15 lb (7 kg) per capita fertilizer consumption to the 128-lb (58-kg) per capita consumption of the developed nations as a step to ward off mass starvation. Biomass for energy has a similar need for fertilizer to achieve high productivities, but economics must be watched very carefully.

About 2.5 percent of natural gas is consumed in the United States for producing ammonia, principally for fertilizer. Moreover, the industrial process is very consumptive of energy—about 42 million BTU (8.4×10^5 kW) per ton of ammonia. The capital costs for the factories and high operating costs drive the price of fertilizer up; it now costs about four times as much as in the late 1960s.

Skinner (1976) has written an excellent overview of nitrogen fixation in which the following prospects are featured:

- Methods for using nitrogen more efficiently.

- Increasing nitrogen fixation by known microorganisms.
- Development of better and perhaps new symbiotic relationships between plants and microorganisms.
- Finding better mutants to utilize organic wastes or sunlight to fix nitrogen.
- Genetic transfer of nitrogen-fixing capability into plants such as corn that now depend heavily on fertilizer.
- Development of industrial catalysts for nitrogen fixation under mild conditions to reduce manufacturing costs.

Evans and Barber (1977) have also emphasized the importance of biological nitrogen fixation as an alternative to expensive fertilizers. From various sources, they have compiled information on nitrogen fixation rates as shown in Table 5.2. Note that the rates for free-living microorganisms are low compared to legumes; thus a simple algal pond or industrial fermentation for nitrogen fixation is not very attractive. In any event, it seems more sensible to fix nitrogen in the fields where it is needed so as to avoid transportation and costs for application. When nitrogen is fixed in the root nodules, the plant can use it efficiently without the losses experienced when the surrounding soil is fertilized artifically.

Table 5.2 Rates of Nitrogen Fixation

Organisms or System	Rate	
	kg/ha · yr	lb/acre · yr
Legumes		
Soybeans	57 to 94	23 to 84
Cowpeas	84	78
Clover	104 to 160	93 to 142
Alfalfa	128 to 600	114 to 534
Lupins	150 to 169	134 to 150
Nodulated nonlegumes		
Alnus	40 to 300	36 to 267
Hippophae	2 to 179	2 to 159
Ceanothus	60	53
Coriaria	150	134
Plant–algal associations		
Gunnera	12 to 21	11 to 19
Azollas	313	279
Lichens	39 to 84	35 to 75
Microorganisms		
Blue–green algae	25	22
Azotobacter	0.3	0.3
Clostridium	0.1 to 0.5	0.1 to 0.4

Genetics

Selection of organisms on a random basis and testing for superiority over the parent strains is tedious and haphazard. Knowledge about biosynthetic pathways and control mechanisms permits a more rational approach to finding better strains because selective media can be devised that cause different growth characteristics. The medium can favor organisms with a given property, and active colonies are picked for further testing. When the medium is nutritionally incomplete or has inhibitors for mutants of the desired type, tiny colonies with poor growth are chosen.

An example of a rational scheme is based on the observation that half of the carbon dioxide assimilated during photosynthesis may be released by respiration, presumably by synthesis and subsequent oxidation of glycolate. Oliver and Zelitch (1977) assumed that slowing glycolate formation would increase photosynthesis. Others have explained the high photosynthetic efficiency of maize, sorghum, sugarcane, and millet in terms of a slow rate of glycolate synthesis. Addition of glyoxylate, an intermediate of the glycolate pathway, to sections of tobacco leaves was found to inhibit photorespiration, thus doubling net photosynthesis. Glyoxylate is the first product formed from glycolate and thus may be expected to inhibit glycolate synthesis by a simple feedback mechanism.

Thus searching for mutant cells with reduced rate of glycolate synthesis is a rational approach to increasing photosynthesis. This and other ways of shifting relative reaction rates offer considerable hope of developing plants with very high photosynthetic efficiency, which, in turn, means high biomass yields per acre.

6

Anaerobic Digestion

Microbial cells can grow in the absence of free oxygen by employing relatively inefficient biochemical pathways in which some molecules are oxidized whereas others are reduced. Gas evolved contains reduced species such as methane, hydrogen, and hydrogen sulfide plus the main oxidized species, carbon dioxide. The net effect of anaerobic treatment of organic material is to destroy about one-half to two-thirds of the solids content by conversion to soluble molecules, some of which escape as gases. Reduction of sludges to less solid mass is termed anaerobic digestion. A simple digester is shown in Fig. 6.1.

Attractive features of anaerobic digestion are (1) the use of elective culture with no attempt to exclude foreign organisms, (2) relatively inexpensive equipment, (3) very simple recovery of the gaseous products and (4) low operating costs because chemicals are used only in small amounts and there is no need for the air and intense mixing needed in aerobic processes. Furthermore, the product gas has moderate heating value and can be upgraded to very desirable pipeline gas by removing the CO_2.

Digestion in Waste Treatment

Biological processes for the treatment of sewage and industrial wastes have been employed for decades. Most of the decomposition of organic materials is conducted in aerated systems where microbial cells, CO_2, and water are the main products. The microbial cells and sludges constitute a disposal problem. Anaerobic digestion is used to reduce the amount of solid mass by converting it to gases plus a smaller amount of sludge that can be collected for ultimate disposal. The reduction of cell mass is more important to waste disposal than is anaerobic production of methane gas, but rate of methane production is a good index of performance.

Anaerobic digestion to produce methane for fuel must achieve high yields of methane and is subject to restraints different from those for waste treatment. For example, concentrated slurries of biomass can be fed, whereas waste treatment processes must accept a fairly dilute feed stream.

Much of the research on methane from biomass is being conducted by people with experience in digestion for waste treatment; thus the terminology is the

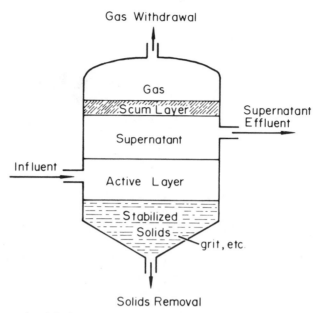

Figure 6.1 Conventional single-stage digester: stratification of highly liquid wastes. Standard operation characteristics: unheated; loading rate 0.64 to 1.60 kg volatile solids/$m^3 \cdot$ day; detention time 30 to 60 days; intermittent feeding and withdrawal; stratification. Courtesy NTIS.

same. Common terms are listed in Table 6.1. These terms reflect the types of analysis practical for ill-defined substances in the reactor. Both BOD and COD are indices of the amount of pollutional material present, but the values seldom agree. Compounds that are metabolized in the time period of the assay are measured by BOD, and this is usually a lower value than that for COD, which measures even those compounds that are resistant to microbial attack. Occassionally, a compound is encountered that is not oxidized by the reagents of the COD test but is metabolized by microorganisms. There is a trend to the use of analyses for total carbon in which CO_2 is measured following combustion of the sample. There is seldom an attempt to measure individual compounds because of the complexity of sewage. A biomass feedstock such as corn stover or a tree species has better defined composition; thus research should progress more rapidly with the use of specific assays instead of BOD and COD.

The VSS measurement is crude but easy; it is a convenient index of organic matter for day-to-day performance of the process. Again, the use of a nonspecific assay makes it difficult to interpret research results. There are fairly good methods for determination of cellulose, of hemicellulose, and of lignin, and excellent chromatographic analyses have been developed that resolve and quantify individual sugars in mixtures. Better characterization of both the dissolved and particulate materials would allow a more scientific approach to anaerobic digestion.

Many years of experience with anaerobic digestion of sewage sludges should translate to digestion of biomass for energy. Methane from sewage sludge is often

Table 6.1   Waste Treatment Technology

Acronym	Term	Definition
BOD	Biochemical oxygen demand	Oxygen equivalent of organic material for microbial metabolism
COD	Chemical oxygen demand	Oxygen equivalent of organic material when oxidized with chemicals
MLSS	Mixed liquor suspended solids	Dry weight of filterable solids in the culture fluid
VSS	Volatile suspended solids	Weight of material that can be driven off dry solids by raising the temperature

used to power operations at the waste treatment plant, and very large plants have been able to supply the gas needs of closely adjacent private users. In a few instances, as with the largest cattle feed lots, there is sufficient manure to supply gas to nearby regions. Nonetheless, municipal and agricultural wastes could supply only 10 to 20 percent of our total gas needs, even if all the wastes could be collected and transported to digesters. Individual digesters for homes would supply negligible power, but units for farms might be economical if fed with animal and plant wastes.

Digestion of municipal solid wastes to methane has been nicely analyzed by Ghosh, Klass, et al. (1978). They recommend dry shredding of the feedstock because wet shredding leads to storage instability, more bulk, and less control of particle size. Various means have been used to remove dirt, metals, glass, and ceramics from refuse, but no single operation can provide satisfactory cleanup. A combination of magnets, screens, and air separators is recomended. Laboratory experiments showed reasonable yields of methane from samples of actual refuse, and provisional designs for a larger plant were provided. Digestion seems worth consideration as an alternative to other means of treating solid wastes, but there can be little impact on national energy needs because not enough municipal refuse is available.

Anaerobic digestion of waste materials and crops to provide methane for energy was reviewed comprehensively by De Renzo (1977). This book includes brief summaries of most of the government-funded research projects on digestion as well as reports from recent meetings and conferences. Although the orientation is biased toward sewage and discussions are mostly noncritical, the book is recommended for the completeness of its coverage.

Almost all plant materials digest well, but often a newly introduced feedstock is not well accepted by a digester. However, after a period of acclimatization, digestion of the new feedstock usually proceeds smoothly. Aquatic plants digest particularly well; algae, Irish moss (*Chrondus crispus*), giant kelp, and the like can be digested efficiently, leaving relatively little unreacted sludge.

Particle size, ease of mixing, and wettability affect rate of digestion. Pfeffer (1978) reports very low conversion efficiency for anaerobic digestion for corn stover. There are no known inhibitors present, so physical factors may be responsible for the poor results.

Economics of a cheap gas such as methane leave little margin for purchasing the feedstock. It probably is not reasonable to produce methane by anaerobic digestion of wood or field crops because their value exceeds that of the methane that could be realized. Aquatic growth, such as water hyacinth or algae, is desirable for bioconversion through digestion, because alternate processes are prejudiced by the excessive moisture present.

Mechanisms of Digestion

Digestion of municipal sewage sludges has traditionally been beset with process upsets and failure to produce methane. This may result from natural fluctuations in the microbial population, inhibitory materials in the waste, operator error, or design inadequacies with respect to process stability. The trend to short detention times in digesters reduces the safety factor so that any imbalance in composition of waste or in process conditions can lead to failure.

In the course of digestion, organic materials are decomposed to small molecules, mostly organic acids, by a variety of anaerobic bacteria. Further reaction to methane is carried out by relatively few organisms, and these are quite fastidious in their tolerance of pH and of concentrations of organics. As outlined in Fig. 6.2, macromolecules lead to soluble organic compounds that are metabolized to organic acids. These acids are attacked two carbon units at a time to yield acetate, and acetate gives methane and carbon dioxide. With the use of radioactive tracers, it has been shown that acid chains longer than acetic are converted stepwise to acetate. Various microorganisms with preferences for different substrates interact synergistically to carry out this scheme.

The acid-producing bacteria are robust and not susceptible to failure. In doing their work, they lower the pH and accumulate acids and salts of organic acids. If the methane-forming organisms are not rapidly attacking these products and thus tending to raise pH and prevent accumulation of acids, conditions are established that are highly adverse to methane formers. Furthermore, the methane organisms are relatively slow in growth and in metabolism, so that heavy organic loading produces acids faster than they can keep up. For large-scale production of methane from biomass, it is essential to maintain dense, active populations of methane-forming organisms, or else detention time and equipment sizes must be very large.

The principal components of cellulosic biomass are cellulose, hemicellulose, and lignin. Anaerobic cultures attack hemicellulose and cellulose but have little effect on lignin, so that the final sludge is high in lignin and other recalcitrant organic molecules and in insoluble inorganic materials. Such sludge is not rich fertilizer but is useful for providing structure to soils, especially to dense, impermeable clays.

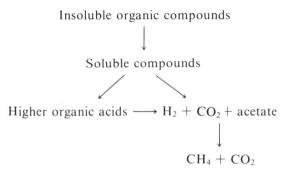

Figure 6.2 Pathways of digestion.

Although anaerobic digestion is usually considered a two-stage process, Bryant (1978) believes that three stages can be delineated: (1) hydrolysis of polymers accompanied by some fermentation, (2) action of hydrogen-producing acetogenic bacteria (proton-reducing acetogenic organisms) and (3) methanogenesis powered by hydrogen. It is hypothesized that hydrogen utilization in stage 3 prevents feedback inhibition of stage 2 because addition of hydrogen has been shown to shut down stage 2. This means that hydrogen partial pressure is a controlling factor in anaerobic digestion.

There are two distinct types of practical anaerobic digestion that depend on temperature. Mesophilic organisms operate in the range of 30 to 37° C, whereas thermophilic organisms function best from 49 to 51° C. Although a thermophilic digestion is faster and more efficient, energy is required to bring the fluids to temperature and to overcome losses. Many digestion processes for sewage operate in the thermophilic range and burn digester gas to supply the energy. However, an economic trade-off may show that mesophilic digestion is better if the gas is to be sold.

There are indications that methane-forming organisms are more temperature sensitive than are acid-producing organisms. At low temperature, the methane organisms have difficulty keeping up with the rate of acid production and there is a tendency for low pH and failure. At higher temperatures, the rate of methane formation is potentially higher than the rate at which acids are produced; thus failure due to acid accumulation is much less likely.

The methanogenic bacteria function in a complicated mixed culture in which microbial interactions are particularly important. Relative abundance depends on nutrition, competition, stimulation or inhibition, predation, and parasitism as well as the physical–chemical conditions. The current status of this fertile area was reviewed by Zeikus (1977).

Kaspar and Wuhrmann (1978) studied the kinetics of some of the main reactions of sludge digestion. Acetate-splitting systems were about half saturated whereas propionate-degrading systems were saturated only to 10 to 15 percent. Hydrogen removal was less than 1 percent of its maximum rate. Their conclusions were that acetate splitting is the rate-limiting step, and capacity for hydrogen consumption was largely unused, with the result that fatty acid oxidation was rapid. There is an implication that natural digestion could be improved dramatically by manipulating the microbiology and conditions of digestion.

A review by Zehnder (1978) includes discussion of the role of hydrogen in anaerobic metabolism. Decomposition of fatty acids involves acetyl coenzyme A as an electron acceptor as the acetyl group is reduced to ethanol. The ethanol is reoxidized by microorganisms with the reaction

$$CH_3CH_2OH + H_2O \rightleftharpoons CH_3COO^- + 2H_2 + H^+$$

$$\Delta G^\circ = 1.39 \text{ kcal/mole}$$

However, the partial pressure of hydrogen in the gas phase does not exceed 10^{-4} atmospheres because the methane bacteria carry out the reaction

$$CO_2 + 4H_2 \rightleftharpoons CH_4 + H_2O \qquad \Delta G^\circ = -33.2 \text{ kcal/mole}$$

The energetics of this reaction greatly favor the formation of methane. Several other anaerobic reactions release hydrogen.

Balch, Fox, et al. (1979) have reevaluated the methanogenic organisms in terms of identification, taxonomy, pathways, and culturing techniques. A major new advance in classification is determination of nucleotide sequences in ribosomal ribonucleic acid. Listings have been compiled of the occurrences of sequences of as small as five nucleotide bases and up to 24 nucleotide bases in various organisms. By comparing the sequences found in an unknown organism with the listing, the choice for identification can be narrowed to one or to a few species. Supplementing this information with morphological examination, biochemical testing, or immunological characterization provides positive identification. Prior to using sequencing, identification was very difficult because methanogens are anaerobic organisms with variable appearance and properties.

Some interesting insight into anaerobic digestion is provided by the studies by Wise, Cooney, et al. (1978) of methane production from the coal gasification products CO_2, H_2, and CO. The reactions are

$$CO + H_2O \longrightarrow H_2 + CO_2 \qquad (6.1)$$

$$CO_2 + 4H_2 \longrightarrow CH_4 + 2H_2O \qquad (6.2)$$

Overall: $\qquad 4CO + H_2O \longrightarrow CH_4 + 3CO_2 \qquad (6.3)$

When excess H_2 is present, all the CO is consumed. An apparatus was devised for anaerobic digestion at high pressure to drive reaction (6.2), where five molecules of gas react to form water and one molecule of gaseous product. The reactor was started with sewage sludge and fed medium plus CO_2 and H_2. Adaptation over a period of days led to vigorous fermentation at 30 atmospheres. The production of methane reached values up to 4 l/l of digester fluid (STP) per hour, which is many times the rate of methane production from sewage sludge. Feeding CO demonstrated the conversion to CO_2 and H_2. Although the intent of this research is quite distinct from biomass fuels, it demonstrates that the biological systems are capable of functioning at rates much higher than those encountered in conventional digestion.

Algal–bacterial mats that grow in alkaline hot springs develop thermophilic methane-producing bacteria (Ward, 1978). The optimum methanogenesis was near 45° C, but the organisms were present at 68° C. By using these organisms, the temperature of thermophilic anaerobic digestion might be extended upward to give faster rates and allow smaller reactors.

Smith and Mah (1978) have found a methanogenic bacterium that decarboxylates acetate to methane at a rate faster than any previously reported. They feel that this step also provides sufficient energy for growth, a view that is sometimes disputed. Cells grow faster on H_2 and CO_2 than on acetate; thus H_2 partial pressure affects the rate of acetate utilization.

Digestion of Manure and Municipal Wastes

Lizdas (1978) has reported on methane production with a pilot unit at a dirt feedlot in Colorado. The system having a 1500-gal (5700-l) digester is outlined in Fig. 6.3. Residue preparation uses two tanks, a mixing pump, and a loading pump. A "V"-bottomed mix–degritting tank is used to prepare the initial slurry of cattle residue and to allow sand and gravel to settle out. A load–metering tank allows the measurement of the total weight of material loaded. A variable-speed, positive-displacement, progressing-cavity-type pump is used to meter the waste load into the system.

A trailer contains all the additional supporting equipment for the system. Fermenter recirculation comes from a centrifugal pump, displacing 200 gal/min (760 l/min). A hot-water heater, a heat exchanger and water heat-transfer loop, and a temperature-control valve are also parts of the system.

Initial startup attempts using only aged pen residue and no fresh manure were unsuccessful. After successful stabilization had been achieved with fresh dairy residue, the substrate was changed to material collected from a "typical" feedlot pen. This substrate, while initially causing a drop in performance, subsequently stabilized, yielding 2.5 ft^3 of methane per pound of volatile solids fed (160 l/kg).

It appears that aged waste, low levels of residual pharmaceuticals, and low levels of various anions and cations all act separately or in concert to exert some additional stress on the biological system. However, it is possible to achieve a stable fermentation and reasonable gas yields from dirt lot residues if a sufficient adaptation period is provided.

One dry ton of organics can yield 8 million BTU of gas and 280 lb (127 kg) of crude protein. Crude protein so produced is as digestible as cottonseed meal (a standard protein supplement). Based on natural gas prices of $2 per million BTU and cottonseed meal prices of $150 per ton, this represents product values of $16 worth of fuel and $51 worth of protein per dry ton of organics. This ratio may be high, but other studies show that the main benefit of anaerobic digestion at cattle feedlots is the production of feed protein and that methane alone would not be the basis of an economically attractive process.

Fresh manure digested very well in terms of stable performance and reason-

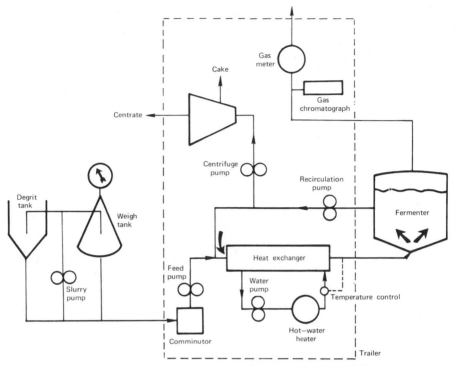

Figure 6.3 Diagram of digestion pilot plant. Courtesy NTIS.

able yields of 2.75 ft^3/lb (171 l/kg) of volatile matter, although these yields are over 20 percent too low based on experience with similar wastes. Pen residue was very low in volatile matter. This was attributed to mixing with large amounts of dirt by hoof action in wet weather. Furthermore, decomposition probably occurred on the ground. Digestion of pen residue gave poor yields of methane (0.5 ft^3/lb, 31 l/kg of volatile matter). Hard water with appreciable concentrations of metal ions in the western part of the United States may pose problems for digestion.

Many details of anaerobic digestion have been investigated by Pfeffer's group (Geisser and Pfeffer, 1977). A single-stage system is more efficient than a multistage system but is much more susceptible to failure if stressed by the processes changes that are so common to anaerobic digestion. With paper as the main biomass feedstock, methanogenesis could easily be upset; manure as a supplemental feedstock improved performance, suggesting that some unidentified nutrients may have a stabilizing influence.

A 50,000-gal (190,000-l) digester for manure has been operated for several years at the Monroe State Dairy Farm in Washington. Typical data are:

pH	7.3
Alkalinity	8250 mg/l

Volatile acids	under 600 mg/l
Gas production	155.7 m³/day
Volatile solids reduction	27 percent
0.24 m³/kg volatile solids added	
0.93 m³/kg volatile solids destroyed	
Electrical consumption	20 kWh/day

When tested on a feed schedule of load every 2 or 3 days, performance was erratic with fluctuations in rate of gas production and percentage of CO_2 (Wise, Ashare, et al., 1979). The digester can serve a large dairy herd, but reduction to 180 head of cattle provided insufficient manure for good gas production. Furthermore, the cattle produced less manure during the winter months.

A digester with feed composed of manure and crop residues is being operated at the U.S. Meat Animal Research Center, Clay Center, Nebraska (Wise, Ashare, et al., 1979). The production rate of 4.65 l/l of medium day is claimed to be the highest reported using animal wastes. This is attributed to avoidance of the antibiotics and growth stimulants used by commercial livestock enterprises.

Attempts to digest manure from a hog farm had little success because the antibiotic lincomycin fed to the hogs was present in the manure (Turnacliff, Custer, et al. (1979). No antibiotic is effective against all microorganisms, and it seems probable that strains could be found to digest manure containing lincomycin. However, several groups have had poor results with manure from animals fed antibiotics.

Several pilot plants or demonstration units for anaerobic digestion have been funded by the U.S. government. A plant at Pompano Beach, Florida converts urban solid wastes to methane-rich gas. The cost was $3.6 million, and capacity is 100 tons/day of waste to give 300,000 ft³ (8.5 million l) of methane. There will be a 2- to 4-yr evaluation of commericial feasibility.

Another plant to be built near Lamar, Colorado at a large cattle feedlot will be financed by a $14.2 million FHA loan. It will process 350 tons/day of manure and produce 1.2 million ft³/day (34 million l) of methane. This is expected to meet 40% of the energy needs of the 10,000 people in the area. Again cattle refeed credits will be important to economic success.

Under a $1.27 million contract from the U.S. Department of Energy, a manure-to-methane plant has been constructed at Bartow, Florida by Hamilton Standard and Kaplan Industries (Lizdas and Coe, 1979). The gas is sent to a nearby packing plant and used to fire steam boilers. Thermophilic digestion with about 10 days of detention works well. A key feature is dewatering the digester solids and using them as supplemental cattle feed. Wastes from 10,000 cattle (25 tons of manure per day) are processed. The initial cost estimate is $5 per million BTU in the gas, but process development could halve the cost. About 23 percent of the fuel gas produced is needed to power the process, but increased methane yields and more efficient operation could reduce this to 7 percent. At this point the value of the supplemental feed is about twice the value of the methane; thus the project is more successful for agriculture than for the fuel produced.

Manure from 130,000 cattle on Oklahoma feedlots is digested to methane by Thermonetics, Inc. (*Biomass Digest*, June 1979). Built at a cost of about $5 million, the plant produces 1.6 million ft^3 (45 million l) of gas per day.

Costs of producing methane are summarized by De Renzo (1977). Methane would cost 31¢/million BTU just for processing. This translates to a total cost with municipal wastes and with converted crops of 39¢ and $2.61/million BTU, respectively.

Anaerobic Digestion of Aquatic Plants

Plants that grow on land present problems for anaerobic digestion because of toughness and lignin, which does not react. Aquatic plants have carbohydrates that are often polymers other than cellulose or hemicellulose. Marine algae contain little cellulose, and their biodegradable fraction consists mainly of carboxylated or sulfonated chains composed of galactose or mannose or mixed sugars. These materials are rapidly and thoroughly digested anaerobically (Sanderson, Wise, et al., 1978).

The marine species *Chondrus crispus* or Irish moss is very readily converted to organic acids. With no pH control, accumulaiton of acid severely inhibits the conversion rate. If organic acids were continuously removed, 98 percent breakdown of the fermentable fraction would occur in 5 days. However, various plants have significant amounts of nonfermentable solids, some of which is inorganic. Table 6.2 compares several plants; hydrilla and duckweed are freshwater plants whereas the others are marine macroalgae. This points out clearly that total dry weight is a poor indicator of potential yields of fuel products.

Giant kelp (*Macrocystis pyrifera*), such seaweeds as *Ascophyllum nodosom*, and water hyacinth have also been shown to digest well on the basis of carbon content. However, there is often confusion between total solids and fermentable solids when advocates are presenting the economics for their pet biomass candidate.

Pretreatment for Anaerobic Digestion

The structure of plants retards microbial attack. Although it is important in nature to have all organic matter ultimately decomposed and recycled, structural elements must resist easy attack. This is accomplished by the structure and biochemistry detailed in Chapter 7. Chemical pretreatment to loosen the structure and to partially degrade the biochemicals can increase rates and product yields for bioconversion.

Anaerobic digestion is exceedingly slow, and the sludge contains unreacted or partially reacted biomass. McCarty, Young, et al. (1978) have reviewed progress of their group in devising effective pretreatment processes. The biodegradability

Table 6.2   Estimated Conversion Percentages

Plant	Percentage of Carbon	Maximum Conversion Yield
Hydrilla	33	83
Gracilaria	17	43
Duckweed	34	85
Ulva lactuca	17	43
Chondrus crispus	18	45

Source: Sanderson, Wise, et al. (1978).

of refuse, which is predominantly lignocellulose, was increased significantly by heat treatment at both high and low pH. Alkaline conditions were most effective. The optimum temperature for increasing biodegradability was 200°C under alkaline conditions and 135°C under acid conditions. Temperatures above these optima gave decreased biodegradability. Soluble products resulting from heat treatment of lignin were inhibitory to the methanogenic culture. Too intense a pretreatment seems to produce toxic substances that more than counterbalance the increased breakdown of the structure.

With newsprint, the pretreatment increased conversion to methane from 25 percent up to a total of 47 percent. The quantity of solids requiring subsequent disposal was reduced by about 30 percent. Feeding alkaline material would assist in pH control of heavily loaded digesters, but excess alkali would have to be neutralized. Acid pretreatment would definitely incur costs for neutralization prior to feeding to the digester. This pretreatment research has interest and theoretical value, but methane is far too cheap a product to justify much additional processing costs.

Plug-Flow Digestion

There is sufficient mixing in conventional anaerobic digestion to inoculate the feed with active organisms. In plug flow, an element moves from inlet to exit with little mixing with adjacent elements. It is necessary to recycle some of the digested stream back to the inlet to inoculate with vigorous, well-adapted organisms. Jewell (1978a) advocates plug flow for anaerobic digestion to lower mixing costs and to minimize construction costs. Flow characteristics of fibrous manure from dairy cattle were too poor for a pilot plant long-tube reactor. A large, full-scale tube might work fairly well, so Jewell's group used four mixed vessels in series as an approximation of a small plug-flow reactor. A 1500-gal (5600 l) plug-flow reactor was fed 7.8 percent solids with a 30-day residence time at 35°C. When digestion was proceeding well, manure as collected was fed at a solids concentration ranging from 10 to 12 percent. Removal of volatile solids by digestion was 25 percent efficient, and methane production was one volume per volume of feed per day. This was thought to be reasonable for a digester with no mechanical mixing.

Jewell (1978a) has also tested a novel digester in which a packed bed of inert solids is coated with the anaerobic cultures. Upflow operation expands the bed slightly and aids distribution of the feed. This reactor was found to operate more rapidly and more efficiently than any other reported using a feed of cattle manure. Scale up to full size reactors might be difficult, and this design requires pumping the feed.

Augenstein, Wise, et al. (1976/77) have also worked with digestion in packed beds. Their system included the reactor, a recirculation pump, and a carbon column to remove potentially toxic materials. Operating with a mixture of shredded newsprint and sewage sludge, plus powdered calcium carbonate for buffering, methane was produced over a 10-month period. The methane yield was 2.66 ft^3/lb (166 l/kg) of substrate. Yields were better with the carbon column in the recycle loop. Municipal solid waste was also tested, showing a yield of 1.52 ft^3/lb (95 l)/kg. After a short time the effluent stream from any run became clear, indicating that microbial solids were well retained in the packed bed.

A similar reactor was used for producing hydrocarbons from biomass (Sanderson, Wise, et al., 1978). Algae were digested in a packed-bed anaerobic digester with loading rates such that organic acids rather than methane were the major products. Following separation of the acids by solvent extraction or a membrane process, Kolbe electrolysis produces hydrocarbons. The reaction with acetate as an example is

$$2CH_3COO^- = CH_3CH_2CH_3 + CO_2$$

A serious drawback is the use of electricity, a prime form of energy for production of combustion gas. However, the overall economics appear reasonable. In a variation of this process, organic acids could be the final products instead of hydrocarbons, and special engines could use them as fuels.

Recovery of Organic Acids

Organic acids are priced considerably higher than methane; thus it is worthwhile to explore their recovery from anaerobic digestion. Jeffries, Olmsteadt, et al. (1978a) and Gregor (1978) have operated laboratory digesters with very high loadings to favor the acid production phase and have pumped the liquor through an ultrafiltration membrane unit. The ultrafiltrate goes through a novel membrane system that is impregnated with a solvent; on the other side this same solvent is used to extract and concentrate the acid. Removal of the acids from the digester provides control so that acid formation is a diversion of the extra substrate and can be costed independently from the methane process.

Gregor (personal communications) has developed polymeric membranes of a sulfonic acid structure that are not fouled by the nonpolar compounds that play havoc with conventional membranes such as those made from cellulose acetate. A membrane life of 2 yr in commercial service seems justified by long runs in the laboratory. Ultrafiltration of actual effluent streams from a large digester fed manure produced a concentrate that was a heavy paste of 12 percent solids. The permeate had low solids content but contained inorganic salts, including ammonium and nitrate. At the present time, the cost is estimated at $1/1000 gal of effluent treated, which seems high. However, the value of the proteinaceous concentrate for cattle feeding justifies the cost. Furthermore, process development could lead to significant cost reductions. There are electrodialysis procedures for recovering ammonia from the permeate, but it might be better to use this stream directly for combined irrigation and fertilization of crops.

Upgrading Methane from Digesters

Gas from a digester is usually 60 to 70 percent methane. To meet pipeline specifications, carbon dioxide, hydrogen sulfide, and moisture must be removed. The practical methods for CO_2 removal are absorption into a liquid, adsorption on a solid, cryogenic separation, permeation, and chemical conversion. Ashare, Augenstein, et al. (1978) have surveyed technology for upgrading digester gas to pipeline fuel. Packed-bed, plate, or spray towers can be used to absorb CO_2 in solvent. Elevated pressure increases solubility. Solvents that have been used are tributyl phosphate, propylene carbonate, N-methylpyrrolidone, methanol, propylene glycol dimethyl ether, and water. Low vapor pressure of the solvent is an asset because losses to the gas phase can be costly.

Water scrubbing is an old process for CO_2 collection. The water is regenerated by lowering the pressure to let gases escape. This process is seldom used for the following reasons: (1) nonselectivity creates intolerable losses of desired gaseous constituents; (2) pumping costs are high; (3) CO_2 removal efficiency is low; and (4) recovered CO_2 is impure.

Chemical absorption of CO_2 into amine solutions or aqueous alkali forms carbonate or bicarbonate ions. Removal efficiency and selectivity for CO_2 are excellent, but regeneration requires heat energy to break bonds. Addition of acid to evolve CO_2 is far too costly and does not regenerate the absorbent. Amine scrubbing is the most widely used method for CO_2 collection.

Adsorption of gas on a solid surface involves weak van der Waals forces, and not much energy is needed for regeneration. Proper choice of the adsorbent provides some selectivity for a given gas. Most gas purification and dehydration adsorbents contain silica, alumina, activated carbon, or silicates that are known as *molecular sieves*. Use of two or more columns such that one is adsorbing while another is regenerating permits continuous operation. Although adsorption is practical, there are problems of high initial cost, poisoning of the surfaces, and plugging.

Cryogenic purification allows fractional condensation and fractional distillation at low temperatures. Pure components may be obtained in convenient liquid form. Commercial trials of cryogenic purification of digester gas have not been very successful because of high capital costs and low thermal efficiency.

When CO_2 concentrations are quite low, it is cheaper to react CO_2 than to remove it. Catalytic hydrogenation gives methane and water. The reaction is highly exothermic; thus cooling is required. Hydrogenation is unlikely to be used for digester gas.

Ashare, Augenstein, et al. (1978) have evaluated a wide variety of systems for removal of CO_2 from digester gas. Although there are many commercial processes based on proprietary absorbents, it was surprising that none were economically attractive for digester gas. The reason is the relatively small scale of digester operations, which does not justify the complexity of most commercial processes, high energy for regeneration, and the high capital costs. A possible exception is the Fluor solvent process based on propylene carbonate as the absorbent; the economics seem favorable, but there is high uncertainty because of insufficient data for this application. Comparisons were facilitated by a computer program for economic estimates. Water scrubbing and membrane separation appeared most economical for upgrading digester gas to meet pipeline specifications. Again this was unexpected because water scrubbing is a very poor choice for industrial gases because poor efficiency means large volumes of water. When recovery of CO_2 is not needed, the spent water can be held in a lagoon for escape of CO_2 to the atmosphere. This eliminates equipment and energy for solvent regeneration and saves money. For very large volumes of gases at higher pressures and with recovery of CO_2 for sale, water scrubbing would not be considered. If CO_2 has sufficient value, alternatives to water scrubbing should be found; the by-product credits can be quite significant.

The best membrane process would probably use superpermeable membranes such as those developed by the General Electric Company. Carbon dioxide passes through much faster than does methane; thus considerable enrichment is possible. The assumed membrane life was 3 yr, and a longer-life membrane could give the process even more favorable economics.

Gardner, Crane, et al. (1977) feel that permeation through membranes will become a very common method for separating gas mixtures. Transfer coefficients are specific to the membrane employed, but certain gases tend to permeate much more rapidly than do others. Values noted by these investigators are given in Table 6.3. The complicated units for the coefficient are of little concern; the important point is the much greater rate of CO_2 than for methane. It should be relatively easy to remove most of the CO_2 with little loss of methane from digester gas. As commercialization of membrane permeation proceeds, there is a good chance of application to digester gas.

Digester Sludge

Digested sludge is biomass of very low value. Economic analyses that claim credits for sludge as a high-grade fertilizer must be questioned. At best, the value of

Table 6.3 Permeation
Coefficients

Gas	Coefficient[a]
H_2	12,000 to 27,000
CO_2	4,000
CO	210
N_2	210
CH_4	160

[a] cm^3 (STP), cm $\times$ 10^{-12}/sec, cm^2, cm Hg.

sludge may be sufficient to pay its processing cost; thus sludge is a loss of biomass and a waste.

Sewage sludge can be filtered or centrifuged. The capital and operating costs of centrifuges are far too high to permit consideration as a major step in a process for producing cheap fuels. Industrial filtration equipment such as the rotary drum filter does poorly with digester sludge in that rates are slow and the filter cake is very wet. Addition of conditioning chemicals such as alum, lime, or polymeric flocculants produces a cake that filters faster and drains well. Of course, these chemicals add to the operating cost.

Solids are filtered from the liquid by a porous medium such as cloth, steel mesh, or tight coil springs. Often the pore size is larger than some of the particles, but a cake of coarser particles builds up and becomes the filter medium. The filtrate from sewage sludges is high in soluble constituents and is recycled to an aerobic treatment step. A factory for direct digestion of wastes would also need a step for treating effluents, and the microbial sludges produced should be quite suitable for supplementing the solids fed to the digester. The need for waste treatment may not be obvious, but such untreated, rich effluent from the digester could not be discharged directly. A digester integrated with an energy farm could very well use digester effluent for irrigation; it would have desirable levels of nitrogen and phosphorous.

The usual disposal method for digested sewage sludges is drying on sand beds, followed by either dumping or use as a soil conditioner. Even for a small sewage plant, sand beds occupy considerable area. The land required for sand beds for a very large plant producing methane from biomass is not likely to be available at reasonable cost. If the sludge is considered as a waste, perhaps the sludge should not be separated from digester effluent to be used for irrigation.

Digesters for Individual Farms

Skrinde (1978) inspected units in India and other countries and reviewed the status of small anaerobic digesters to supply family needs for cooking, lighting, and heating. He noted that countries that have abundant forests show little success with digesters for small farms. However, when labor is very cheap and fuel is very expensive, methane by anaerobic digestion can be important.

Critical energy shortages during the World War II period led Germany to operate military equipment on alcohol. England, France, and Germany during the same period developed small farm digestion plants for gas used for heating and for trials in tractors and trucks. These digesters quickly fell into disuse because of poor reliability, inconvenience, and the return of plentiful supplies of petroleum after the war.

A program of the United Nations adopted the Colombo Declaration in April 1974, resolving that one of the most urgent priorities is energy. A regional project for the development of anaerobic digestion throughout Asia was approved in November 1974. The term "biogas" signifies small, diversified digestion operations. A mission sent to India, Japan, Korea, Pakistan, the Philippines, and Thailand reported extremely favorably on the potential of biogas plants, suggesting additional use of effluents for the growing of algae and the raising of fish, ducks, and crops.

India

India imports two-thirds of its energy requirements. For 30 years there has been development of anaerobic fermentation of cow manure, known as *gobar* in Hindi. The first biogas plants began in 1951; by 1978 the number had reached 50,000, with an ultimate target of 100,000 units. The Government of India attaches great importance to this program because of the energy crisis and shortage of feed and fertilizer.

The development of biogas plants in India has resulted in preparation of standard designs utilizing concrete tanks, concrete inlet and outlet basins, and steel covers serving as floating gas holders. The expense of the steel cover is a major cost. One year of a family's income may equal the purchase cost of a digester. These digesters have no pumps, motors, mixing devices, or other moving parts, and digestion takes place at ambient temperature. Retention time is generally about 50 days. The digesters contain a baffle in the center, much like a normal septic tank. A schematic diagram representative of the present Indian standard plans is shown in Fig. 6.4.

Even with government encouragement, the biogas program in India is not progressing well. With high interest rates and the cost of maintaining the steel gas holder, all for a return of energy equivalent to only 0.5 to 0.8 gal (2 to 3 l) of kerosene a day, there is some reluctance on the part of the farmer to invest in a biogas plant. Daily labor requirements, a lack of reliability, and inconvenience are other factors. Even though the digester solids are a good soil fertilizer, this is apparently not sufficient incentive to create a continuing national movement to small biogas plants. A number of biogas plants have been discontinued.

Japan

In Japan, biogas generation in small plants seems to have been common through World War II, and many original designs of digesters and gasholders were devel-

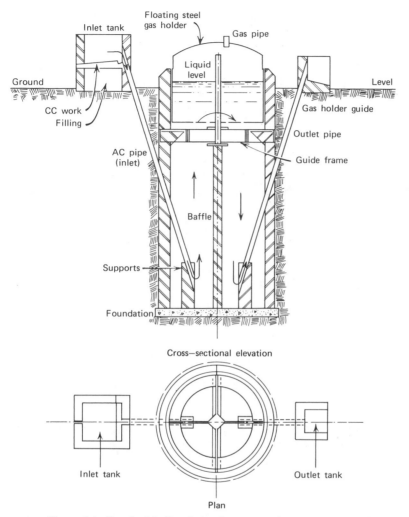

Figure 6.4 Standard Indian design for a small digester. Courtesy NTIS.

oped in that country. Currently the process is seldom used in the villages, however, because of greater affluence and the easy availability of petroleum fuel and chemical fertilizers.

Republic of Korea

In the Republic of Korea, 25,000 digesters have been installed. The severe Korean winter and the lack of a cattle-based economy prejudice the program. Premature felling of trees has denuded the land, and most of the straw from rice and barley crops is used as fuel. A major part of the construction has been done by the farmers themselves, but to accelerate the program the government had been subsidizing 33 to 50 percent of the cost. This subsidy has been discontinued, and the number of new units being constructed has declined. Shortage of animal waste, difficulty in collecting wastes, and the rapid urbanization of rural areas has resulted in a decline in the number of small digesters.

Pakistan

In Pakistan, nearly 100 biogas plants have been built, and there is growing interest in the biogas industry. The government has been supplying the gasholders free, provided the farmers build their own digesters. The main problems in extension of the program have been low temperatures from November to March and the high cost of steel imported for fabrication of gasholders.

The Philippines

Fuel is not a major problem in Philippine villages because of the ready availability of firewood. There is a small biogas program for fuel, and pollution control is one of the motivating factors.

Thailand

Farmers in Thailand have an abundance of charcoal and wood, and this factor, coupled with problems of animal manure availability and collection, has inhibited a rapid development of biogas plants. In the southern region, however, considerable hog manure is available. The present emphasis on biogas plants is mainly for treatment of human wastes to reduce breeding of house flies and to minimize infectious diseases. Anaerobic digestion is also being developed to treat industrial wastes, but here again the emphasis is on waste treatment rather than on energy production.

Taiwan

Nearly 7500 biogas units are in operation in Taiwan, and most family-sized units were designed to handle household wastes plus the wastes from several hogs. A unique feature of the design is manual agitation by pulling and releasing a rope fastened to a mixer made of plastic pipe.

   The use of bag digesters may have originated in Taiwan. A sausage-shaped bag is provided with a polyvinyl chloride (PVC) inlet and outlet. Advantages of bag digesters are reported to be low-cost, mass-production capabilities, and easy transportability. Bag digesters are available in both circular and rectangular configurations and in sizes ranging from 180 to 7000 ft^3 (5 to 200 m^3).

   Huang, Won, et al. (1978) have designed and operated anaerobic digesters for swine waste. Taiwan raises 5 to 7 million hogs per year, most of which are on family farms. Before World War II, the Japanese generated methane on Taiwan, but the technology faded away. The authors propose a system for small farms based on their research, in which two vessels in series are used. Loading should be about 0.68 g of volatile solids per liter per day with a detention time of 16.7 days.

Indonesia

The availability of firewood is not a problem in most parts of Indonesia; there-fore, biogas development there has been slow. Acceptability of biogas may be hampered also by religious convictions and certain prejudices. For example, a biogas unit set up by the Indonesian Board of Voluntary Services had to be closed down because of opposition from the Muslim villages for using hog ma-nure. In another case the use of digested night soil as fertilizer was discontinued when a local witch doctor in Bali attributed sickness to the consumption of prod-ucts grown using the digested slurry as fertilizer. Approximately 12 biogas plants are operating throughout Indonesia, with numerous experimental units being evaluated to reduce construction cost.

People's Republic of China

The People's Republic of China has for many years digested human waste with other organic matter such as garbage, animal waste, and plant materials to pro-duce biogas and fertilizer. Several hundred thousand plants are reported to have been built.

A significant feature of the People's Republic of China design is incorporation of the gas storage compartment within the fermentation vessel by means of baf-fles. This allows construction of a fixed cover facility, obviating the need of costly steel floating covers commonly used in other countries. The gas displaces liquid in the gas holding compartment, and a simple water-filled U-shaped tube in the effluent gas line serves as a safety valve, releasing gas if the pressure becomes ex-cessive. It should be noted, however, that few people have visited biogas plants in the People's Republic of China to verify the enthusiastic reports. Biogas plants are not numerous in large cities and more affluent eastern areas, but may be more common in poorer western areas.

The National Academy of Sciences (1977) of the United States has issued a report on methane generation that includes a review of practices in various coun-tries. Table 6.4, taken from the report, shows gas yields from digestion from sev-eral different feedstocks.

In the most rural areas of the poorest countries where the cost of labor is low and the kerosene equivalent of biogas is expensive and electricity is nonexistent, a very simple biogas system producing a substantial portion of the family's needs may be feasible. Even in these cases the present biogas plant designs are too costly, and the value of the biogas produced is hardly equal to interest on the construction cost. Where human waste is incorporated into the process, such as in the People's Republic of China, and where the plants are extremely low in construction and operating cost, the biogas system can be a good method of incorporating sanitation with fertilizer enhancement and energy production.

United States

Jewell is a leading American proponent of anaerobic digesters for individual farms. He points out that the maximum energy value is about $5/ton of fresh, wet manure and that there is an additional $5 to $10/ton worth of nutrients and refeedable fiber. More than 80 percent of all U.S. farms are less than 500 acres (200 ha) in size, and the total of farms is about 2.3 million. Manure has too little value to justify hauling to central processing plants; thus digestion near the source of manure must be considered. With a 10-yr payback on the investment, Jewell estimates that the maximum allocation to construct a digester for a small herd of 10 animals is $2400 and $24,000 for 100 animals. As this 10-yr payback is acceptable only to utilities and monopolies, a much lower cost would be reasonable for private financing to pay back in 2 to 3 yr. Nevertheless, manure is a nuisance and a waste to be disposed of. As Jewell's research is finding it difficult to meet his goals with utility financing, it is highly questionable whether methane from individual farm digesters can be economic.

In the harsh winters of most of the United States the digestion rates will fall to almost nothing unless the units are well insulated. A liner laid on insulating material in an excavation is probably the cheapest construction technique. An inflatable bag would be a low-cost collector, but the gas might not have sufficient pressure to serve as fuel. Digesters occasionally explode, so leak prevention and safety devices are important and can add significant costs. When methane is a marginal financial proposition at large feedlots and is dependent on refeeding credits that are two times the value of the methane, the problems for small farms are so formidable that the wisdom of performing much R&D is questionable.

Jewell (1978b) reported on a digester operated with 16-day retention time at 35°C using 10 to 12 percent dairy cattle manure. Gas production was 1.4 l/l day when volatile solids reduction was 28 percent. To avoid metal components and their corrosion problems, PVC elements were used in the heat exchangers. However, pumping of thick slurries led to hydraulic shocks that damaged the plastic. Another problem was encountered in the flexible plastic cover for the digester because leakage occurred. The original liner at $10.75/m² has been replaced with a polyurethane–ether liner costing $21.52/m² (B. F. Goodrich Company). These problems should serve as a warning to anyone planning a homemade digester. A team of skilled, experienced investigators has difficulty in designing a digester.

Methane from Landfills

Sanitary landfills are dumps of solid wastes sandwiched between layers of soil. After a number of years the landfill site will stabilize so that it is suitable for construction of airports, light buildings, and the like. However, the odor of methane

Table 6.4 Yield of Biogas from Various Wast Materials[a]

Raw Materials	Biogas Production per Unit Weight of Dry Solids		Temperature		CH₄ Content in Gas (Percent)	Fermentation Time, days
	ft³/lb	m³/kg	°F	°C		
Cow dung	5.3	0.33	—	—	—	—
Cattle manure	5	0.31	—	—	—	—
Cattle manure (India)	3.6 to 8.0	0.23 to 0.50	52 to 88	11.1 to 31.1	—	—
Cattle manure (Germany)	3.1 to 4.7	0.20 to 0.29	60 to 63	15.5 to 17.3	—	—
Beef manure	13.7[b]	0.86[b]	95	34.6	58	10
Beef manure	17.7	1.11	95	34.6	57	10
Chicken manure	5.0[c]	0.31[c]	99	37.3	60	30
Poultry manure	7.3 to 8.6[f]	0.46 to 0.54[f]	90.5	32.6	58	10 to 15
Poultry manure	8.9[c]	0.56[c]	123	50.6	69	9
Swine manure	11.1 to 12.2	0.69 to 0.76	90.5	32.6	58 to 60	10 to 15
Swine manure[d,e]	7.9	0.49	91	32.9	61	10
Swine manure	16.3	1.02	95	34.6	68	20
Sheep manure[d]	5.9 to 9.7	0.37 to 0.61	—	—	64	20
Forage leaves	8	0.5	—	—	—	29
Sugar beet leaves	8	0.5	—	—	—	14
Algae	5.1	0.32	113 to 122	45 to 50	—	11 to 20
Night soil	6	0.38	68 to 79	20 to 26.2	—	21

[a] Some figures have been rounded to the nearest tenth.
[b] Based on total solids.
[c] Based on volatile solids fed.
[d] Includes both feces and urine.
[e] Animals on growing and finishing rations.
[f] Based on volatile solids destroyed.

183

is recognized from landfills throughout the stabilization period and for years afterward. The gas produced and trapped in a landfill may be tapped by perforated pipes. As with gas from anaerobic digestion, there is roughly 60 percent methane and 40 percent carbon dioxide.

The status of landfills as sources of methane was briefly reviewed (*Chemical Engineering News,* August 28, 1978). Since 1970, methane has been collected from a landfill at a Palos Verdes near Los Angeles, California. A purification plant was constructed, and gas was sold to a pipeline company in 1975. There have been operating problems, and corrosion was troublesome until better materials were employed. A slight vacuum in the collection system draws gas from the landfill, and molecular sieves are used to remove carbon dioxide. Production is about 500,000 ft^3/day (18,000 m^3/day).

Although only large landfills constitute much of an economic resource, this approach seems very sensible for recovering energy from past accumulations of biomass. The useful lifetime for generation of methane is unknown, but the Palos Verdes plant seems to be maintaining production smoothly. The question may be raised about landfills as an alternative to anaerobic digesters. There are obvious advantages in very low construction costs and no problems of sludge handling or waste disposal, but the efficiency appears to be low, and volumetric productivity is poor compared to a digester. However, where land is cheap and biomass is available, the landfill approach is worth considering. After several years, when the land has been upgraded to a higher value, the plant could be dismantled and moved to another site.

7

Fractionation and Pretreatment

Overview

Although much research has dealt with conversion of cellulose to fermentable sugars, it is becoming obvious that cheap fuels cannot be obtained if noncellulosic components are wasted. Not only are credits for possible products lost, but it is costly to treat the large amounts of biological waste materials left over after using only the cellulose. Biomass typically contains 30 to 40 percent cellulose. Assuming the purchase of biomass at $30/ton, the material cost for only cellulose is about $90/ton. This leaves little margin for profitable hydrolysis to glucose at a price that is attractive for fermentation to ethanol.

The situation brightens considerably when the other biomass constituents are also utilized. Hemicellulose already has some value as a feedstock, and new fermentation processes are being developed for the sugars from hydrolyzed hemicellulose. Lignin has more carbon and hydrogen in proportion to oxygen than do cellulose and hemicellulose and has the highest heating value. Direct burning can utilize lignin while awaiting the development of other more valuable large volume applications. See discussions of lignin in Chapters 3 and 10.

Biomass Fractionation and Hydrolysis

Native biomass is resistant to microbial attack because of its physical and biochemical structure. Anaerobic digestion is a sort of brute-force bioconversion that accepts crude biomass and converts much of the readily metabolizable constituents to CO_2 and methane while leaving recalcitrant components such as lignin mostly unaffected to be lost in the sludge. Fermentation by controlled cultures can give high yields of selected products, but native biomass is a very poor substrate. To optimize fermentation, it is necessary to fractionate and/or hydrolyze crude biomass. Physical, chemical, and biological steps can be employed.

Structure of Biomass

Lignocellulosic materials are mostly cellulose, hemicellulose, and lignin in ratios of roughly 4:3:3. This can vary considerably as softwood contains typically 42 percent, 25 percent, and 28 percent of cellulose, hemicellulose, and lignin, respectively, whereas corn cobs contain about 40 percent, 36 percent, and 16 percent. Cellulosic biomass contains some protein as do all living cells, and some proposed energy feedstocks contain sufficient protein to justify recovery. In fact, edible protein is much more valuable than the fuels or chemicals designated as the main products. Ideally, biomass can feed the world, provide energy, and supply chemical feedstocks. Matching supply to demand for a multiproduct array is a formidable problem; thus a discussion should focus on one large-volume product and treat others as by-products. Trees, corn stover, and other biomasses composed mainly of structural portions of plants are relatively low in protein, and it is too costly and difficult to extract.

Cellulose is a polymer of glucose rings linked at the 1,4 positions. The 1,4 positions are also linked in starch, which has additional bonds at some of the 1,6 positions to form branches. Whereas the 1,4 links in starch are of the α type and very easily hydrolyzed, cellulose has β 1,4 links resistant to hydrolysis (Fig. 7.1). Starch is a rather dynamic biochemical that serves as a food reserve that may be drawn on quickly and easily. In contrast, cellulose furnishes rigidity to plants and should maintain its integrity until death. Snails are one of the few animals to produce enzymes that attack cellulose. Other animals that feed on cellulose rely on microorganisms in their intestinal tracts to digest cellulose to smaller molecules that can be absorbed in the intestine. Rumen organisms, wood rots, and other microorganisms hydrolyze cellulose. Most are molds, but some bacteria produce cellulase enzymes.

Hemicellulose is a polymer with a linear chain of mostly xylose units joined through 1-4 linkages plus side chains of diverse units such as arabinose, glucuronic acid, mannose, other pentoses, and hexoses (Fig. 7.1). Side chains are usually initiated through 1-3 links to the linear portion. Hemicellulose is much lower in molecular weight than cellulose; roughly 100 to 200 monomers are polymerized. There is no highly ordered crystalline state as with cellulose.

Hydrolysis of hemicellulose to mono- and oligosaccharides is easily accomplished with either acids or enzymes under moderate conditions. Native cellulose resists hydrolysis because of (1) its highly ordered crystalline structure and (2) a physical barrier of lignin surrounding cellulose fibers. The 1,4-β-glucosidic linkage in cellulose may be about as easy to break as the 1,4-α-glucosidic linkage in starch if the cellulose molecules are fully hydrated and exposed and free of lignin. In other words, the primary linkage of the cellulosic polymeric chain may not be as important in causing slow and incomplete hydrolysis of cellulose as are secondary and tertiary structures of cellulosic materials.

Degree of polymerization (DP) is the number of glucose units in an average chain. Depending on its source, the cellulose DP ranges from about 1000 (cellulose in newsprint) to as high as 10,000 (in cotton). In hydrolysis with acids, the

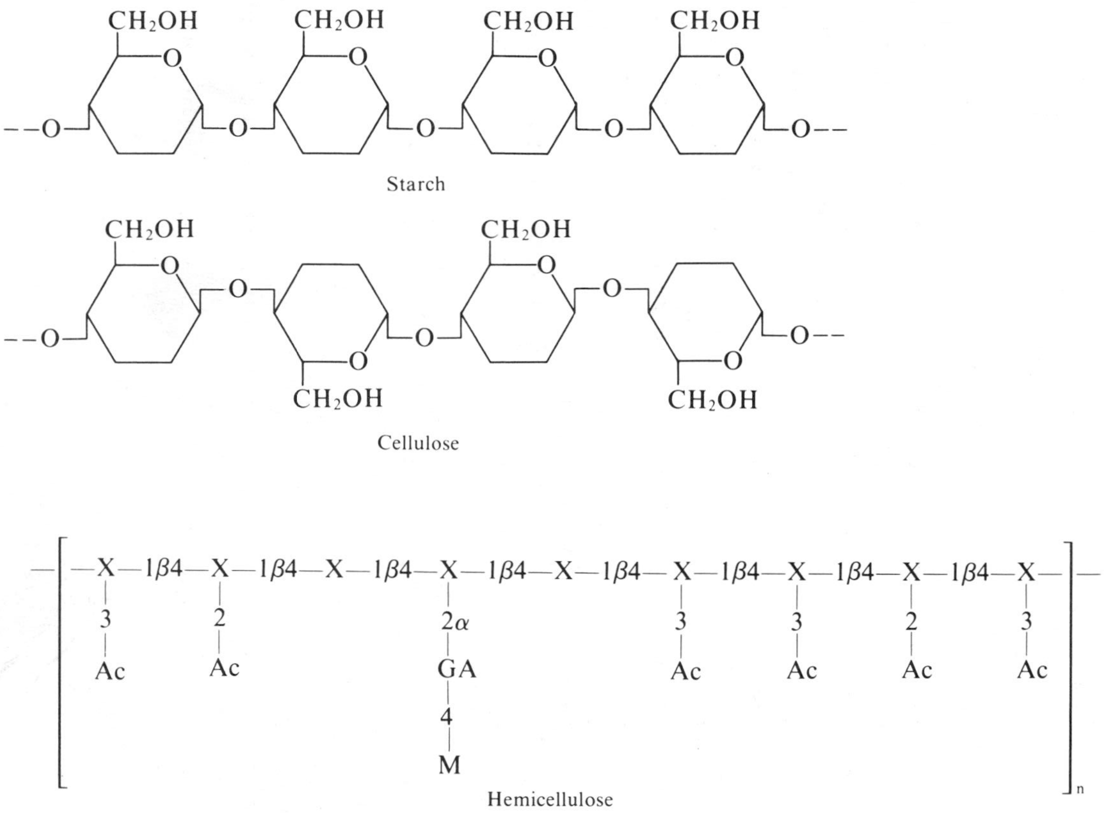

Figure 7.1 Structures of starch, cellulose, and hemicellulose [hemicellulose based on Schury (1978)].

DP drops quickly and levels off at a more or less constant value of 100 to 200, corresponding to a length of 500 to 1000 Å. The easily hydrolyzable portion of cellulose is often referred to as the "amorphous" region and the resistant residue, the "crystalline cellulose." On average, cellulose is 15 percent amorphous and 85 percent crystalline. To hydrolyze crystalline cellulose, a strong acid at a high temperature is needed; chemical and equipment costs can be prohibitively high. The strongly acidic conditions degrade some of the glucose into undesirable by-products.

Once the amorphous cellulose is removed, rod-shaped particles of crystalline cellulose can be obtained. In the case of cotton, the particles are about 400 Å long and 100 Å wide. Further hydrolysis with strong acid can reduce particle weight by as much as 80 percent and yet the crystalline order remains.

The crystalline structure of cellulose is still a subject of considerable controversy. Two major schools of thought exist. Based on the observation of chain folding in linear synthetic polymers, models of cellulose fibrils involving folded chains have been suggested. On the other hand, models based on extended cellulose molecules are also found in the literature.

The anhydroglucose units are linked together through 1,4-β-glucosidic bonds, most of which exist in a configuration known as *Hermans form*. Through repeated Hermans β linkages, linear polymeric cellulose molecules can be built up. Variations of the normal Hermans form probably exist. With two or three such successive deflected β bonds, a loop can be formed in a cellulose polymer to produce a 180° U turn essential for chain folding. These exposed, deflected β linkages are likely to be more susceptible to hydrolytic cleavage. Therefore, the so-called amorphous regions in a cellulose fibril could be zones rich in loop bonds containing many deflected β-glucosidic linkages. After hydrolysis by a dilute acid that removes "amorphous" regions, cellulose left is of an increased "crystallinity" and is very resistant to further acid hydrolysis.

Native cellulose fibers exist in the cell walls of trees and other vegetative mate-

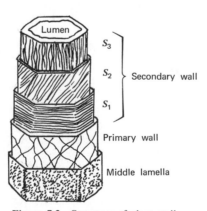

Figure 7.2 Structure of plant stalks.

rials. The cell wall architecture is schematically depicted in Fig. 7.2, which suggests the cementing role of lignin and the hindrance it creates to cellulose hydrolysis. The lumen represents space once occupied by the cytoplasm of the living cell.

Surrounding the fiber is a region called the *middle lamella,* which contains mostly lignin. The outermost layer of the fiber is called the *primary wall,* which was formed on cell division. The secondary wall, formed during the growth and maturation of the cell, is subdivided into the transition lamella (S_1), the main secondary wall (S_2), and the inner secondary wall (S_3). The middle lamella is about 1 to 2 microns thick, amorphous, and generally porous. The primary wall is usually very thin, but the secondary wall (S_1, S_2, and S_3) thickens during cell growth and contains the majority of cellulose. Some molds send their hyphae right into the lumen for efficient digestion.

With the concepts of strong crystalline structure and lignin barrier, the difficulties of cellulose hydrolysis by either acids or enzymes are understandable. Acids are nonspecific catalysts that can attack cellulose and, to a small extent, paper. Only very strong acids can hydrolyze cellulose well, but they also promote the degradation of glucose and reduce its yield.

Cellulases can convert cellulose into glucose with few by-products. Cellulase enzymes cannot easily penetrate through the lignin seal surrounding the cellulose fibers; thus the problem is nonaccessibility of cellulose to the enzymes.

The structure of a piece of cellulosic material is analogous to that of a reinforced concrete pillar with cellulose fibers resembling the metal rods and the lignin the natural cement. To obtain a significant amount of glucose from a piece of cellulosic material, the material is often pretreated. Mechanical grinding is an obvious method. Mechanical milling to reduce a piece of cellulosic material to small particles of an average size of 400 mesh can enhance the sugar production with 48 hr of contact with enzymes from about 50 to about 75 percent. Unfortunately, the cost of milling is too high.

Hemicellulose is insoluble in water, soluble in dilute alkali, and very readily hydrolyzed by dilute acid. There are two types: that found in cellulose fibrils and the spongelike, amorphous material between cells.

Lignin is a very complicated polymer of a wide variety of phenolic compounds (Kringstad, 1978). Figure 7.3 shows a structure for lignin and its principal monomers. Ether linkages between aromatic rings are possible at several positions; thus a three-dimensional structure results. Many –OH and methoxy groups are present. Lignin is slowly decomposed by microorganisms, but the aromatic rings are opened before the ether bonds are broken. Catalytic hydrogenation can cleave the ethers to give valuable aromatic compounds. Biological cleavage of aromatic ethers has not been observed. The resulting phenols would be toxic, so a microbial process to produce them is unlikely.

After extracting hemicellulose, the solid mass could be treated with a solvent for either lignin or cellulose. Lignin is fairly soluble in butanol, xylene, and the like; however, the insoluble cellulose may retain its crystallinity and be resistant

Structure of a portion of lignin

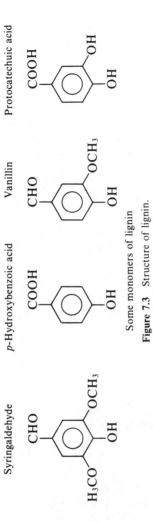

Some monomers of lignin

Figure 7.3   Structure of lignin.

to hydrolysis. Dissolution of the cellulose instead renders it more amorphous and leads to easy hydrolysis. Ladisch, Ladisch, et al. (1978) have tested a number of solvents for cellulose and found several that are effective. It has not been necessary to solubilize the cellulose, and soaking in the solvent destroys enough of the crystallinity to permit ready hydrolysis so that the resulting sugar solution can be washed away from the residue of lignin.

Some perspective on the resistance of lignin to microbial attack is gained from research by Martin and Haider (1979). Radiotracer experiments showed that only 48 percent of the carbon atoms in the aromatic rings was oxidized to CO_2 after 6 months of incubation in fertile sandy-loam soil. Carbon atoms in the side chains were more readily attacked, and 60 percent appeared as CO_2 in the same time period. Although lignin is degraded in soil and thus has some fertilizer value, the rates even with rich mixed cultures are discouragingly slow for consideration as the basis for an industrial process. Addition of easily metabolizable substances has little effect on the rate of lignin degradation, and there are little differences in the rates of attack on lignins from various materials.

Ban, Glanser-Šoljan, et al. (1979) report reasonably rapid degradation of lignin with mixed culture of microorganisms. The substrate was calcium lignosulfonate from pulping of beechwood, and the culture contained at least two yeasts and two bacteria. About one-half of the lignin was destroyed in 24 hr. Several other papers report lignin degradation (Reddy, 1978).

Acid Hydrolysis

Acid hydrolysis of wood is an old technology, and projects during World War II led to the Madison process, which optimized time, temperature, and acid strength. The process is not economical in the United States, but other countries, particularly the Soviet Union, have many plants for hydrolyzing wood to sugars. A few plants produce alcohol by fermenting the sugar, but single-cell protein for animal feed is the most common product. Furfural is sometimes derived from the pentose fraction from hemicellulose.

Acid hydrolysis leads to a sequence of reactions. Hydrolysis is approximately 1000 times faster for hemicellulose than for cellulose. The sugars from each are degraded by acid to resins, polymers, and furfural derivatives. Reaction conditions are thus set for a compromise between hydrolysis and degradation such that the final mixture contains unreacted biomass, unwanted products, and the desired sugars. Since the sugars from hemicellulose are formed early, there is time for considerable degradation leading to major losses. The maximum yield of fermentable sugars is about 55 percent by weight of starting cellulose.

Hemicellulose can be removed by dilute acid treatment with very little effect on the cellulose. There is a time–temperature relationship whereby low acid concentrations take more extreme conditions and longer times to extract the hemicellulose. Although it might seem essential to avoid the cost of pressure vessels by using strong acid and lower temperatures, it turns out that the cost of neutralizing agents wipes out the savings in capital cost. Adequate conditions range from 0.1 percent sulfuric acid at 170° C to 1 percent at 120° C with times up to 1 hr. The pressure vessels for a batch process are not overly expensive because these temperatures do not correspond to very high pressures. However, for a continuous process there might be some problems in pumping slurries of biomass. Sulfur dioxide gas can substitute for sulfuric acid and can produce more concentrated extracts because it dissolves in the moisture of the biomass itself. Lee, Yue, et al. (1976) have found that SO_2 pretreatment also facilitates hydrolysis of cellulose in a subsequent step with sulfuric acid. During hemicellulose removal there is some reduction in the degree of polymerization of cellulose.

Hokkaido Process

The Hokkaido process developed in Japan in 1948 hydrolyzes wood with concentrated sulfuric acid. If xylose is preferred over furfural, wood chips are pretreated with 1.2 to 1.5 percent sulfuric acid at 140 to 150° C. Furfural is produced at 180 to 185° C. Pretreated wood is dried and crushed prior to hydrolysis with 80 percent sulfuric acid at room temperature. To obtain a desired low acid : wood ratio, crushed wood is sprayed with a thin film of acid. After only about 30 sec of contact, filtration under pressure and washing gives a sugar yield greater than 90 percent. The concentration of the spent sulfuric acid from the combined filtrate and washing water is 30 to 40 percent. An ion-exchange membrane process recovers 80 percent of the total sulfuric acid as a solution of 25 to 30 percent that is concentrated by evaporation to 80 percent and reused. Solution retained by the membrane contains 5 to 10 percent sugars and 5 to 15 percent sulfuric acid. Warming to 100° C for about 100 min further hydrolyzes sugar oligomers. Lime for neutralization forms calcium sulfate, which is removed by filtration. The overall yield of crystalline glucose is 83 to 85 percent of theoretical. Direct fermentation of the syrups would save energy and raise yields.

In a process developed by Nippon Mokuzai Kogaku Company, the sulfuric acid is not recovered but converted into gypsum that is marketed as a by-product. Other products are crystalline glucose, crystalline xylose, and refined molasses.

Bergius–Rheinau Process

The first commercial process using hydrochloric acid is the Bergius–Rheinau process. Wood is hydrolyzed with 40 percent hydrochloric acid to yield a 10 percent sugar solution. However, reuse of this solution several times for hydrolysis brings the sugar concentration to about 40 percent. Distillation under reduced pressure at 36° C recovers hydrogen chloride gas in about 80 percent yield (Sieman, 1975). The sugar solution is dried to remove residual hydrochloric acid.

The dried residue is converted to glucose and di, tri, and tetramer sugars with 41 percent hydrochloric acid. As before, hydrochloric acid is recovered by vacuum distillation. The resultant solution is 60 to 65 percent sugar, of which approximately one-half is dimer and trimer sugars that are hydrolyzed by reheating. With a modified process the hydrochloric acid losses were reduced from over 18 percent in the Bergius–Rheinau process to 5 to 6 percent.

Hydrogen Chloride Gas Processes

Hydrogen chloride gas shifts the hydrolysis reaction into the interior of the wood particles. The wood is wetted with a small amount of concentrated hydrochloric acid and then hydrogen chloride gas is permeated into the wood. The acid can be recovered relatively simply by heating the wood.

The Chisso Corporation of Japan has devised a flash saccharification process in which dry sawdust, hydrogen chloride gas, and some hydrochloric acid are heated for 90 min (*Chemical Engineering News,* April 3, 1978). The glucose yield is 95 percent of theoretical. If HCl can be recovered and recycled economically and if corrosion can be controlled, this could be a good way of producing cheap fermentable sugars. Several variations of HCl processes are reviewed by O'Neil, Colcord, et al. (1978).

There is a Battelle-Geneva process for acid hydrolysis of biomass, but the details are proprietary. It seems that concentrated hydrochloric acid is used near room temperature to get excellent yields of both glucose from cellulose and pentoses from hemicellulose. The hydrochloric acid is recovered during evaporation and recycled. Fermentation of mixed sugars to ethanol has presented serious problems, but new fermentation process for ethanol or other products are being developed.

Gregor and Jeffries (1979) have commented on hydrolysis of wood with HCl as a prelude to the proposed use of membranes for acid recovery. A prehydrolysis step with 35 percent HCl for 30 min at 20° C removes the hemicellulose as its constituent sugars. The acid concentration is raised to 41 percent, and mild temperature is sufficient to solubilize cellulose by conversion to shorter polymers of glucose. Acid is removed until the concentration is 0.5 to 1.0 percent so that raising of the temperature to 100 to 130° C completes the production of glucose.

Evaporation of the 41 percent acid leads to a condensate that is 24 percent HCl. Older technology used expensive concentration steps, but Gregor and Jeffries propose electrodialysis that would use much less energy. Membranes that could process concentrated HCl are not available commercially, although it is claimed that known technology can handle the synthesis of suitable polymers.

Hosaka and Suzuki (1960) found that about $\frac{2}{3}$ of the pentosans in hardwood are readily hydrolyzed with hot acid, but the rate is slower for the remaining $\frac{1}{3}$. Selective removal of the hemicellulose fraction of hardwoods by acid hydrolysis has been investigated by Lee, Yue, et al. (1976). The removal is about 94 percent in 1 to 2 hr with 0.1 to 0.2 percent sulfuric acid at 170° C. Higher acid concentration such as 0.4 percent increases removal but ruins much of the xylose in the hydrolysate. Particle size reduction of the wood is desirable, and grinding is facilitated by acid weakening of the structure. For example, a grinding time of 480 min for untreated controls was reduced to 180 min by treatment with hot 0.1 percent sulfuric acid. Only 6 min of grinding to obtain fine chips was required when the wood was impregnated with sulfur dioxide at 1 atmosphere. The gas has a heat of solution, but additional heat was required to reach 170°C.

Hydrolysis conditions were adjusted to give a solution that was about 90 percent xylose. A preliminary cost estimate including neither cost of wood nor by-product credits showed 2.9¢/lb (6¢/kg) of pentose. Steam was a major expense to heat the reactor and for evaporation to a 7 percent solution deemed necessary for fermentation of xylose to useful products.

The implication of this work is that an inexpensive pentose sugar by-product would be available from a process aimed at producing glucose from cellulosic biomass. As steam is a major cost, there is a fine opportunity to integrate this plant with some industry that has a large amount of low-grade waste steam. Furthermore, development of highly energy efficient means for concentration could lower costs significantly.

Chambers, Herendeen, et al. (1979) have hydrolyzed southern red oak hardwood in a batch reactor at temperatures of 140 to 180°C using dilute sulfuric acid. The best yields have been about 83 percent of the hemicellulose; conditions were 150°, 0.2 percent sulfuric acid, and 90 min of residence time. The product mix is shown in Table 7.1. The glucose results from partial hydrolysis of cellulose.

The author was privileged to tour a Soviet plant for hydrolysis of wood to produce several products, including fermentation alcohol. The following description is taken from a private report by H. E. Grethlein, who was a member of our touring group.

There are about 30 hydrolysis plants throughout the Soviet Union. The Leningrad plant was built in 1935 as a production unit and a full-scale pilot plant to test process modifications. Products are ethanol, yeast, carbon dioxide as dry ice and furfural. Other hydrolysis plants in the Soviet Union produce mainly yeast or furfural. No continuous acid hydrolysis is in commercial operation.

The Leningrad plant has four 4200-gal (16-m^3) hydrolyzers. Various-sized wood pieces are used from sawdust to chips. To ensure good percolation, an initial charge of large chips is laid down at the bottom of the hydrolyzer before the general mix, which may include fines.

After live steam is injected to heat the chips to 170°C, 0.5 percent sulfuric acid is percolated through the bed. The residence time of the liquid phase is of the order of 10 min, while the chips remain in the reactor for 3 hr. About 13 to 15 parts of aqueous phase are collected for each part of dry wood charged.

Flash cooling of the hydrolysate to 100°C removes a major part of the furfural. Condensate is sent to a furfural distillation section. To complete the hydrolysis of soluble oligomers, the hydrolysate is held about 8 hr at 100°C. In newer plants the Soviets complete the inversion at 130°C with a residence time of about 30 min.

Hydrolysate is neutralized with lime; calcium sulfate is removed by sedimentation. The overflow contains about 3 percent total reducing sugar of which 70 percent is hexose and 30 percent is pentose. After the wood chips

Table 7.1   Products from Mild Hydrolysis of Hardwood

Product	Percentage of Soluble Solids
Xylose	50
Glucose	26
Oligomers	10
Acetic acid	10
Furfural	3
Methanol	1

have been extracted for 3 hr, the lignin residue containing about 60 percent water is conveyed to the power plant boiler for combustion.

Five lead and trail tanks are used in a continuous fermentation, each with a volume of 26,000 gal (100 m³). The total residence time is about 7 hr at 34°C. Centrifugal separation is used to recover the *Saccharomyces* yeast, and a part is recycled to the head fermenter. About 5 percent fresh yeast is added to the recycle. With a pH of about 4, no sterilization is required since most potential contaminants grow poorly. The inlet sugar concentration is only 3 percent; thus the ethanol is a very dilute 1.5 percent, which means a large load on the ethanol recovery still. The carbon dioxide from the fermentation is dried and recovered as dry ice.

The bottoms from the first ethanol column is an aqueous solution of pentose that is sent to the protein plant; the overhead is about 20 percent ethanol. In subsequent columns it is purified to the composition of the azeotrope.

Protein production is carried out with the pentose and residual organic fractions. An unsterilized continuous aerated tank process is used with the yeast *Candida scottei*, although some other species are also present. Centrifuged yeast concentrate is dried on a drum drier and packed in sacks for animal fodder. During the seige of Leningrad during World War II this plant played a key role in providing protein for human consumption.

Ethanol from wood is not competitive with that from petrochemical ethylene. It is the special high purity of the ethanol and the semiexperimental status of the Leningrad plant that allows production of ethanol at about $1.70/gal (45¢/l). It is not valid to comment on Soviet costs because the dollar exchange rates are rigged, and the Soviets process wood wastes they consider to have zero value.

The Leningrad plant yields from 1 ton of dry wood are 0.28 tons of hexose and 0.12 tons of pentose for a total of 0.4 tons of total reducing sugar, which gives 0.14 tons of 96.2 percent ethanol.

Capacity is 2.6 million gal (10 million l) of 96.2 percent ethanol per year, 3500 tons of protein, 500 tons of carbon dioxide, and 200 to 300 tons of furfural. This implies a daily input of 1600 tons of oven-dried wood per day. The inputs to other Soviet plants range from 70,000 to 300,000 dry tons/yr.

Saeman (1945) developed a simple kinetic model for the Madison process:

$$\frac{dC_A}{dt} = -k_1 C_A$$

$$\frac{dC_B}{dt} = k_1 C_A - k_2 C_B$$

$$\frac{dC_C}{dt} = k_2 C_B$$

where C_A = cellulose concentration
C_B = glucose concentration
C_C = decomposed glucose concentration
k_1, k_2 = reaction rate coefficients
t = time

By well-known procedures with the Arrhenius equation, activation energies can be found from the changes in reaction rate coefficients with temperature. The coefficients have been determined from experimental data and have been found to depend on the substrate. Additional equations for analyzing the results are

$$k_1 = k_{10} e^{-E_1/RT}$$
$$k_2 = k_{20} e^{-E_2/RT}$$
$$k_{10} = p_1 C_h^m$$
$$k_{20} = p_2 C_h^n$$

where C_h = wt percent of H_2SO_4

E_1, E_2 = are activation energies

R = the gas constant

T = absolute temperature

$p_1, p_2, k_{10}, k_{20}, m, n$ = coefficients

Knappert, Grethlein, et al. (1979a) have plotted the expected yield of sugars from cellulose as a function of time and temperature (Fig. 7.4). Note that best yields are obtained at high temperatures for very short times. However, times less than 0.1 min can be dismissed from consideration because there would be insufficient mixing time for acid solution and biomass. The predicted yield does not exceed 55 percent; thus acid hydrolysis must be improved or replaced with a different technology. There are indications that pretreatment of the feedstock can greatly increase reaction rate coefficient k_1. This could raise the yield of glucose to perhaps 85 percent by reducing the time for the degradation reaction.

Snyder (1958) developed a hydrolysis process buffered at pH of 3.1 to 3.4. Little water is added, and steam at 70 atmospheres gives hydrolysis in 3 min. High yields may result from fewer side reactions with controlled pH.

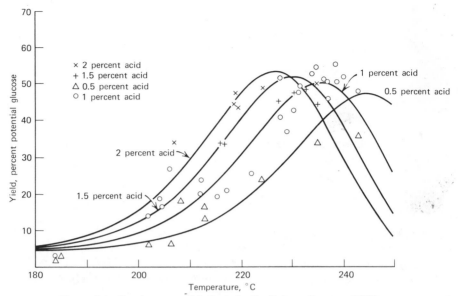

Figure 7.4   Continuous acid hydrolysis of cellulose. Courtesy NTIS.

Continuous acid hydrolysis of wood or paper wastes are demonstrated at a scale of 1 ton/day by Brenner (1979). A key is use of a twin-screw extruder of a type used by the plastics industry, first to mix cellulosic material and then to express the water to form a dense plug that is heated to 450°F (232°C) by steam in a jacket around the exit section. Near the outlet, 0.5 percent sulfuric acid is injected to give a hydrolysis time of about 20 sec. The yield approaches 60 percent glucose from cellulose, as would be predicted from the kinetic studies carried out by Knappert, Grethlein, et al. (1979a). The twin-screw extruder seems very logical for commercializing acid hydrolysis, but the current energy consumption is 1600 BTU/lb (888 cal/g) of cellulose. On the basis of 60 percent yield to glucose and 50 percent fermentation yield to ethanol, this energy consumption is more than the energy required to distill the ethanol and thus could be 10 to 15 percent of the total operating cost.

Conventional Pulping

The sulfite process for making paper pulp involves cooking wood chips in a solution of calcium bisulfite containing excess SO_2. Cooking is typically done at pressures of about 6 atmospheres and about 140°C for 8 hr. Hemicellulose is hydrolyzed to pentose and hexose sugars; lignin is converted to lignosulfonic acid and is extracted as the soluble calcium salt. The pulp, rich in cellulose, is used for making paper products. Waste sulfite liquor contains both calcium lignosulfonite and hemicellulose sugars with typically total solids of 11.0 percent, which are 51.6 percent lignin and 16.9 percent sugars.

Sulfite waste liquor has been fermented to such products as single-cell protein or ethanol, but the economics have been marginal. Pulping is nonselective, and the waste liquor contains intense pigmentation, partially reacted sugars, and ill-defined lignin derivatives. This method of pulping is unlikely to have a major role in a fuels-from-biomass program because the value of side streams has been sacrificed for the sake of the fiber. Other pulping methods may be developed to produce good fiber and side fractions acceptable to processes for conversion to fuels and valuable chemicals. Fiber is more valuable than energy, and a high-value product would make an energy by-product even more economically attractive.

The approximate cost for conventional pulping is $340/ton of wood, which is far too high for consideration as a treatment step in a fuels from biomass process. The largest pulp mill handles 4000 tons of wood per day; thus 5000 tons per day would seem to be an upper limit for one biomass fuels factory based on wood.

Treatment of wood chips with steam at 185°C is a first stage of a variation of the kraft process for paper pulp. Organic acids liberated from acetyl groups in the wood cause hydrolysis of hemicellulose and some breakdown of lignin that aids its removal in the pulping step. Treatment at 185°C is much the same as a dilute acid step being developed by Knappert, Grethlein, et al. (1979b). Lora and Wayman (1978) have found that steaming at 175 to 220°C increases the amount of lignin that is extractable from wood. Wayman and Lora (1978) report that lignin polymerizes to reduce its solubility, but several compounds retard polymerization. Addition of 2-naphthol before heating results in increased amounts of extractable lignin. Protection of lignin might be helpful in other processes proposed for fractionation of biomass.

Chambers, Lee, et al. (1979) have considered the use of paper pulping machinery for carrying out the extraction of hemicellulose from biomass. There are several pulping plants in North America and Europe that have quite new equipment for contacting wood chips with alkaline pulping reagents. Rakes move the chips for countercurrent contacting with the solutions in large-diameter columns that are 200 to 300 ft high (60 to 90 ms). Washing is conducted on a comparable scale. With proper attention to materials of construction resistant to hot, dilute sulfuric acid, the designs for this equipment could be adapted easily to processing of biomass for fuels.

Recovery of Dilute Acids and Bases

Residual acids from pretreatment of biomass or acid hydrolysis will require neutralization with caustic. Although some acid may be recovered and recycled, it will be impossible to remove all acid from the biomass. Dilute acid is not worth recovering by any conventional method. If neutralization of sulfuric acid is carried out with CaO, $Ca(OH)_2$, or $CaCO_3$, there are problems of solubility, of precipitates coating the neutralizing agent, and of disposal of copious amounts of fine, wet $CaSO_4$ formed in the reaction. Sludges of $CaSO_4$ have no value and are usually allowed to fill a settling pond that can later be abandoned* or its sludge may be scraped away and buried.

Gregor and Gregor (1978) have reviewed membrane technology and have suggested a way to recover both acid and base from soluble salts. Figure 7.5 shows a membrane device for this purpose. Ion-exchange membranes are used that are permeable to ions of one charge only. In the center compartment, water ionizes to hydrogen ions and hydroxyl ions. The hydrogen ions under the influence of an imposed electrical field move through the ion-exchange membrane, which passes only positive ions, while the hydroxyl ions pass the selective membrane for negative ions. Consider the case where the dilute salt is sodium chloride. Adjacent compartments enclosed by additional ion-exchange membranes permit selective entry of sodium ions to partner the hydroxyl ions and chloride ions to partner the hydrogen ions. Once into these compartments, the direction of flow of ions in the imposed electrical field and the selective membranes prevent escape; thus the acid and base can become quite concentrated. This arrangement is termed a "water-splitting cell," and the process is much different from electrolysis of water whereby hydrogen is evolved at one electrode and oxygen is evolved at the other. Of course, with the water-

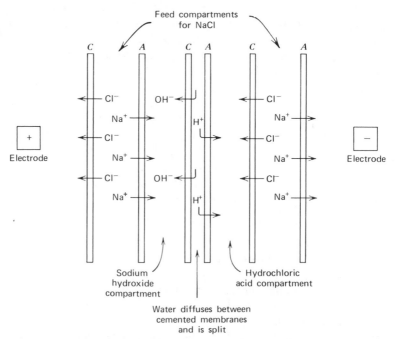

Figure 7.5 Water-splitting membranes for recycling reagents. *C* designates a membrane that repels positive ions. *A* designates a membrane that repels negative ions.

splitting cell, complicated electrolysis reactions can occur in the electrode compartments. By stacking cells in series, it is possible to minimize the number of electrode compartments.

Water-splitting cells are not commercially available, but new membrane materials show promise. There is a problem of desiccation of the membranes themselves as water within them is split. Furthermore, high current densities for economical operation require careful design of membrane properties.

The energy for this scheme is much less than that required for electrolysis because ions are only changing partners and being concentrated. Not only is this important to fuels from biomass, but many energy-saving applications are possible throughout the chemical process industries.

Enzymatic Hydrolysis

All microorganisms that grow well on cellulose must produce active cellulases. Active enzyme preparations are seldom obtained, probably because cellulase producing cultures work synergistically and few organisms in pure culture secrete adequate levels of all members of the enzyme complex. Of the bacteria that digest cellulose, few have been found to be good producers of enzymes that could be isolated from a fermentation process. However, a number of molds produce cellulase enzymes in submerged culture or in surface culture, and recovery is not difficult. Elwyn Reese has worked on biological decomposition of cellulose for over 30 yr, mostly with the mold formerly known as *Trichoderma viride*. In his honor, the name has been officially changed to *Trichoderma reesei*. Such a changeover takes time, and both names can be found in the current literature.

Commercial use requires enzymes that are inexpensive, stable, and pure enough for the intended purpose. Production of cellulases has advanced to the point where fermentation titers are sufficiently high that the broth can be diluted for use and need not be purified or concentrated. For cellulose hydrolysis, enzyme tolerance to 50°C for several days is needed. Fortunately, commercial preparations have adequate stability in the pH range of optimum activity.

Enzymatic impurities can catalyze recombination of glucose units so that the product is contaminated with small amounts of oligosaccharides. For hydrolysis of native cellulose, the proportions of various cellulases in an enzyme preparation may be unsuited to a particular biomass. Analytical procedures have been developed for different cellulases, and some understanding has been gained of factors that shift proportions of cellulase enzymes in a fermentation for their production.

There are many thermophilic organisms that attack cellulose, but those with highest activity are actinomycetes, clostridiae, and sporocytophagas (Bellamy, 1979). Typical conditions for optimal growth are pH 7 to 8 and temperature 50 to 60°C. Rate of cellulose hydrolysis is low unless the feedstock has been pretreated to disrupt the attachment of lignin or to reduce the crystallinity of the cellulose. The enzymes from thermophilic organisms are expected to be relatively stable, and warm temperature favors rapid reaction kinetics. Cellulases from the thermophilic soil fungus, *Thielatia terrestris*, have been studied by SRI International (*Chemical Engineering News,* August 7, 1978). The organism may be cultured from 60 to 90°C and gives enzymes that are functional between 60 and 70°C, which means a faster reaction rate and little risk of contamination. *Clostridium thermocellum* is a bacterial candidate for supplying cellulases. This organism thrives at 60°C and completes its fermentation in about 2 days, whereas *T. reesei* takes from 1 to 2 weeks. The enzymes from *C. thermocellum* are slightly different in activity, and a blend or enzymes from various sources might have superior properties.

The proportions of isoenzyme and enzymatic activity vary for different organisms. For a practical hydrolysis process, it may be well to select a cellulase-producing organism best suited to a given substrate. Bisaria and Ghose (1978) compared hydrolysis of bagasse using cellulase from a single organism with use of mixtures from two organisms and found faster rates with the mixture. As the substrate properties are very strongly dependent on the method of pretreatment, the problems of matching enzyme to substrate and of optimizing the overall process are very complicated. Blending of cellulases from different microbial sources to obtain ideal activities would be expensive because two or more cellulase fermentations would be needed. It would probably be more cost effective to use a mutation–selection program or genetic engineering of several cultures to get one organism with a good balance of the required activities.

In Japan, research on the production of cellulases by organisms growing on solid media such as moistened bran has led to a scheme whereby enzymes are flushed out with water (Toyama, indirect communication). Such a simple system may be suited for use on small farms where hydrolyzed biomass could be fed to nonruminant animals.

Induction of Enzymes

There are two types of enzyme: constitutive, which are produced by the organism under nearly all conditions of growth, and induced, which are produced only when an "inducer" is present in the medium. The constitutive enzyme of some species may, however, be induced enzymes in other species. An inducer is usually the substrate of the enzyme or a variation of the substrate, but when the substrate is a large molecule that cannot enter the cell, another inducer small enough to enter is required. For cellulases, this turns out to be the soluble products obtained by enzymatic hydrolysis of the polymer. Induction by soluble products may be masked because products or intermediates reduce enzyme activity. The inducer at high concentrations may act as a repressor.

In the case of *Trichoderma reesei*, Binder and Ghose (1978) have shown cellulase production to be triggered by adsorption of cellulose to the cell's surface. In other organisms, cellulases are excreted without contact with the substrate. The close proximity of organism and substrate should provide a competitive advantage for the cellulase producers over foreign organisms.

Catabolite repression is a phenomenon whereby the accumulation of compounds from enzyme reactions decreases the rate of synthesis of the enzymes. This exercises feedback control so that excess product is not formed. Mutants can be isolated in which the control mechanisms are damaged so that repression is weakened or inoperative. Often mutants that hyperproduce enzymes are still subject to catabolite repression. Further mutation of these hyperproducing strains to obtain less catabolite repression can give higher yields of enzymes.

Inhibition of cellulase action by accumulation of cellobiose from hydrolysis is well known. However, through analysis of kinetic data, Howell and Mangat (1978) have concluded that part of the decline in cellulase activity results from a process other than enzyme inhibition. They have modeled the data quite well by considering inhibition by cellobiose and first-order deactivation of the enzyme–substrate complex.

The saccharification of cellulose with enzymes can take many days if no pretreatment is used. With mild pH and slightly elevated temperature, contaminants can thrive on the sugars that are formed. Antiseptics or antibiotics can be added to reduce contamination, but cellulase activity may be impaired (Spano, 1976). Removal or destruction of the protective agent may be needed prior to the fermentation of sugar to ethanol. Even if the fermentation culture is unaffected by the protective agent, there is a pollution problem if a toxic agent is present in the final effluent. It seems advisable to omit these agents and carry out saccharification quickly so that contaminants have insufficient time to reach troublesome concentrations.

Cellulases

Cellulolytic enzymes from fungi were recently reviewed by Enari and Markkanen (1977) and by Eveleigh and Montenecourt (1977). It is well established that at least three distinct types of activity exist: exo-β-1,4-glucanase, endo-β-1,4-glucanase, and β-glucosidase. Specific enzymes attack cellulose by cutting the long chains into smaller fragments. It is thought that the bonds where the chains fold back are more easily attacked than those in the main chains. The exo enzymes work from the ends of a chain, whereas endo enzymes attack along the polymers. Some glucose is formed, but the enzyme cellobiohydrolase splits from a terminus to give cellobiose (a disaccharide of two glucose units). Other enzymes, cellobiase or β-glucosidase, split cellobiose to glucose. There may be a transition of crystalline to amorphous cellulose, but there is probably just a rate difference where the same enzymes attack the amorphous form more rapidly. There are five or more isoenzymes of the endo enzyme, two of β-glucosidase, and still others may be found. *Trichoderma reesei* cellulases have been studied most extensively, but many other organisms also contain these enzymes. Some properties of cellulolytic enzymes are listed in Table 7.2

The initial mechanism of cellulose degradation is not completely clear. Solutions of reconstituted mixtures of purified enzymes have less activity toward cellulose than do unpurified mixtures with equal amounts of the known enzymes, suggesting that there may be important but as yet unknown activities. Furthermore, oxidative enzymes seem to play a role, perhaps by breaking hydrogen bonds between chains.

Most of the research on cellulose enzymology uses purified, crystalline cellulose. Native cellulose has amorphous and crystalline portions in a matrix of hemicellulose and lignin. This raises the question of how to translate theory to actual commercial practice.

Although many organisms release cellulases from their cells, others must make direct contact with the substrate. It may be that some cellulases are most efficient when bound to the cell. If energy is spent on enzymes released to the medium, contaminants with no cellulases of their own can intercept glucose and can compete effectively. Bound enzymes release glucose very close to the producing organism, thus providing little for contaminants. It has often been observed that two organisms that grow poorly, if at all, in pure culture on cellulose may thrive together. This may indicate that whereas some enzymes are bound to the cells, other enzymes are diffusing to the other organism.

Cellulase is an inducible enzyme complex that requires the presence of cellulose, lactose, or cellobiose to stimulate yields. Product inhibition is severe; cellobiose greatly reduces the rate of cellulose hydrolysis. Glucose inhibits the action of β-glucosidase.

Ryu, Andreotti, et al. (1979) have operated two-stage continuous cultures of *T. reesei* for production of cellulases. The first stage was for rapid cell growth; lactose was the carbon source and served as an inducer for cellulase

Table 7.2 **Properties of Some Enzymes from *Trichoderma reesei***

Name	Molecular Weight	Isoelectric Point
Exo-β-1,4-glucanase	42,000	3.79
Endo-β-1,4-glucanase (I)	12,500	4.60
Endo-β-1,4-glucanase (II)	50,000	3.39
β-glucosidase	47,000	5.74

formation. Cellulase productivity was best when the second-stage dilution rate was 0.026 to 0.028 hr^{-1}. This is roughly equivalent to $1\frac{1}{2}$ days of fermentation and is significant because the batch fermentation can take 1 to 2 weeks.

Eriksson (1978), who has contributed much to the understanding of cellulase from *Trichoderma reesei*, has reviewed the enzyme mechanisms for cellulose hydrolysis by wood rot fungi. There are somewhat different pathways for *T. reesei* and *Sporotrichum pulverulentum;* the latter can split both glucose and cellobiose from the nonreducing end. An oxidative enzyme important for both cellulose and for lignin attack is found in wood rot fungi, but not in *T. reesei*.

The University of Pennsylvania team effort with the General Electric Company has been using an organism identified as *Thermoactinomyces* sp. Further testing plus confirmation by workers at Rutgers University has corrected the identification, and the proper designation is *Thermonospora* sp. (perhaps *T. alba*). There have been a number of reports with the incorrect name *Thermoactinomyces* sp. (Hägerdal, Ferchak, et al., 1978). At 55°C this organism elaborates active cellulases; yields are nearly comparable to those of good mutants of *T. reesei*. Mutation should lead to higher yields, but continued improvement of other species that produce cellulases means that comparisons must continually be updated. The *Thermoactinomyces* enzymes offer different proportions of the various activities and should be stable at fairly high temperatures. The β-glucosidase activity is associated with culture solids while the cellulases are released to the medium. This could be an advantage if fractionation is desired or a disadvantage by requiring a step for releasing β-glucosidase if an enzyme mixture is being prepared. The β-glucosidase from *Thermoactinomyces* is unusual in that there is very little inhibition by glucose. Glucose syrups approaching 20 percent concentration were made from cellobiose using only *Thermoactinomyces* cells (Pye and Humphrey, 1979).

Comparisons of several strains of *T. reesei* by an independent group (Wilke and Blanch, 1979c) showed the following:

Rutgers C-30	14.4 units/ml cellulase,	26 units/ml β-glucosidase
Natick QM9414	6.1 units/ml cellulase,	1 units/ml β-glucosidase

Attempts were made to optimize the conditions for each strain; thus the validity is better than for tests where a new strain was evaluated under the conditions established for the old strain. Superior performance of the Rutgers strain appears to result not only from more cellulase activity, but from many times the β-glucosidase activity.

Smith and Gold (1979) have characterized the β-glucosidase from the wood white rot *Phanerochaete chrosoporium*. Several forms of this enzyme are produced, and it is not clear why there is an evolutionary advantage for doing so.

Grove and Bracker (1970) have shown that various fungi have at their hyphal tips the secretory vesicles for synthesis and transport of extracellular enzymes. Young tips may be most active in secretion, but some workers hold the opposing view that old cells release cellulases when autolysis weakens the cell membranes.

This fits the observation that cellulase in the culture medium peaks late in the fermentation when aged cells are predominant. In young cells the enzymes may be tightly bound and functional with a minimal amount being secreted. Hyper-producing mutants have abnormal cell walls, which may account for release of more enzyme to the medium.

Mutation can lead to improved titers of enzymes. Programs at Natick Laboratories and at Rutgers University have led to much higher yields of cellulase activity and to enzymes less inhibited by accumulation of products.

Methods

Improved analytical procedures can greatly accelerate the pace of research. The determination of cellulase activity is particularly troublesome because the term "cellulase" is applied to a mixture of at least four or five enzymes with differing actions on a spectrum of substrates. Most experts develop assays that are best suited for the questions to be answered by their own research. However, there is a need for comparison of samples from different laboratories. For routine measurement of gross cellulase activity, most workers determine how much sugar is released by action of the enzyme sample on filter paper. Montenecourt, Eveleigh, et al. (1978) have an improved method in which a disk of Whatman #1 filter paper is incubated with various dilutions of the enzyme sample and the concentration of reducing sugar is analyzed at a given time interval.

Other procedures are based on release of color from dyed cellulose. Leisola and Kauppinen (1978) have reported on automatic analysis of cellulase activity in a continuous stream from a fermentation to produce the enzyme. A Technicon Autoanalyzer is used to meter the sample and the reagents and to develop, measure, and record the color.

Practical Hydrolysis of Cellulosic Materials

Linko (1977) has recently reviewed enzymatic hydrolysis of cellulosic materials. As noted by others, he points out that enzyme cost and pretreatment to increase susceptibility of cellulose to hydrolysis are the key factors. Well-known pretreatment schemes such as alkali swelling, steaming at 160 to 170° C, or hot inorganic acids can improve hydrolysis yields, but much more effective methods have recently been reported.

Except for microbial biomass, which is already well dispersed, biomass for hydrolysis requires subdivision. If shredding is the sole method of pretreatment, hydrolysis yields are poor. It might be possible to produce a fine powder to achieve good hydrolysis, but the power costs for grinding become totally impractical. The newer pretreatment processes will accept coarse chips or roughly shredded materials that are readily prepared with conventional equipment with quite low power costs.

It is very important to obtain high sugar concentrations for the fermentation step so that products are not too dilute in the broth. Recovery by distillation of dilute solutions means that excessive water would be heated, vaporized, and condensed. Evaporation of the hydrolysate is feasible, but it, too, is costly; thus it is best to strive for high sugar concentrations directly. The product inhibition of cellulases as previously mentioned causes a lowering of hydrolysis rate; high sugar concentrations can be achieved by using excess enzyme or allowing a prolonged detention time. As enzyme is a major expense, large excesses are intolerable. However, reuse of recovered enzymes is possible; the hydrolysate may be rich in enzymes that can be recovered by well-known methods. A problem arises from the very tight binding of cellulases to cellulose. To obtain high sugar concentrations, the feed concentrations of cellulose must be high. This leads to considerable unreacted cellulose and adsorbed enzymes. Agents such as urea that weaken hydrogen bonds can desorb enzymes from cellulose and increase recovery yields by a factor of 2 or more. Unfortunately, enzyme recovery is expensive. If urea or some other agent is to be used, it, too, must be recovered and reused.

It seems sensible to partially hydrolyze with acid to solubilize cellulose to shorter fragments and to complete the hydrolysis with enzymes. Prolonged acid hydrolysis decomposes simple sugars, but partial acid hydrolysis should have negligible by-products. With no solids present, enzymes could not be adsorbed and losses would be minimized.

One excellent approach to lowering costs is enzyme immobilization, so that the enzyme can be reused many times. Cellulase enzymes immobilized in collagen have been used in a fluidized-bed reactor to hydrolyze prepared cellulose (Avicel) (Karube, Tanaka, et al., 1977). The kinetics are poor for a particulate substrate and a tethered enzyme, but high yields of glucose were noted at prolonged times. The half-life of native cellulase was about 30 hr, but immobilization increased half-life to 21 days. Much faster reaction would be expected if cellose dissolved in a special solvent could be hydrolyzed with an immobilized enzyme stable in that solvent.

Immobilization of β-glucosidase for splitting cellobiose to glucose makes a great deal of sense because this enzyme is usually present in insufficient proportions in natural cellulases. In nature, sugars from cellulose do not tend to accumulate because feedback control turns off the enzymes producing them. Small levels of β-glucosidase are adequate for cellular metabolism that uses sugars as they are produced. Supplemental β-glucosidase works well *in vitro* when immobilized since its substrate is a soluble, relatively small molecule.

Isaacs and Wilke (1978) have immobilized β-glucosidase from *Aspergillus phoenicius* on phenol–formaldehyde resins by coupling with glutaraldehyde. Up to 80 percent of starting soluble enzyme activity was retained. When columns with immobilized enzyme were operated in conjunction with hydrolysis of cellulose by cellulases, there was little difference in the rate at which reducing sugars were formed. However, cellobiose was split to give a higher yield of glucose that is acceptable to most yeasts whereas cellobiose is not fermented. A group at the University of Connecticut has also had good success in immobilizing β-glucosidase and improving hydrolysis of various cellulosic materials (*Biomass Refining Newsletter,* August 1979).

Sachdev and Ghose (1978) have tried several different methods for immobilizing β-glucosidase. With cyanogen bromide for coupling the enzyme to special glass beads with controlled pore size, 33 percent of the starting activity was bound, and $\frac{2}{3}$ of the activity was retained for 19 days at room temperature. Use of the immobilized enzyme at 4°C indicated that half of the activity would be present at 264 hr.

Ghose and Sahai (1979) have operated continuous cultures of *T. reesei* at a 5-1 scale and obtained cellulase productivity of 30 IU/1 · hr. The optimum dilution rate was 0.025 hr^{-1}. A common problem for research in this area is handling of concentrated slurries of cellulose; fine particles used in this study were easy to dispense but remained suspended in the effluent stream. A sedimentation vessel with 20- to 25-min detention time retained almost all of the mold cells but passed the cellulose fines. The collected mold cells were fed back to the fermenter at a recycle ratio of 1:2. The actual enzyme titer of 1.2 IU/ml is comparable to batch culture yields with this strain but well under the yields with some of the very new mutant strains. However, it is encouraging that continuous culture performs well; thus high priority should be given to testing new strains and to increasing the dilution rate.

The 10-1 fermenters used at Natick laboratories for the production of cellulases are shown in Fig. 7.6. There are also units of 200-1 capacity in their pilot plant. Typical data are presented in Fig. 7.7. This is for *T. reesei* strain QM9414 growing on 1 percent cellulose. After 35 hr the pH falls to pH 3.5, where it is controlled at the optimum for enzyme production. By 50 hr most of the cellulose has been consumed, but little cellulase is detectable. Cellulase, other enzymes, and protein begin to be secreted, and by 120 hr maximum levels are reached. Extracellular soluble protein (enzyme) reaches about 2 mg/ml, equivalent to 20 percent of the cellulose consumed (10 mg/ml); or, alternatively, to 40 percent of the maximum cell weight (5 mg/ml). Higher cellulase yields require richer media, and since cellulase production is repressed by soluble substrates, this means that cellulose levels must be increased. In shake flasks, enzyme yields decline at cellulose concentrations greater than 1 percent because low pH develops. In fermentors with pH control, cellulose concentrations of 12 percent give greatly increased cellulase yields.

Peitersen and Ross (1979) have developed a mathematical model for enzymatic hydrolysis and fermentation of cellulose. The model is quite complicated and accounts for the crystallinity of the cellulose; thus applicability of the model to pretreated substrates may be questioned. Nevertheless, this article is recommended for its discussion of mechanisms and for the compilation of reasonable rate constants for the various reactions. Sasaki, Tanaki, et al. (1979) have developed a crystallinity index for cellulose based on X-ray diffraction. Pretreatment by various means greatly lowers the crystallinity index, but little difference was noted in the yield because saccharification reached 93 to 94 percent of the theoretical yield following any of four good pretreatment methods. Incorporating crystallinity index into a mathematical model could be a significant advance.

Figure 7.6 Fermenters at U.S. Army Natick Laboratories. Report from U.S. government laboratories.

Countercurrent contacting of cellulosic biomass with enzyme should be effective and efficient because easily hydrolyzed components of fresh biomass are contacted with weakened enzyme whereas the most potent enzyme contacts biomass containing resistant substrate. Lindsey and Wilke (1978) tried a four-stage countercurrent system and found the surprising result that equivalent hydrolysis was obtained in a single stage with the same total contacting time. Failure of countercurrent contacting to outperform a batch system is not logical; there may have been losses during the solid–liquid separations between steps. Another possible explanation is adsorption of incoming enzyme to the exiting biomass so that much of the activity is lost.

There has been emphasis on hydrolyzing cellulose to obtain concentrated glucose syrups that do not need to be evaporated prior to fermentation (Pye and Humphrey, 1979). With cellulose pretreated by swelling in phosphoric acid, hydrolysis by cellulases supplemented with β-glucosidase gave syrups of 20 percent glucose by weight. The hydrolysis took 7 days, and all components were made up with minimum water. Nevertheless, this laboratory experiment clearly shows the potential for eliminating or minimizing the need for costly evaporation.

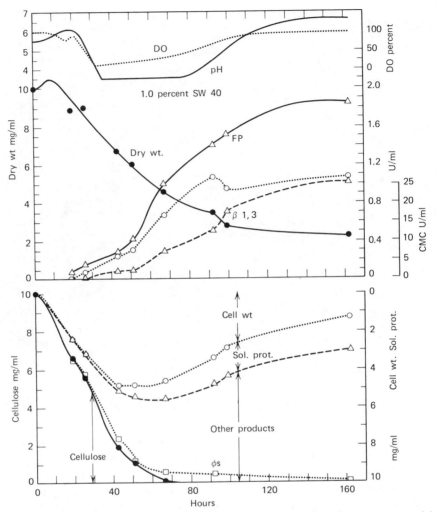

Figure 7.7 Time course of cellulase production: DO = dissolved oxygen; FP = cellulase activity using filter paper assay; CMC = endo-β 1,4-glucanase activity using carboxymethyl cellulose; β 1,3 = β 1,3 glucanase. The growth substrate was Solka Floc (SW40); the organism *T. reesei* QM9414. Data courtesy of R. Andreotti.

Dissolution or Swelling of Cellulose

Ladisch, Ladisch, et al. (1978) have reported almost quantitative yields of glucose from enzymatic hydrolysis of cellulose following pretreatment with solvents. The key finding was that cellulose need not be dissolved but merely soaked or permeated with solvent to disorganize the crystalline structure. This could prove a breakthrough in cellulose hydrolysis since much less solvent is needed than would be required to dissolve the cellulose. Furthermore, solutions of cellulose kept concentrated to keep the costs for solvent reasonable are so highly viscous that handling creates problems, but slurries in the solvent for soaking are not difficult to handle. However, cost considerations do dictate high solvent recovery yields.

Cadoxen or tris ethylenediamine cadmium hydroxide is a clear, colorless, stable liquid. Typical composition is 25 to 30 percent of ethylenediamine in water plus 4.5 to 5.2 percent of cadmium (added as oxide or hydroxide) based on the total liquid weight. At room temperature cadoxen can dissolve some 10 percent by weight of cellulose, which is reasonably high. Although cadoxen is probably the best available solvent for cellulose, it has little chance of commercial use because cadmium is highly toxic. When excess water is added to a cellulose–Cadoxen solution, cellulose will reprecipitate in a soft floc. On standing, this cellulose may recrystallize and again become resistant to hydrolytic attack, but when the cellulose is still in the form of soft floc, it can be easily hydrolyzed with either cellulase enzymes or acids to give high yields of glucose. *Trichoderma reesei* cellulases were found to be active in the presence of the Cadoxen–water solution.

After hemicellulose is removed from biomass with acid, cellulose and lignin are left in a wet state. Thorough washing to remove acid is essential for minimizing reaction with alkaline Cadoxen. If the hemicellulose extraction is done with basic agents, washing is less critical. In fact, addition of NaOH to Cadoxen can increase the solubility of cellulose by several percent.

No special precautions are needed for an iron-based cellulose solvent with the cryptic designation "CMCS" because all its ingredients are generally recognized as safe. A typical formulation for CMCS is 20 parts of sodium tartrate, 15.5 parts of ferric chloride, and 14.5 parts of sodium sulfite dissolved in 1000 parts of 5 percent sodium hydroxide solution. Stock material can be prepared in either liquid or dry form. A green powder of the other constituents can be dissolved in 5 percent caustic. This opens the possibility of manufacturing the "solvent" in a central location for shipment dry to individual plants utilizing cellulosic materials. There would be an overall savings in capital and operating costs since individual plants need not invest in the equipment required to produce the powder. When CMCS is dissolved in 5 percent caustic, it will readily dissolve up to 4 percent of cellulose at room conditions. Cellulose is reprecipitated on addition of methanol or water.

The Purdue group feels that the most practical solvent for cellulose is concentrated sulfuric acid. Concentrations of 60 to 70 percent sulfuric acid dissolve cellulose fairly well at mild temperature, and it can be reprecipitated by adding methanol or ethanol. This separates most of the cellulose and sulfuric acid. The small amount of acid wetting the precipitated cellulose can be used for partial acid hydrolysis by adding water and warming to near boiling. Either methanol or ethanol can be distilled easily from the concentrated acid stream so that it can be recycled for dissolving more cellulose. In conjunction with a process for ethanol by fermentation, ethanol is a more logical precipitant for cellulose than is methanol, although the latter can be purchased more cheaply. This process appears quite attractive, but larger-scale testing is needed to determine the permissible recycle of sulfuric acid without building up troublesome impurities. Yields appear quite high, and the keys to process economics will be reagent costs and acid recovery costs.

The saccharification step is not yet ready for a decision between acid, enzymes, or a combination of acid and enzymatic hydrolysis. With removal of hemicellulose under mild conditions and decrystallization of cellulose with solvents, acid hydrolysis of cellulose is faster whereas there is less time for decomposition, although the rate of destruction of sugars is unchanged. Experiments are still in progress, but the Purdue developments may increase yields of glucose from cellulose to over 95 percent.

Wilke and Blanch (1978) have been exploring a Soviet process by Shnairer, Shpultova, et al. (1960) for cellulose hydrolysis with 85 percent sulfuric acid using high shear. The overall process has considerable similarity to those being developed at Purdue because the hemicellulose fraction is first removed with dilute acid. The intimate contact from high shear mixing permits a reduction in the amount of acid to an acid:wood ratio of about 0.6:0.62 by weight. This works out to a sulfuric acid:total fermentable sugars weight ratio of 2.4. Since the acid must be recovered for the process to become economical, it is quite important to reduce the amount of acid. Acidic washings may be used elsewhere in the process as, for example, removal of hemicellulose.

Low-temperature steam treatment of Aspen chips has been reported by MacDonald and Mathews (1979). The condition was 190°C for 7.5 min. Hydrolysis of treated wood with cellulase enzymes gave about five times the free sugar noted with untreated controls. However, the total sugar was less than 30 percent of the dry weight of the wood. Other pretreatments can lead to much better hydrolysis yields; thus steaming at moderate pressure appears inferior to the same conditions, except with acid present.

Tsao, Gong, et al. (1979) and Chang and Tsao (1979) have observed some very interesting properties of treated cellulose. Native cellulose has folds in the crystallin portions where chains turn back over each other. These folds complex iodine to produce a brownish color. Starch has a helical structure that accepts iodine to give a very well known deep blue color. After solvent treatment of cellulose, a very similar blue color is obtained with iodine; this supports the idea of a spiral cellulose structure with β linkages. Furthermore, treated cellulose is hydrolyzed fairly well with amylases that were thought to be specific for the α linkages in starch. Not only does treated cellulose acquire features closely resembling starch, but its rate of enzymatic hydrolysis becomes rapid and more characteristic of starch than of native cellulose. This work provides a basis for interpreting the profound effects of pretreating cellulose.

Dunning and Lathrop (1945) observed that removal of pentosans from corn cobs in a prior step led to better hydrolysis of cellulose to glucose. Their explanation was increased porosity of the cobs. This might be expected with a biomass very rich in pentosans. However, Tsao's group at Purdue has found removal of pentosans to improve subsequent cellulose acid hydrolysis in corn stover, a material rich in cellulose. Knappert, Grethlein, et al. (1979b) have found that pentosan removal aids enzymatic hydrolysis of cellulose.

Explosive Decompression

It has long been known that pressurizing wood with steam and then suddenly releasing to atmospheric pressure would shatter the structure in a popcornlike effect. Such shattered biomass is more amenable to acid or to enzymatic hydrolysis. A modification of the process as operated by the Canadian Iotech Company uses 45 to 52 atmospheres of steam in a batch reactor and releases through dies that disintegrate the biomass. Many feedstocks have been tried, and wood chips work quite well.

Enzymatic hydrolysis of exploded biomass can give over 80 percent of the theoretical amount of glucose. Hemicellulose does not fare well in this process; yields of pentose sugars are less than 30 percent, and there is an odor of decomposed material. Removal of hemicellulose prior to explosion could give a major boost in overall yield of reducing sugars. The processing characteristics for exploding biomass after hemicellulose removal are unknown at present.

Lignin is little affected and can be recovered in nearly native form. For use in lignochemicals, it is far superior to lignin from paper pulping. Lignin from exploded biomass could command a premium price for special markets, but these are much too limited to figure heavily in a fuels from biomass program. Nevertheless, credits for lignin could greatly boost the profits for the first factories; some very exciting new uses for lignin are in plastics and as a plywood binder.

Solvent Delignification

A joint project of the University of Pennsylvania and General Electric Company has been investigating alcohol extraction of lignin as a pretreatment of hydrolysis of cellulose. The conditions are 150°C, 5 atmospheres, and pH 8.2 with a carbonate buffer, using a ratio of 10 parts solvent to 1 part wood chips. A 50 percent aqueous butanol solution extracts about half of the lignin and changes the wood structure sufficiently that about 80 to 90 percent of the cellulose is readily hydrolyzable with enzymes. Ethanol may be substituted for butanol, but it is not as good a solvent. Lignin precipitates during solvent recovery, but there is the option of using lignin solutions or slurries as diesel fuel.

Solvent delignification is also the basis for some processes for producing paper pulp. Katzen, Frederickson, et al. (1980) have described a process for unbleached pulp using ethanol and wood chips. Important features are effective contacting so that the wood need not be pulverized and modern, energy-efficient alcohol recovery. Experience with alcohol pulping should be invaluable in developing a solvent pretreatment process for hydrolysis to fermentable sugars, but pulping economics can be different because the product has relatively high value.

Biological Delignification

Microbial degradation of lignin could make the cellulose in biomass easier to hydrolyze. Brooks, Bellamy, et al. (1978) proposed the use of the mold *Chryso-*

sporium pruinosum (recently identified as an imperfect state of *Phanerochaete chrysosporium*) for aerobic attack on lignin. An ingenious reactor with spiral mixers, entry ports for air, and spray addition of nutrients was to be tested if laboratory results were successful. Attack on lignin is relatively slow, and over 40 percent remains after 30 days. Cellulose and hemicellulose are more readily utilized, and it seems that it would be difficult or impossible to find an organism that has no attack on cellulose or hemicellulose. In an aerobic process, solubilized materials are quickly converted to carbon dioxide, water, and cell mass, which are losses from the point of view of fuels. An anaerobic removal of lignin would be desirable, but no appreciable anaerobic metabolism of lignin has been demonstrated.

Acid Pretreatment

Knappert, Grethlein, et al. (1979b) have found that conditions inadequate for acid hydrolysis of cellulose can still be very effective as pretreatment for enzymatic hydrolysis. In a flow reactor, 0.2 percent sulfuric acid at 180°C for 30 sec gives little glucose, but the degree of polymerization of cellulose is greatly reduced. Subsequent hydrolysis with cellulase gave over 90 percent of the theoretical amount of glucose. Runs with oak are shown in Fig. 7.8. A control with no pretreatment gave only 20 percent of the theoretical glucose in 48 hr. Processing without acid increased the yield to over 40 percent, and dilute acid gave about 90 percent hydrolysis yields with *T. reesei* cellulase.

A similar process has been developed at the University of California Forest

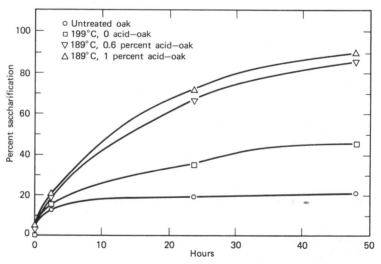

Figure 7.8 Enzymatic hydrolysis following dilute acid pretreatment of oak. Courtesy of NTIS.

Products Laboratory, using batch rather than continuous operation. A slurry of about 23 percent wood in water is treated for 45 minutes at high mixing shear with 0.1 percent sulfuric acid and 180°C. The treated material is reported to be easily hydrolyzed to high yields of sugar.

This approach has excellent prospects because dilute acid is cheap and yields are outstanding. Although continuous processing must overcome problems in pumping slurries and designing good heat exchanges, higher productivity should reduce capital costs for large factories.

Another proposed pretreatment has been found to have insufficient effect (Linden and Murphy, 1978). Although ethylene gas causes extreme swelling of the cell walls of various plant structures, the hydrolysis of biomasses treated with ethylene was little better than the controls. The best yields were less than 10 percent of the theoretical amount of glucose with enzymatic hydrolysis, and other pretreatment steps do much better.

Additional Reading

For additional reading see Bailey, Enari, et al. (1975); Gaden, Mandels, et al. (1976); Ghose (1978); Jahn (1971); Timell (1976); Turbak (1975); and Wilke (1975).

8

Fermentation

Fermentation Technology

The most highly developed fermentations are those of the pharmaceutical industry. Starting with penicillin in the 1940s, more and more antibiotic or vitamin fermentations have been developed. Conversions of one steroid to another by microorganisms are also important processes. Common to all these processes is the high value of the products. This means that costs of steam, nutrients, electricity, and labor are not of overriding concern. Although costs savings are not to be ignored, the dominant factor in process improvement is achieving higher yields of expensive products. Fermentations for fuels and industrial chemicals can make use of some of the refinements of pharmaceutical fermentations, but much more attention must be paid to cheap materials of construction, prices of ingredients, steam, and product recovery.

Industrial fermentations for compounds such as citric or lactic acid have an economic philosophy more akin to that for fuels in which raw material costs are predominant. For example, citric acid is produced from cheap molasses for which a major expense is treatment to remove heavy metals that are toxic to the fermentation organisms.

Production of beverage alcohol must be concerned with product quality. A beer must have a characteristic and consistent taste from batch to batch to satisfy loyal customers. Whiskey is often blended; thus some batch variations are acceptable if within a given range. Although the outlook differs from that for fuel fermentations, beverage alcohol processes can provide significant leads. Centuries of experience with fermentation in open wooden vats prove that strict asepsis and stainless-steel pressure vessels are not essential.

Still another group of fermentations, those for foods such as pickles, cheeses, and vinegar, also have some relevance. Several of these can be run in crude equipment, and the cultures are often mixed species.

Different types of fermentation will be examined in more detail to see where the technology is suitable for producing fuels and petrochemical substitutes. It must be emphasized that no fermentations have ever been run on a scale approaching that needed for making a significant contribution to U.S. energy supplies.

Fermentations for Pharmaceuticals

It is interesting to trace the history of commercial production of penicillin. This antibiotic was discovered on a petri dish and studied for several years in test tubes and flasks. When commercialized, early production runs were with large flasks or milk bottles. Dairy equipment was purchased for washing bottles on a large scale. One badly contaminated bottle could contain sufficient penicillinase enzyme to destroy most of the penicillin in a batch pooled from the contents of thousands of bottles. Titers were less than 1 μg of antibiotic per milliliter. Steel pressure vessels of about 50-gal capacity were adapted for fermentation and were found to be far superior to milk bottles. Titers increased to several micrograms per milliliter. Little technology existed for well-mixed, highly aerated fermenters; thus invention was required throughout the stages of scaleup. Literally, air filters for tanks were sized on the basis of the cotton plugs used for shaken flasks. About two dozen companies embarked on penicillin fermentation using various sizes and designs and quite crude purification processes. Oversupply ruined prices, and few companies still manufacture penicillin. Tanks of several hundred gallon's capacity gave way to those of several thousand gallons, and 15,000 to 25,000 gal became a typical size. In more recent years, a few fermenters in the 80,000- to 100,000-gal size have been tried, but contamination, although rare, is a serious risk if so much material is spoiled. A new plant built today would probably select 40,000- to 60,000-gal (75,000 to 250,000 l) fermenters.

Biologists have made much more spectacular increases in penicillin yields than have engineers. Whereas chemists and engineers have improved recovery from 10 percent to over 70 percent and have seen some respectable fermentation increases due to better agitation and aeration, microbiologists and geneticists, through strain selection and mutation, have improved titers by several orders of magnitude. The exact numbers are closely kept industrial secrets, but titers of 20 g/l of penicillin have been rumored for several years now.

Some of the steel fermenters that were installed 35 yr ago are patched and repaired but still in service. Most fermenters, however, are stainless steel with stainless-steel coils through which water or steam passes. Cooling water is needed to control temperatures during a run, and steam is used in the coils and in the fermenter during sterilization. Jackets are inadequate for cooling a very large fermenter from sterilizing temperature down to operating temperature because of insufficient heat-transfer area. Coils inside the tank can offer a large amount of surface.

Most of these processes are aerobic, so good aeration and agitation are required. Enormous air compressors are used for a plant, usually in pairs to allow one to carry the load during maintenance of the other. Aeration leads to foaming, which is usually controlled by surface-active agents such as lard oil or synthetics of silicone or polyglycol types. Pumps are uncommon in a fermentation plant because it is easy to use air pressure to force liquid from one tank to another or to harvesting equipment.

Fermentation solids are removed by large rotary vacuum filters or by centrifuges prior to product recovery. The mainstays of purification are ion exchange, solvent extraction, adsorption, and crystallization. Penicillin is an organic acid, and at low pH has its ionization is suppressed, and thus it strongly favors nonpolar solvents. Its instability at any pH not close to neutrality dictates contact with a water-immiscible solvent immediately after adding acid.

Acetone–Butanol

The feasibility of an acetone–butanol fermentation was recognized around 1910, and a basis for an industrial process was attained by Weizman's group by 1914. During and after World War I the process was important in several countries. Although rapid wartime development was aimed at acetone for manufacture of explosives, butanol became the more desirable peacetime product. It is of interest to note that butanol was not in demand, but its availability created new markets. As by-products become available from factories producing fuels from biomass, new uses will be found to spur demand and to create a price structure that will make the overall process more economic. Acetone and butanol are good solvents, as are their derivatives, butyl acetate and 2-propanol. All have uses as intermediates in chemical synthesis.

Intense economic competition from petroleum-based compounds led to a decline in acetone–butanol fermentation starting in the late 1940s and continuing to the present. One of the few remaining installations is in South Africa, where special circumstances permit operation. This fermentation is discussed by Bu'Lock (1975).

Anaerobic fermentations using strains of *Clostridium acetolbutylicum* were reviewed in detail (Prescott and Dunn, 1959; Underkoffler and Hickey, 1954, Cassida, 1968). There has been only primitive strain selection and no genetic engineering of these cultures. The metabolic pathway intercepts trioses from the breakdown of sugars and generates hydrogen, carbon dioxide, and an acetaldehyde complex that condenses to acetoacetyl through acetyl coenzyme A. Part is decarboxylated to acetone, and other molecules are reduced to butanol. If reducing power is insufficient, as would occur during incomplete anaerobiosis, unwanted butyric acid is formed. Other unwanted products resulting from faulty operation are ethanol and 2-propanol. Acetone:butanol ratios vary with the strain used and with fermentation temperature. Potentially useful by-products are hydrogen and cell biomass, both of which tend to have little value at present. The cells are rich in riboflavin and could be valuable.

The range of raw materials that has been used for the acetone–butanol fermentation is wide, including both high- and low-grade maize, blackstrap and high-test cane molasses, beet molasses, wheat, rice, horsechestnuts, Jerusalem artichoke, cassava, starch and sugar wastes, oat hulls, corn cobs, wood hydrolysate, sulfite waste liquor, and whey. Some of these were little more than pilot projects under wartime conditions, and others reflect local or temporary surpluses, but it is clear that many carbohydrate sources could be considered. This is not, however, to say that all are equally convenient for running the fermentation, and a body of data regarding the special characteristics of the major substrates is available. Unless the product tolerance of the organisms can be improved, these carbon sources need to be employed at relatively low concentrations.

Wang, Cooney, et al. (1978a) have tried the acetone–butanol fermentation in modern equipment. Even with pH control, the solvent yields are about the same as for industrial fermentations of 30 yr ago. It was confirmed that acetone is not toxic to *Clostridium acetobutylicum* at over 3.5 percent concentration, but butanol shows some toxicity at 1.2 percent and is totally inhibitory at 2 percent concentration. A typical yield from sucrose is shown in Table 8.1.

One of the few acetone/butanol plants in use is in South Africa. Its operation has been described by Lurie (1975). In the South African sugar belt every molasses is different, and availability changes from month to month. The fermentation is sensitive to these changes, and the characteristics that lead to good yields in short times have been investigated. Batches of molasses are segregated and blended according to properties. Initial sugar concentration in the fermentation is 6 to 6.5 percent. Corn steep water is absolutely essential to high yields, and the fermentation plant is adjacent to a cornstarch factory. When the value of corn rose, corn steep water became too high priced for fermentation, and amount in the fermentation was reduced by 90 percent and replaced by ammonia. However, the remaining 10 percent cannot be cut. Ammonia is added to adjust pH and to supply nitrogen. A typical batch run lasts 48 hr. Continuous fermentation has been described by Hospodka (1966). Several stages were run to obtain a total solvent yield of 40 percent, half of which was butanol.

The operation is aseptic, and the inoculum is only 2 gal to a 20,000-gal fermenter. This means that contamination must be avoided and that conditions must be correct to avoid a poor start and low yields. Traces of copper are disastrous; thus no copper or brass pipe or fittings must contact the makeup water or the ingredients. Yields are about 2 percent total solvents, and the yearly output is 1800 tons/yr of butanol and 900 tons/yr of acetone. The steam cost is high, and the effluent has to be evaporated as well because no degradation of local natural waters by industrial pollution is permitted. Waste solids have some value as animal feed, and fermenter gas is purified for salable carbon dioxide and a waste gas contains hydrogen that is burned to supply heat.

Attempts to find cultures more tolerant to butanol have been unsuccessful, and lipid structure of membranes may be damaged at about 2 percent. A mutational program for improvement might do well to emphasize developing altered cell membranes. Recovery of solvents by distillation from so dilute a fermentation medium is expensive. However, Abrams (1975) has pointed out that the distillation temperature is well matched to low-grade heat. Waste steam from other factories or heat from nuclear processing would be very cheap; thus an acetone–butanol factory might be an exceptionally good adjunct to another type of factory even if not economical independently.

In South Africa, where acetone–butanol fermentation uses molasses, waste disposal has been a key difficulty; presently all the residue is concentrated and spray dried. Attempts at using waste for irrigation were unsuccessful because of odors and runoff that polluted streams and caused the death of cattle. Initially

**Table 8.1 Products from 100 tons of Blackstrap Molasses
(57 tons of Sugar)**

	Yield, tons	
Butanol	11.5	⎫
Acetone	4.9	⎬ Total solvents 16.9 tons (28 percent)
Ethanol	0.5	⎭
Carbon dioxide	32.1	
Hydrogen	0.8	
Dried stillage	28.6 (protein 6 tons)	

there was little demand for the spray-dried powder, but now the total output goes into feed for ruminant animals.

The People's Republic of China also has some acetone and butanol produced by fermentation.

Butane Diol

This product was ready for commercialization 20 yr ago, but petrochemical routes became more attractive. A strong technological base is available for fermentation, and extensive pilot scale work has been performed. A key biochemical intermediate is acetoin, which is reduced to butan-2,3-diol. Theoretical yield from hexose is 50 percent. Different stereoisomers of the diol are produced in various proportions by microbial strains. The organism of choice seems to be *Aerobacter aerogenes,* which will give about 85 percent of the theoretical yield of diol plus a small additional amount of acetoin and ethanol. No cheap way of recovering the diol has been presented; its vapor pressure is far too low for simple distillation.

Ethanol

Archaeological evidence indicates that alcoholic beverage fermentation is 10,000 yr old (Greenshields, 1975). A temple at Thebes dating from 6000 B.C. shows in strip pictures how wine and beer were made. Christian monasteries in Europe distilled fermented liquor by about 800 A.D.

With older technology, about 12 percent sugar represents an optimum for good ethanol yield in a reasonable time. A final ethanol concentration of 15 percent can be obtained by fermenting concentrated syrups, but the production rate becomes uneconomically slow far below this level. Temperature from 25 to 40° C may be used, and times range from 36 to 72 hr. The principal cost is raw materials.

The two main feedstocks for alcohol factories are corn and molasses. Juices from the extraction of sugarcane are sometimes fermented directly, but evaporation of these juices causes crystallization of sucrose of high value for food. Further evaporation gives additional crops of crystals, but the syrups become concentrated in carbohydrates other than sucrose and in ions of heavy metals.

Concentrated syrup or "blackstrap molasses" is a relatively cheap feedstock for fermentation, although the price mirrors the erratic course of world sugar prices. The approximate composition of molasses is 55 percent sugar, of which 35 to 40 percent is sucrose and 15 to 20 percent is invert sugar (equal amounts of glucose and fructose from hydrolysis of sucrose). There is also a high-test molasses from which no sugar has been crystallized. It is 70 to 80 percent sugars and has been hydrolyzed with acid to give mostly invert sugar, which is sweeter than sucrose. Some molasses from sugar beets is available in the United States; its use for making fuel alcohol has unfavorable economics because beet sugar farming would die out without a large price support by the government.

The process for ethanol from molasses (Fig. 8.1) was developed in the days of cheap energy and must be considered obsolete. It is a batch operation in which the molasses is diluted to 10 to 15 percent sugars and sterilized with steam. Blackstrap molasses may not need additional nutrients, but high-test molasses or sugar juices are supplemented with a nitrogen source, usually ammonium sulfate, and small amounts of phosphate. Yeast cultured aseptically is inoculated into wooden or steel fermentation tanks where cooling is necessary because metabolic heat is generated. Scrubbing of the evolved carbon dioxide with fermenter make-up water reduces losses of ethanol in the gas stream. After 28 to 72 hr of fermentation, the alcohol is recovered by distillation. Energy is wasted by (1) dilution of molasses, which has an energy investment for evaporation, (2) sterilization, which has been proven unnecessary, (3) no reuse of yeast (thus sugar is needed for new growth), and (4) inefficient distillation. Alternatives are discussed later.

The carbohydrate feedstocks for Brazilian ethanol plants are varied. Molasses or sugar mill concentrated streams account for 47 percent of the total with 44

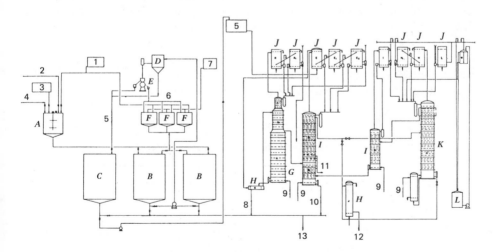

Figure 8.1 Flowchart for alcohol manufacture of ethanol from molasses, sugarcane juice, and syrup: (1) water, (2) mixed juice, (3) molasses, (4) nutrients, (5) beer, (6) yeast, (7) dilute acid, (8) waste, (9) steam, (10) fuel oil, (11) anhydrous alcohol, (12) stillage, (13) waste; (A) dilution tank, (B) fermenter, (C) collection tank for beer, (D) decanter, (E) centrifuge, (F) yeast tank, (G) beer still, (H) heat exchanger, (I) fuel oil column, (J) condenser, (K) dehydration column, (L) benzene reservoir. *Source:* modified from *Guide Book,* ISSCT XVIth Congress, International Society of Sugar Cane Technologies, Sao Paulo, Brazil (1977), p. 139.

percent from sugarcane juices and 9 percent from mandioca (Yand and Trindade, 1979). One project for the use of babassu has been approved. A sugarcane and a mandioca distillery are compared in Table 8.2. The processes are quite comparable, except that amylase enzymes are added to mandioca to hydrolyze the starch to glucose. One important difference is water consumption; sugarcane processors use 200 volumes of water per volume of alcohol whereas manioca processors use 43.3 parts. Stillage is high in salts and organic material, which would pollute rivers if discharged directly. The most widespread treatment practice is lagooning to allow microbial decomposition of organic materials, but some processors spray raw stillage on the sugarcane fields for irrigation and fertilization. Other treatments of stillage being considered are purification to produce animal feed, anaerobic generation of methane, and cultivation of single-cell protein for animal feed.

The Brazilian plants collect the yeast, perform an acid wash to retard bacterial contaminants, and recycle the yeast to the fermenters. This gives a high cell density so that the fermentation time is shortened to about 12 hr. Furthermore, less sugar is diverted to new cell growth, and thus percentage conversion to ethanol is higher.

Guatemala has followed the lead of Brazil for ethanol as supplemental motor fuel. Espinosa, Cojulun, et al. (1978) have reported some features of the Guatemalan process, including recycle of yeast, continuous fermentation, collection of excess yeast for animal feeding trials, direct sugarcane fermentation, bagasse burning to power distillation and evaporation, and water to scrub ethanol from the evolved carbon dioxide. Their feed is 130 g/l of sucrose to obtain over 9 percent ethanol v/v.

Rolz, De Cabrera, et al. (1979) of ICAITI in Guatemala have been developing a process termed *Ex-Ferm* in which chopped pieces of sugarcane are fermented directly. Both fresh and stored dry pieces can serve as the feedstock. This eliminates the expensive step for extraction of fermentable sugars from the cane. The final broth had material that gave a positive test for sugar, but this is more likely to be unfermentable oligosaccharides than residual sucrose. A packed bed rather than a stirred fermenter is recommended because thick suspensions of sugarcane

Table 8.2 Typical Materials in Brazilian Fermentation[a]

Material	Amount	Material	Amount
Sugarcane	1910 tons	Mandioca roots	860 tons
Bagasse	480	Enzymes	0.6
Chemicals	5.85	Chemicals	6.99
Fusel oil	0.61	Fusel oil	0.61
Stillage	1580	Stillage	1330
CO_2	90	CO_2	90
		Solid wastes	50

[a] Basis: 100 metric tons of absolute ethanol.

pieces are difficult to mix. Good drainage to avoid product losses with the pieces is essential.

The conventional process for making ethanol from corn grain includes a variation of wet milling. Ground corn grain is treated hot with amylase enzymes to extract and hydrolyze the starch. Fermentation of this material to ethanol leaves a stillage very rich in corn protein. In contrast, the usual wet milling of corn grain is a slow process with fractionation of the components (Fig. 8.2). Corn oil, a valuable product, is found in the germ. Gluten has most of the corn protein. Starch is an important article of commerce for many uses in foods, in paper manufacturing, as textile sizing, and the like. A part of total starch production is hydrolyzed to syrup that may be isomerized to a mixture of glucose and fructose; this mixture is a valuable, excellent sweetener. It is evident that fermentable sugars from corn grain cannot be cheap because there are competing uses for starch and investment costs are high for constructing the factory.

Sheppard (1978) has presented one of several schemes proposed for integrating wet milling with ethanol manufacture. The idea is to sacrifice yields in starch extraction while taking the easily removable starch under mild conditions. The remaining starch plus gluten would go to the fermentation step. The advantages are lower plant investment, short detention time, corn oil and starch as valuable coproducts, and bypassing purified starch as an intermediate before fermentation. Tsao (private communication) proposes much the same scheme, except that the "easy" starch would go to fermentation and the residue would be processed to food such as corn chips.

Fermentation produces mostly ethanol, but about 3 percent of the carbohydrate feedstock goes to glycerol. Succinate and acetate represent less than 1 percent of the feedstock. All these unwanted side products have low vapor pressures and tend to remain in the stillage. However, traces of volatile acids are carried to the first distillation column, and stainless steel is used to minimize corrosion.

Saccharomyces cerevisiae is not the only organism that is used commercially for ethanol fermentation. Swings and DeLey (1977) have reviewed *Zymomonas* bacteria, which are used throughout the world to produce ciders and beverages such as pulque. Under ideal conditions the molar yield can be 1.9 moles of ethanol/mole of glucose, about as good as yeast can do. The alcohol concentration seems to be at most 5 to 6 percent, which is not as high as that obtained with yeast. Bacteria have potential advantages of very rapid growth rates and thus reduced opportunity for contamination. Other yeasts of various genera are sometimes used commercially. A densely flocculating yeast has been mentioned as being used in Europe. Its advantage is easy retention in the fermenter to get very high organism concentration and thus a high rate of ethanol production.

Continuous Fermentation

Fermenters require sizable investments; higher productivity per unit can mean significant savings in capital, particularly for the massive factories required for

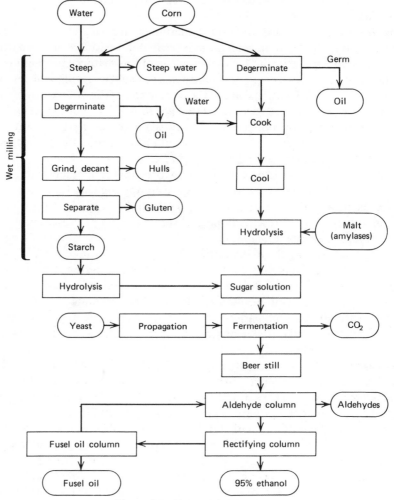

Figure 8.2   Ethanol from corn.

fuel production. A typical batch fermentation such as that for beverage alcohol may take 48 hr. Part of this time is required for growth of the organisms; product formation starts, increases, and declines until it is best to halt this batch and to start another. In contrast, a continuous fermentation maintains vigorous cells in a dynamic equilibrium with constant feed converted to product at a high rate. Yeasts grow quite rapidly and offer the possibility of short detention times. Table 8.3 shows the potential advantages in productivity of continuous culture.

A conventional fermenter often serves a dual purpose as a sterilization vessel; a continuous fermenter must be furnished with sterile media, or at least with feed streams relatively free of active contaminating organisms. Batch sterilization is

Table 8.3 Comparison of Batch
and Continuous Culture

Detention Time	Relative Productivity
Batch (48 hr)	1
24	2
12	4
6	8
4	12
3	16
2	24

probably out of the question for fuel fermentations because steam is used for heating and its energy is lost to the cooling water needed to bring the medium back to fermentation temperature. A well-designed continuous sterilizer exchanges the heat of the hot, sterile medium with the feed stream. A small amount of steam is required to complete the heating, and some cooling water accomplishes the final cooling. Thermal efficiency can be quite high.

One of the largest fermenters in the world is at the Billingham plant of Imperial Chemicals Industries, Ltd. It measures 200 ft (61 m) long by 25 ft (8 m) in diameter and will produce microbial protein for feeding livestock. Rated capacity is 70,000 tons/yr of protein (*Chemical Engineering News,* April 10, 1978, p. 86).

Continuous fermentation to produce ethanol was reviewed by Prince and McCann, (1978), who point out that 95 percent of organic products from the sugar in batch fermentation are ethanol and CO_2 but that 3.3 percent goes to glycerol and 0.6 percent goes to succinate. Although of minor importance to beverage alcohol because they have no taste or odor, these by-products to be found in the stillage are a loss for fuel alcohol processes and add expense for waste treatment. Recovery from such dilute solutions is not appealing. Loss of ethanol in the CO_2 stream is mentioned as the reason for operating the fermentation at 28 to $32°C$. At higher temperatures, increased vapor pressure of ethanol would require recovery from the CO_2. A conventional scrubbing system would be expensive, but bubbling the CO_2 stream through makeup water for the fermentation step should be simple and efficient. Of course, there are costs for ductwork and either a tank system for dispensing the CO_2 into fine bubbles or an absorption tower. Suggestions for using thermophilic organisms and elevated temperatures to increase productivity or to operate a vacuum fermentation at milder pressures must consider the problems of ethanol recovery from the gas phase.

"Fill and draw" is the term for harvesting part of a fermenter and replacing with fresh medium. This can reduce the time to reach peak cell growth because of the heavy inoculation of vigorous organisms. An intermittent process is usually inferior to a truly continuous fermentation in which an optimum steady state is established. A commercial continuous fermentation plant in New Zealand has several stirred tanks in series followed by a separation step for yeast that are recycled to the tanks. Total residence time for producing ale continuously is 3.5 hr, and some runs have lasted 12 months.

Single-stage continuous fermentation can do as well as or better than batch fermentation in approaching the theoretical yield of ethanol. Englebart and Dellweg (1978) obtained 98 percent of theory versus 93 percent for the usual batch process. Glucose solutions or starchy mashes from wheat or potato or manioc were used. The high yields may be the result of utilizing carbohydrates other than those containing glucose. The substrate had a major effect on ethanol tolerance in that runs with glucose or beet sugar solutions showed good growth of yeast as the alcohol concentration passed 8 percent. Molasses and starch could have fermentations in which the yeast were inhibited by 5 percent alcohol, and too rigorous cooking of starch gave worse inhibition. Caramel compounds in molasses and in starch hydrolysate were thought to potentiate the inhibition by ethanol. Results from a 1-1 fermenter translated nicely to a 250-liter fermenter; thus scale-up should be easy.

Vacuum Fermentation

Finn and Ramalingam (1977) showed that operation of ethanol fermentation at low pressure allowed removal of ethanol as it was formed; thus its concentration could be maintained low enough to avoid inhibition of the yeast. A sketch is shown in Fig. 8.3. Note that the gases must be compressed because the low partial pressure of ethanol in the vapor requires very cold water or brine for condensation, and this is too expensive compared to normal cooling water. Cysewski and Wilke (1977) optimized vacuum fermentation of alcohol and incorporated cell recycle to obtain very high cell densities in the fermenter. With fermentation at 35°C, it is common to use cooling water to remove metabolic heat. Vacuum evaporation removes this heat and represents about a 7 percent saving in cooling water. There is an obvious improvement in vacuum fermentation by eliminating the decompression and recompression of CO_2. The method is also shown in Fig. 8.3. Atmospheric pressure exists in the fermenter, and CO_2 escapes normally. Little CO_2 reaches the pot to waste pumping energy by diluting the vapor. Also, condensation is easier because the partial pressure of ethanol is higher.

There is general agreement that vacuum fermentation at 35°C is a marginal or losing proposition because of the extra capital cost and the energy required for recompression. At 55°C the fermentation is more rapid with thermophilic cultures, and the vapor pressure is so much higher that less vacuum is needed. The metabolic heat evolution matches the heat of evaporation quite well, and vacuum operation may be practical.

Some nutritional adjustments are necessary to achieve good performance. Furthermore, the fermentation must not be highly anaerobic; thus a very small input of air or oxygen is required. With the external vacuum pot of the modified vacuum fermentation, the main tank is operated normally with traces of air preventing too low an oxidation–reduction potential.

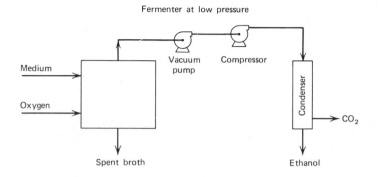

Figure 8.3   Vacuum fermentation. Courtesy NTIS.

Tower Fermentation

Continuous production of alcohol in tower fermenters is a well-established practice in Europe (Blanken, 1974; Greenshields and Smith, 1971). The APV Company in Crawley, Sussex, England markets the process shown in Fig. 8.4. A heavily flocculated yeast culture with over 15 percent solids is retained in the fermenter; thus there is no need for external collection and recycle. Detention time to reach about 8 percent ethanol is a remarkably short 4 hours, with the result that small units representing low capital cost are as productive as much larger conventional fermenters. It is also surprising that the temperature is 15 to 17°C; a higher temperature might be better for fuel alcohol where quality control is not as important.

A change in the yeast can affect its flocculating properties and lead to losses. Both air and CO_2 are admitted to the bottom of the column; air in

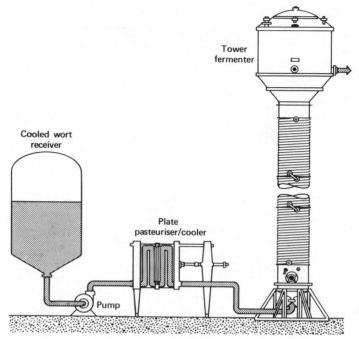

Figure 8.4 Tower fermentation. From trade literature.

small amounts avoids strongly anaerobic conditions, and CO_2 aids mixing. The lower portion of the column has a stable agglomeration of yeast. Higher up, the yeast is less well flocculated. In the upper section and in the settling cone, the flocs are small and well mixed. As medium flows up the column, sugar concentration decreases while alcohol and CO_2 are formed. This produces concentration gradients and specific gravity gradients, both of which are important to the metabolic state and sedimentation properties of the cells. Feed rate can be controlled to maintain gradients that have been found to work well.

Páca and Grégr (1979) found that the partial pressure of oxygen was important for production of ethanol by *Candida utilis* in a tower fermentation. Some oxygen is necessary for good productivity of ethanol, but an excess causes a decline in ethanol production while concentration of acetate increases.

Greenshields and Smith (1971) have reported tower fermentations in the laboratory at 20°C that gave 95 percent utilization of sugars in 4 to 8 hr with 10 percent sugar from molasses. A small sacrifice in percent utilization results from faster dilution or higher sugar concentration. Such short residence times would minimize the effects of contaminants.

Lager beer is produced by tower fermentation of pasteurized wort (nutrients). Some runs persisted for 6 to 10 months of continuous operation. When contamination becomes apparent, it is sometimes possible to increase the flow rate and flush small bacteria from the fermenter while the heavily flocculated yeast are retained. The top of the fermenter is a truncated, conical separator that allows escape of gas and collection of cells that sediment well. Yeast in excess of that to fill the column will go out in the effluent, or portions can be withdrawn directly from the column.

The reason that a tower fermentation works so well is the very high numbers of organisms present. Although alcohol inhibits the fermentation rate of individual cells, the total number of cells is so great that low activity per cell is overcome. The tower fermenter has a thick floc that allows cell retention, but other means of achieving high cell density are possible. Ordinary yeast cells can be collected and returned to the fermenter; this requires a separate step, whereas a collection area is built into the tower fermenter. Del Rosario, Lee, et al. (1979) have used a heavily flocculating strain of *Saccharomyces carlsbergensis* to pack very high cell concentrations into a fermenter. There were about 3×10^9 cells per milliliter for a total cell weight of 83 g/l, which is almost a paste. Batch fermentation gave about 7 percent by weight of ethanol in 5 hr. Continuous culture with cell recycle could give over 8 percent ethanol with residence times of 3 hr or more, but yields fell off for shorter residence times. Tsao's group at Purdue University has devised a packed cell reactor using *Rhizopus*, a filamentous mold that has strands of cells that can be strained from the medium and packed into a column. Preliminary experiments have been run with a small column through which mixed sugars and oligomers are passed. With 20 percent by weight of fermentable sugars, the product is about 10 percent ethanol, using a detention time of a few hours. An even more remarkable unit has been described by Chibata (1979). Cells of *Saccharomyces carlsbergensis* are immobilized in a gel in which cells can continue to multiply. With a 1-hr detention time, 15 percent ethanol has been reached. The preparation of the gel particles containing yeast is described by Tosa, Sato, et al. (1979).

It is difficult to justify the research in the United States to develop a highly productive fermenter by continuous removal of the ethanol to reduce its inhibition on the culture. Tower fermentation already has high productivity and is commercially proven. Fermenters with packed or immobilized cells show great promise. Unless vacuum fermentation can achieve some unanticipated energy savings, these alternate approaches will be far superior in productivity and in operating cost.

Ethanol from Cellulose

Glucose from cellulose hydrolysis should be the same as glucose from starch. However, Wilke and Blanch (1978) have found the enzyme hydrolysates are not always readily fermentable. Sugar solutions produced with cellulases from *T. reesei* QM9414 from Natick Laboratories were fermented very poorly by yeast. The inhibitory material could be removed by carbon adsorption to give sugar solutions that fermented nicely. This *Trichoderma* strain produces a yellow pigment, whereas Rutgers strain C-30 does not. Sugar solutions produced with Rutgers cellulase showed no inhibition of the ethanol fermentation; thus it would be of interest to seek a correlation between pigmentation and inhibition.

Ghose and Tyagi (1979a) encountered no toxicity for the fermentation of sugars obtained by enzymatic hydrolysis of bagasse; again the cellulases came from *T. reesei* QM9414. Perhaps the enzymatic activity was purified, but no such step was mentioned. Both batch and continuous fermentations using *S. cerevisiae* were rapid; cell recycle for continuous fermentation led to 9.7 percent ethanol at a 5-hr detention time. Ghose and Tyagi (1979b) studied inhibition of the fermentation by high glucose concentration or high concentrations of ethanol. A multistage continuous process is recommended to provide for fermentation of the high levels of residual sugar leaving a first stage that has a high rate of ethanol fomation. Equations were derived for optimizing a multistage process. Two stages, each with residence times of 9 hr, were equivalent to a single stage of the same total volume, but with a 42.4-hr detention time.

Sitton, Foutch, et al. (1979) have developed a process for manufacturing ethanol from cornstalks and have analyzed the costs. The steps are:

1. Hydrolysis of pentosans and removal of sugars in the filtrate. Conditions are 100° C, 4.4 percent sulfuric acid, and 50 min contact time.
2. Hydrolysis of cellulose with 8 percent sulfuric acid at 100° C for 10 min.
3. Recovery of acid from the sugar solutions by electrodialysis.
4. Fermentation of sugars to ethanol in columns filled with support material coated by microbial cells. Alcohol from glucose by *Saccharomyces cerevisiae* in such a fixed-film reactor is based on extensive research. Use of *Fusarium oxysporum* with mixed sugars from pentosans should work but has not yet been reported for a film reactor.
5. Recovery of ethanol by distillation.

Excellent hydrolysis yields have been demonstrated, and the projected economics for ethanol production are quite favorable. However, the major single cost was for cornstalks, and the assumed price of $15/ton is low compared to $30/ton assumed by other authors. Electrodialysis is also costly as it constitutes 50 percent of the plant investment and 40 percent of operating cost with associated items such as power and chemicals for neutralization.

This process is very attractive in terms of excellent yields amd clever technology. Some features need improvement; for example, ethanol concentration is only 1.5 percent in the fermentation. Questionable accumptions such as the use of one distillation column for ethanol create doubt about the cost analysis. Nevertheless, this process or a hybrid process incorporating the best features of several processes could lead to early commercialization of ethanol from cellulosic materials.

Simultaneous Hydrolysis and Ethanol Fermentation

One way to eliminate product inhibition that slows the rate of cellulose hydrolysis is to remove glucose as rapidly as it is formed. A yeast fermentation to alcohol has been shown to operate simultaneously with degradation of cellulose to glucose by cellulases from *Trichoderma reesei* (Blotcamp, Takagi, et al., 1978; Meyers, 1978; Takagi, Abe, et al. 1978). Purified cellulose from two sources was tested: Avicel 105 (American Viscose, Marcus Hook, Pa.) and Solk-Floc, (Brown Co., Berlin, N. H.). A complication arises because the optimum fermentation temperature is 28 to 32°C, whereas saccharification is best at 45 to 50°C. The compromise temperature selected is usually about 40°C.

Inoculum size was found to have a minor effect in that increased cell concentration gave only slight improvements in ethanol yield. Lower enzyme concentrations suffice for the simultaneous system because end-product inhibition is essentially eliminated. The glucose concentration with yeast present tended to remain below 0.5 mg/ml, and cellobiose was less than 0.4 mg/ml. Ethanol concentrations of 5 percent in simple saccharification were not detrimental, and 10 percent ethanol may provide a small enhancement of β-glucosidase activity.

The simultaneous process was tested with cellulose slurries from 2 to 18 percent by weight, and increases above about 6 percent gave no additional ethanol. This amount of cellulose saturates the available enzyme, but addition of more enzyme gives a low percentage conversion of cellulose under the experimental conditions.

A program for producing chemicals by fermentation is described by Mooney (1977). Cellulosic materials have been converted to ethanol at a pilot plant scale. Another process with simultaneous cellulose hydrolysis and ethanol fermentation was reported by Savarese and Young (1978). A variation being explored at the University of Pennsylvania is based on a thermophilic organism, *Clostridium thermocellum*, which produces ethanol plus acetate (Pye, 1978). High-temperature fermentation allows a much better match with the temperature optimum of the cellulase enzymes and gives a higher vapor pressure; thus milder vacuum removes ethanol as it is formed. High-temperature cellulases from *Thermoactinomyces* can be used with the *Clostridium* at 60°C and pH 6.6. There has been difficulty in raising ethanol concentration to 2.5 percent in the broth; recovery costs are unattractive until the concentration is at least twice this amount. Hägerdal, Harris, et al. (1979) have found that the cellulase complex produced by *Thermoactinomyces* is incomplete because β-glucosidase activity is almost nil. This enzyme is bound the cell wall and not released to the culture filtrate; thus supplementation from another source of β-glucosidase may be required.

Although simultaneous hydrolysis and fermentation solves the problem of product inhibition, much flexibility is lost for process development of the individual steps. Temperature, pH, and other conditions may have to depart significantly from their optima for the individual steps and be set at compromise levels. Furthermore, biological kinetics probably will not match; the ethanol fermentation probably will be rate limited because liberation of glucose from cellulose is slow. Working with a solid substrate is also a headache for fermentation development because of the need to handle slurries and to remove or recover unreacted solids. The simultaneous process is thus an important concept with marked potential, but its problems leave much room for alternate approaches.

Mixed Culture Fermentation of Cellulose

Separate hydrolysis of cellulose has been avoided by combining enzyme treatment with ethanol fermentation as mentioned previously. A possible advance over this technique would be to use a mixed culture in which some organisms are highly cellulolytic and others ferment the released sugars to ethanol. Brooks, Bellamy, et al. (1978) reported the use of a thermophilic *Sporocytophaga* mold that produces very active cellulase anaerobically. A search was made for thermophilic bacteria to use in association with the mold to produce ethanol. Ethanol tolerance of the organisms and of selected mutant cultures was not very good, and 4 percent ethanol was quite inhibitory. Mixed cultures gave only traces of ethanol, and acetic acid was the major product. The concept merits further study, but defined mixed cultures are more difficult to study than are pure cultures because of the greater number of interacting parameters. Thick slurries of biomass should be fed to processes to avoid excessive cost for concentrating product streams. However, some slurries may prove to be extremely difficult to handle (Wise, Ashare, et al., 1979). At 12 to 13 percent of straw in water, flow almost ceases and mixing is prohibitively expensive. A 10 percent slurry of straw could not be pumped, and at 3 percent flow was still poor. To handle such streams, it may be best to heat treat or to hydrolyze partially with acid, base, or enzymes.

Wang, Cooney, et al. (1979) have selected mutant cultures of *Clostridium thermocellum* and *C. thermosaccharolyticum* for direct fermentation of cellulose or of cellulose with hemicellulose. Strains resistant to inhibition by ethanol have been found; growth is 50 percent of the control growth rate up to 6 percent ethanol, whereas parent strains grew poorly in 1 percent ethanol. With these strains, lactic acid tends to accumulate as the cells age, and it appears to depress the rate of ethanol fermentation. The parent strain produced considerable lactic acid and about as much acetate as ethanol. The mutants produce relatively little acetate and less lactate. *Clostridium thermocellum* mutants attack cellulose and hemicellulose but accumulate the sugars from hemicellulose. *Clostridium thermosaccharolyticum* can be added later in the fermentation because it ferments the hemicellulose sugars. With pure xylose, this mutant gives 2.6 percent ethanol, a concentration approaching that needed for a commercial operation. Mixtures of the two mutants yielded 3 percent ethanol from solka floc (a treated cellulosic material containing hemicellulose). This research has achieved an order of magnitude improvement in ethanol concentration in a short time. If the rate of progress can be sustained, there will quickly be a strong basis for direct fermentation of biomass without a costly hydrolysis step.

Fermentation of Hydrolyzed Hemicellulose

Research groups that have fermented pure sugars such as xylose as a prelude to studies with actual hydrolysate have been surprised that the latter does not work as predicted. At Auburn University (McCasky, private communication) 25 strains were selected for fermentation of xylose. Only four of these did well on hydrolysate, and the yields were markedly lower than with pure xylose. A group at Purdue University (Flickinger, private communication) has found that toxic materials produced during hydrolysis can be removed or their concentrations can be lowered by adjusting hydrolysis conditions; the fermentation then proceeds well. They have also found sequential use of the different sugars in the hydrolysate. It is well known that a sugar such as glucose, which is in the main portion of a metabolic pathway, controls the enzymes for the metabolism of less common sugars. There is no need for the cell to synthesize special enzymes for these uncommon sugars until the glucose is nearly exhausted. Growth on mixtures of two sugars often shows a drop in rate as the common sugar is depleted and a resumption of rapid growth after the enzymes for the uncommon sugar are activated. This is termed *biphasic growth* or *diauxie*. The Purdue group followed concentrations of various sugars in hydrolyzed hemicellulose and found quite sharp sequential use. Although xylose made up about 85 percent of total sugar, it was fourth in sequence of use. The implication is that data on fermentation of pure xylose may be unreliable for predicting performance with hydrolysate.

Although the Purdue group has a strain of *Aeromonas hydrophilia* that has been adapted to high xylose concentrations and produces 0.52 moles of ethanol per mole of xylose consumed, this research is being deemphasized. A more promising process is based on a recent discovery that xylose can be converted to glucose by yeast if an enzyme is added to the system (G. T. Tsao, personal communication). This enzyme is identical to glucose isomerase, which is very widely used commercially for the production of glucose-fructose sweetener from glucose. The implications are that the best yeast strains for making alcohol can now be used with the mixed sugars from cellulose hydrolysis, and the yield of alcohol from lignocellulosic biomass could increase by 60 to 80 percent. Furthermore, there are good prospects for inserting the genes for synthesizing glucose isomerase into yeast. This would eliminate the cost of added enzyme.

By-products more valuable than ethanol may be produced from hemicellulose, but their markets will be small compared to those of alcohol fuels. It is highly likely that the economics of alcohol factories will benefit significantly by recent improvements in fermenting the sugars from hemicellulose.

Acetic Acid

Groups at both the Dynatech Company and MIT are developing processes for acetic acid by fermentation. Acetic acid is a valuable chemical in itself, and it could be a clean liquid fuel. Rapid, high-yield fermentations to acetic acid are well known. The MIT group ferments sugars with *Clostridium thermoaceticum* (Wang, Fleishchaker, et al., 1978b). The Dynatech group is converting marine algal biomass to acetic acid with mixed cultures.

Acetic and other organic acids can be electrochemically converted to hydrocarbons via the Kolbe reaction:

$$R - COOH \longrightarrow R - R + 2CO_2$$

Ethane from acetic acid would be excellent as pipeline gas, and higher acids give liquid hydrocarbons for motor fuel.

Vinegar generation is an old art. Even though acetic acid is easily synthesized from petroleum, acetic acid from fermentation has a superior flavor because of impurities. The usual fermenter for vinegar generation is packed with wood shavings. Microorganisms attached to the shavings are retained in the reactor and used over and over. Lai and Wang (1978) have described very rapid and inexpensive production of vinegar in a reactor consisting of two barrels. This fermenter, suited to small operations, can turn out roughly 25 gal (100 l) of 4 percent acetic acid per day.

Conversion of glucose to acetic acid does not require evolution of carbon dioxide; thus weight yields can be high. About 0.9 to 0.92 g of acetic acid are obtained from 1 g of glucose. Of course, the heat of combustion of acetic acid suffers because one of the carbon atoms is already highly oxidized.

Single-Cell Protein

Litchfield's review (1978) notes several large-scale fermentations for producing microbial protein. One plant produces 3.35 tons/hr of yeast paste (50% solids) from sulfite waste of a paper pulping plant. Another factory produces 5000 tons/yr of Torula yeast. A British process for baker's yeast from molasses maintains a yeast concentration of 1.9 percent in the fermenter using continuous culture at a dilution rate of 0.14. Plants for producing 60 to 100,000 tons/yr of yeast from petroleum have been built in Sardinia (Italy), Romania, and the Soviet Union, but questions have arisen concerning hydrocarbon residues in the product.

Algae are popular food in the Orient, and the Japanese have elaborate, glassed-in culture facilities for commercial production. Food value justifies expenditures that cannot be considered for algae processes directed toward cheap fuels. Near Mexico City, the alga *Spirulina maxima* is cultivated on natural saline ponds for sale to the Japanese. Algae also grow on organic substrates, but protein extraction may bring along pigment that would threaten acceptance as human food.

The Swedish Symba process develops microbial protein from carbohydrate wastes such as those from potatoes. A mixed culture system of *Endomycopsis fibuligera* and *Candida utilis* is employed. Humphrey, Moreira, et al. (1977) have experimented with production of single-cell protein from cellulose with *Thermoactinomyces* sp. and obtained yields up to 0.45 g of cells per gram of cellulose.

Goals for microbial protein as food are high productivity, high proportion of cell mass as protein, a good profile of desirable amino acids, good performance in feeding livestock, and no toxic or carcinogenic components. For human food, there must be the additional requirements of no objectionable tastes or odors, and nucleic acids must be removed because the levels associated with microbial protein are toxic to humans.

Lignin might be converted to useful small molecules or to cell mass by microbial cultures. In most such studies, molds are grown on cellulosic biomass in which cellulose and hemicellulose are metabolized more rapidly than lignin. Any attempt at biological delignification has the potential drawback of unacceptable losses of carbohydrates. An important application of lignin fermentation is upgrading of straws for use as animal feeds. Lindenfelser, Detroy, et al. (1979) reported an example of this in the use of mushrooms to increase the protein content and digestibility of wheat straw.

Alcohol fuel production from lignocellulosic biomass will have large amounts of relatively pure lignin left over. Some can be used for lignochemicals, but the remainder may have to be burned for disposal and for energy credits. A better use would be as a fermentation feedstock. However, no direct fermentation to a valuable product has been found. Single-cell protein would be a desirable product if good yields were demonstrated, but toxic, phenolic compounds tend to slow the fermentation and suppress yields. The MIT group has an interesting approach in which mixed cultures are employed with some species degrading lignin to small molecules serving as the substrates for a directed fermentation to a selected product. This reduces the accumulation of intermediates, thus permitting more rapid rates.

It is disconcerting when promoting methanol and ethanol as potential cheap fuels to note that both are so inexpensive as to be considered seriously as substrates for the production of single-cell protein. Cooney and Makiguchi (1977) have assessed single-cell protein from methanol-grown yeast, and Laskin (1977) has reported on growing *Acinetobacter calcoaceticus* on ethanol.

Upgrading of straw for animal feed has been the dream of agricultural research for many years. Grant, Anderson, et al. (1977) have made cost estimates for two fermentation processes that increase the feed value of grass straw: (1) hydrolysis with dilute sulfuric acid followed by yeast fermentation and (2) alkali treatment and then fermentation with cellulolytic bacteria. Production costs were estimated to be roughly $80 to $90/ton by either method, but the capital cost was much less for the alkali processes for a plant handling 100 tons/day of dry straw. Expensive, corrosion-resistant pressure vessels were responsible for the high capital cost of the acid process. The fermented straw contains protein and fat and properties of digestibility that would command a premium, but it was concluded that costs will not become competitive until prices of current feed constituents increase.

Fixed Nitrogen by Fermentation

Nitrogen fixation is important to biomass production in several ways. The large crops of biomass needed for energy would quickly deplete nitrogen from soil; thus some means for replenishment is absolutely essential. Leguminous plants are associated with symbiotic microorganisms in root nodules that fix atmospheric nitrogen. Such plants might have no requirement or greatly reduced need for nitrogen from fertilizer. However, economics may dictate that crops other than legumes be grown because high productivity outweighs cost of fertilizer. In other words, land acquisition, site preparation, cultivation, and harvesting costs may demand productivity unreachable with legumes. This raises the question of how best to reduce the price of fertilizer.

For several years, there have been suggestions for an industrial process for microbial nitrogen fixation leading to ammonia or protein. Ammonia has a variety of uses in the world economy, including explosives, synthetic organics, and fertilizers, with the latter accounting for the major usage. Because the industrial process for ammonia is heavily dependent on fossil fuels, chemical nitrogen fixation can accurately be viewed as a contributor to the present energy crisis.

Laboratory systems have been developed for nitrogen fixation based on NIF mutants. This designation means that the gene location for nitrogenase enzyme has been effected. In normal cells, ammonia or an intermediate in the nitrogen pathway represses synthesis of nitrogenase. These NIF mutants are derepressed and produce excess nitrogenase, which speeds product formation. Glutamine is an expensive, essential requirement in the culture medium. The organisms of current interest are *Klebsiella pneumoniae*, an anerobe that evolves hydrogen while excreting ammonium compounds, and *Azotobacter vinelandii*, an aerobe that excretes little fixed nitrogen but incorporates it in cell mass.

Wallace and Stokes (1978) have analyzed the costs of industrial processes based on laboratory systems for microbial nitrogen fixation. Glutamine costs immediately ruin any chance for profitability, so the assumption was made that glutamine could be ignored because means might be found for its elimination. Even so, the product concentration is too low (about 0.1 percent) for economical recovery. The most favorable estimate is a cost of $2/lb of ammonia by an industrial microbial process. This is far from the present value of 6 or 7¢/lb for ammonia. Perhaps direct application of dilute ammonia solutions to fields of crops could serve for fertilization and irrigation. However, the logistical problems of producing ammonia very near the fields to keep pumping costs low could be formidable.

Stillage

Gregor and Jeffries (1979) have studied membrane processes for concentration and desalting of the aqueous residue from ethanol distillation. The stillage from grain alcohol is relatively low in salts and high in corn proteins and other proteins. This stillage has long been evaporated to dryness and sold as cattle feed. Stillage from ethanolic fermentation of molasses is too high in salts to be used as cattle feed because the cattle may develop bloating or diarrhea. The removal of salts by a membrane process produces material that is probably equivalent to stillage from grain alcohol, but much more testing as cattle feed is required.

Hydrolysis of cellulose in biomass frees very little protein; thus the stillage from ethanol fermentation and recovery should be worth little as cattle feed. There will be a trace of biomass protein plus enzymes and other proteins released from the fermentation culture. Membrane processes might be worthwhile for water recovery for recycle, but there is more logic to using these streams for irrigation and fertilization of growing areas near the factory.

Hertzmark and Gould (1978) point out that over 50 percent of the protein in feed rations for cattle comes from soybeans. Meals from other oilseeds, animal protein, and grain protein account for the rest. Distillers dried grains play a small role, but drastic expansion of processing corn to alcohol will overwhelm the cattle-feeding market with dire effects on the prices of all protein sources, especially soybeans. Lipinsky, Scantland, et al. (1979) fear major dislocations of agriculture in the United States if great excesses of distillers dried grains become available. Farmers with crops other than corn may voice violent opposition to alcohol fuels, thus weakening the present enthusiastic support from agricultural lobbies.

Solvent Extraction of Products

Removal of products to eliminate their inhibition of the fermentation organisms might be accomplished by liquid–liquid extraction. Distribution coefficient is the ratio of the concentration of the desired product in one solvent to its concentration in a second solvent at equilibrium. A favorable distribution coefficient means that solvent extraction may be performed with relatively small volumes of extractant.

All liquids have some solubility in water, but this can be very small for nonpolar liquids. However, fermentation products such as ethanol, acetone, and butanol are themselves fairly polar and thus tend to have a greater affinity for water than for nonpolar organic liquids. Higher alcohols, such as decanol, are good extractants but are very toxic to the fermentation organisms. The MIT group found paraffins and silicone oils to be immiscible with water and nontoxic, but the distribution coefficients were poor. Lipids such as corn oil are not particularly good extractants and tend to coat the equipment and attach to clumps of cells. However, oil lost in the aqueous phase is a useful, although costly, nutrient.

Extraction of organic acids should be a better recovery method than distillation because their vapor pressures are low, and water is distilled from organic acids rather than distilling acids from water. Continuous extraction from fermentation broth that is recycled to the fermenter poses serious restrictions on the solvent not to be inhibitory. Simple recovery from final broth allows a much wider choice of solvents. Helsel (1977) has reported on using trioctylphosphine oxide (TOPO) for extracting organic acids from dilute waste streams. Its molecular weight is 386; its melting point is 56°C, and it boils at about 460°C. Water solubility is about 1 ppm, and thermal and chemical stability are excellent. These properties are well suited to commercial practice. A step for recovery of acetic acid from a 2 percent solution was estimated to add about 6¢/lb to the selling price. For recovery from a waste stream, this is a reasonable price, but further cost reduction would be necessary before this process can be attractive for fuels that have costs for raw materials and for fermentation.

The author participated in coordination of fermentation projects sponsored by the U.S. Department of Energy from 1976 to 1980. Because of this close involvement and a strong opinion about the importance of fermentation, capsule reviews for the various projects are presented in Appendix B.

Additional Reading

For additional reading see Heperer (1977).

9

Photobiological
Processes

Introduction

There is a hierarchy of sophistication in the processes for harnessing solar energy through photosynthesis. The simplest approach is to grow an ill-defined material termed "biomass" and to burn it for heat energy. Controlled thermochemical conversion of biomass to a mixture of fuels and chemicals is more complicated but gives products of more value than just heat energy. Bioconversion makes use of enzymes and pathways whose understanding and improvement must be based on a very high level of scientific achievement. Processes for bioconversion need a knowledge of the composition of biomass because sugars, starches, cellulose, hemicellulose, lignin, and other biochemicals differ greatly in their reactions. Direct production in the plant of gum naval stores, rubber precursors, and hydrocarbons has already been discussed. The ultimate in bioconversion would be to intercept energy right at the fundamental steps of photosynthesis and to shunt it to the production of fuels or chemicals. This would circumvent many of the intermediate steps now required to make biomass, to collect it, and to convert it.

Solar energy through photosynthesis generates ATP and NADPH, which do biochemical work in the usual cellular pathways. Short-circuiting of these pathways or diverting the ATP and NADPH to other purposes creates the possibility of solar-powered direct production with the biomass left in place. In theory, any biochemical could be the product, but there are already bases for photobiological manufacturing of hydrogen and fixed nitrogen. Hydrogen, an extremely clean fuel, is evolved naturally by photosynthesizing cells under certain conditions. Atmospheric nitrogen is fixed by microorganisms in the root nodules of leguminous plants, but this is a very indirect use of biochemicals produced relatively far away in the leaves. Some microorganisms have nitrogen fixation more closely coupled to photosynthesis. The industrial process for ammonia uses large amounts of electricity; thus solar-powered nitrogen fixation has important implications to the national energy picture.

There is a recent review by Weaver, Lien, et al. (1979) of photobiological processes from the point of view of commercial production of hydrogen and fixed nitrogen. This review is highly recommended for the large number of original references cited and for the detailed, comprehensive analysis of the subject.

Light is composed of energy bundles called *quanta* that vibrate at frequencies from the ultraviolet (uv) to the infrared (ir) ranges of the electromagnetic spectrum. For solar energy conversion, the wavelengths of interest range from 300 to 1200 nm. The energy of typical chemical bonds is 50 to 90 kcal/mole, whereas light in this range comes in quanta that could excite electrons to 40 to 70 kcal/mole. This provides little margin for converting light energy to chemical bond energy; thus the energy of a quantum is seldom sufficient to break a chemical bond. However, photosynthetic systems have means of collecting energy from several quanta to match the energy needs of a particular biochemical task. Electron bonds in photosynthetic pigments are excited to a higher energy state by light and immediately start to decay spontaneously back to the original or ground state. Those that are lost without doing useful work generate heat energy and represent a loss in efficiency. Some of the excited pigment molecules energize electron transport biochemicals that carry out useful reactions, but the net effect is a small increase in the reducing power.

Milne, Connolly, et al. (1978) classify photochemical processes as homogeneous or heterogeneous and direct or photosensitized. Homogeneous reactions occur in a single phase of gas, liquid, or solid. Although homogeneous gas-phase reactions are very important in air pollution, reactions in liquid solution are of more commercial interest because of compactness and easier control. Some reactants absorb light energy directly and react whereas other systems must be photosensitized, a term that refers to a substance absorbing light energy without reacting and donating its energy to a reactive molecule. The heterogeneous systems have dispersed solids or liquids that participate in the photochemical reactions. Photobiological systems are photosensitized and heterogeneous with enzymes and pigments associated with membranes and cellular apparatus.

Solar energy may be transformed directly to chemical energy by using chemical catalysts or biological systems. Of course, there are also photovoltaic devices for direct conversion of solar energy to electrical energy that could be used for chemical purposes. One scheme for using solar energy is to generate electricity for the electrolysis of water to hydrogen and oxygen, which are relatively easy to store. The hydrogen could be used as a clean fuel, as an industrial chemical, or could be recombined with the oxygen in a fuel cell for electricity. Direct photolysis of water to hydrogen and oxygen is also possible. Other solar devices are being studied for photochemical industrial reactions because of fuel savings and because of the ability to direct the reactions with specific catalysts and with light of a selected wavelength.

Photolysis of Water

The bonds between hydrogen and oxygen in water can be broken by radiant energy, and various inorganic catalysts facilitate this reaction. A typical energy efficiency of producing hydrogen using a very good catalyst might be a few percent. Efficiencies of over 10 percent seem readily achievable, but these figures are very misleading, for these catalysts absorb energy as their chemical bonds are excited at narrow bands of wavelengths of light. Other wavelengths are essentially useless; thus these efficiencies, which are based on monochromatic light of the optimum wavelength, would be much lower based on the total radiant energy of incident sunlight. There are materials problems with catalysts. They must resist corrosion in a wet environment and must not be degraded by long exposure to intense light.

Biophotolysis of water is a different concept of energy farming. The biomass is left in place to produce its product. Photosynthetic pigments have optimum wavelengths as do the nonbiological catalysts, but there is a complex of several interacting pigments that traps energy over a range of wavelengths. The photosynthetic mechanism in the chromatophore was mentioned in Chapter 5. A single quantum of light is unlikely to carry out bond splitting; each quantum striking a chromatophore raises its level of excitation until it exceeds the required energy.

At 25° C the free-energy change for conversion of liquid water to oxygen and hydrogen is 56.7 kcal/mole. Graves and Stramondo (1976) have pointed out that only single photons of short wavelengths can provide this energy. Multiple photon splitting of water greatly broadens the usable spectrum, and the theoretical efficiency is increased. If the required free energy could be obtained by adding contributions from both low and high-energy photons, quantum efficiencies in the range of 60 percent would be possible.

Photosynthesis is accomplished by a sequence of coupled reactions. Such coupling is very common in biochemistry because it allows for stepwise exchange of energy for greater control, and each reaction can have sufficient energy to drive the next. The photosynthetic couples are shown in Fig. 9.1. Research on photosynthesis is continuing, and refinements to our knowledge are almost certain to be forthcoming. Photochemical excitation of chlorophyll a670 produces a strong oxidant, chlorophyl a$^+$670, and reduced plastoquinone. As this strong oxidant is converted back to chlorophyll a670, an ill-defined series called the *Hill reaction* liberates oxygen gas and hydrogen ions from water. Plastoquinone is regenerated by a cytochrome reaction involving ADP so that ATP is formed and can be used for energy in many cellular reactions.

Another photochemical event occurs with the pigment P700 whose excited form contributes an electron to an unknown compound X, thus forming another strong reductant. This reduces ferrodoxin, which reacts with NADP to form NADPH, a storage compound of reducing power for many cellular reactions. Overall the photosynthetic reactions produce oxygen, hydrogen ions, and two biochemicals, NADPH and ATP, which are used to operate the reactions of life.

Hydrogen ions do not accumulate during photosynthesis because they may participate in biochemical reactions. Anabolism is mainly a system of producing reduced molecules, and catabolism is mainly a sequence of oxidations. Compounds in the Hill reaction accept hydrogen that can be liberated by hydrogenase enzymes. These enzymes operate at low oxidation–reduction potential; several anaerobic bacteria, photosynthetic and nonphotosynthetic, evolve hydrogen.

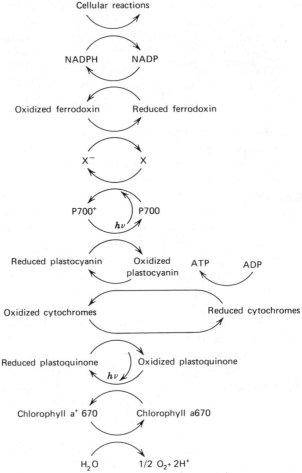

Figure 9.1 Photosynthetic couples.

The immediate precursor for the reduction of protons is usually reduced ferrodoxin, a relatively low molecular weight protein high in iron and sulfur.

A similar species of protein, although smaller in molecular weight, is in the normal pathway of electron transfer in the photosynthetic apparatus. In this latter case the reduced ferredoxin reduces NADP, which, in turn, is the source of electrons for the reduction of carbon dioxide to plant materials. With some anaerobic bacteria it has been demonstrated that reduced NADP can reduce ferredoxin rapidly in spite of the unfavorable thermodynamic barrier. Chloroplasts can photosynthetically reduce NADP by the oxidation of water if ferredoxin is present. Photosynthetic bacteria cannot use water as an electron source and instead use reduced organic or inorganic substances. *Rhodopseudomonas capsulata,* a photosynthetic bacterium, can form copious quantities of hydrogen from malate under anaerobic conditions, provided no elemental nitrogen or fixed forms of inorganic nitrogen are present. In general, the evolution of hydrogen by photosynthetic bacteria depends on nitrogenase and is greater when concentration of free nitrogen is low. Mutants with impaired nitrogenase give off no hydrogen. Nonphotosynthetic bacteria have hydrogenases and can evolve hydrogen when utilizing carbonaceous substrates. These organisms could be good sources of hydrogenases for fabrication of devices for artificial photosynthesis. All photosynthetic bacteria have nitrogenase and can use nitrogen gas as a sole source of nitrogen.

There are two photosystems for algae and leafy plants as shown in Fig. 9.2. One system splits water, generates ATP, and creates a sufficiently low oxidation reduction potential for the other system to develop an even lower potential for driving the reduction of NADP. This is in striking contrast to photosynthetic bacteria in which only one photosystem is present. The single system is coupled to NAD instead of NADP and requires the input of biochemicals for reducing power.

Lien and San Pietro (no date) have surveyed in depth the scientific bases for practical biophotolysis and estimated maximal efficiency from

$$E = \frac{1.25 \, V_e}{2 \cdot h_\eta}$$

where E is the energy efficiency, V_e is the energy per electron volt, and h_η is the energy of the absorbed photons. Based on photons at 680 nm (the approximate

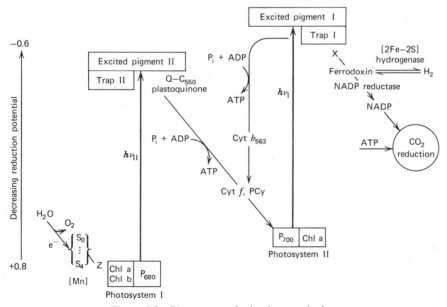

Figure 9.2 Electron transfer in photosynthesis.

maximum for absorption by chlorophyll) and a corresponding energy of about 1.83 eV the absolute maximum energy efficiency of photosynthesis is roughly 30 percent for red light and 26 percent at the mean wavelength of solar radiation. The light harvesting pigments are present in high concentrations and capture nearly all the useful photons that strike them. These photons are 43 to 46 percent of incident sunlight; thus the theoretical efficiency of photosynthesis is the product of energy efficiency times percentage of useful photons, or 10 to 12 percent.

Actual efficiency can be determined from the biomass harvest and the incident sunlight. Eight tons of biomass per acre per year (18 tons/ha) in the United States represents about 0.4 percent photosynthetic efficiency. Superior yields of sugarcane in Texas and Hawaii of 50 tons/acre·year (112 tons/ha) equate to 2.5 percent efficiency. The reasons for low actual efficiencies are limitations other than sunlight. Inadequate water, insufficient nutrients, low temperature, and biological mechanisms not adapted to ideal conditions all contribute to reduction of efficiency. In short-term experiments with an excellent environment for growth, an efficiency of about 4.2 percent has been demonstrated.

Efficiencies for carbon fixation are definitely not the same as those for producing hydrogen. With a few direct reaction steps, it might be possible to approach the 10 to 12 percent photon efficiency. However, natural systems are poorly adapted to high intensity and high productivity. The special architecture of the photosynthetic apparatus has evolved for efficient use of weak light; efficiency falls off rapidly as light intensity increases. After the photosynthetic pigments are saturated, excess photons induce side reactions and can destroy pigments. Success with commercial biophotolysis may come only after extensive genetic manipulation to create cells with greater turnover rates for the photosynthetic apparatus or development of reactors using immobilized catalysts of chemical, biological, or mixed origins.

The review by Lien and San Pietro (no date) covers a variety of experiments on rapid photosynthesis. Kinetics can be studied by flash illumination and by cycling between light reactions and dark reactions. Certain mutants have small photosynthetic structures but unusually high saturation rates. This is encouraging to prospects of improvement through genetic manipulation.

It is interesting to compare nitrogen fixation and hydrogen evolution. The equations are

$$N_2 + 6e^- + 6H^+ + nATP \longrightarrow 2NH_3 + nADP + nP_i$$
$$2H^+ + 2e^- + mATP \longrightarrow H_2 + mADP + mP_i$$

The shorthand symbol for inorganic phosphate is P_i. The fixation of nitrogen is reversible and requires much energy in the form of ATP to drive the reaction. Furthermore, the oxidation–reduction potential of the system affects the energy requirement. When conditions are favorable, the coefficient n is 4 to 5 but can be in the range of 35 to 40 at a poor oxidation–reduction potential. Evolution of hydrogen is not readily reversible, but the coefficient m ranges from 1 to 2 with little change so long as the potential is sufficiently low to prevent inhibition of the hydrogenases.

Systems for Hydrogen Generation

Microalgae can evolve hydrogen under anaerobic conditions, which means that oxygen must be trapped or removed as it forms. There are numerous reports of laboratory-scale hydrogen systems. For example, Miyamoto, Hallenbeck, et al. (1979) have studied hydrogen production by the thermophilic alga *Mastigocladus laminosus*. There was some evolution of hydrogen for several days, but the conversion efficiency for light energy was only 0.1 to 0.2 percent outdoors, compared to 2.7 percent in the laboratory. Furthermore, cultures were protected from oxygen by sweeping with argon plus small amounts of nitrogen and carbon dioxide. As with other proposals for biological hydrogen production, commercial applications are probably far in the future.

Systems using photosynthetic bacteria require organic substrates, which tend to be more valuable than the hydrogen. Nevertheless, research on such systems has value in laying a foundation for inventing cell-free devices. Several groups have demonstrated hydrogen evolution by bacteria. Zurrer and Bachofen (1979) were able to maintain hydrogen production by batch cultures of *Rhodospirillum rubrum* for 80 days. By using cheese whey as a feed, the rate was 6 ml of hydrogen per gram of cell dry weight per hour, in contrast to 20 ml/g·hr (65 ml/hr·l of culture fluid) with lactate as a hydrogen donor. The composition of the evolved gas was 70 to 75 percent hydrogen and 25 to 30 percent carbon dioxide.

Another class of photosynthetic organisms that might seem of interest for energy generation are the *Halobacteria*. These grow in highly saline environments and have a purple membrane powered by light. With almost no exception, conditions are obligately aerobic. However, growth has not yet been demonstrated with light as the sole source of energy, and the main role of the purple membrane seems to be in regulation of ionic concentrations by pumping ions against a gradient. Aspects of the interesting organisms are reviewed by Lanyi (1978).

Blue-Green Algae (Cyanophyta)

Blue–green algae have a very interesting differentiated structure called a *heterocyst* that can fix atmospheric nitrogen. A mature heterocyst has lost the photosystem for oxygen evolution, but the other photosystem is highly active in generating ATP for use in nitrogen fixation. The vegetative cells adjacent to the heterocyst carry out normal photosynthesis of organic molecules that diffuse to the heterocyst to act as reductants. The net effect is to protect nitrogenase enzymes from evolved oxygen while allowing diffusion of needed intermediate biochemicals. Vegetative cells also have the ability to fix atmospheric nitrogen, but only when unusually anaerobic conditions prevail.

There is only about one heterocyst for 15 to 20 vegetative cells, and rates of nitrogen fixation are fairly low. For growth in natural waters, blue–green algae could satisfy their own demand for nitrogen. However, very rapid growth in ponds for biomass production would outstrip the fixed nitrogen supply from heterocysts and supplemental fertilization would be necessary. Blue–green algae thus have no great advantage for producing biomass, but these organisms are very attractive for photoreactions because nitrogenase and hydrogenase activities are closely coupled and good rates of hydrogen evolution have been demonstrated. Green algae are less promising for making hydrogen because evolution occurs only under anaerobic conditions, with low CO_2 tension, and with light intensities well below saturation.

Restriction of the nitrogen supply to blue–green algae promotes hydrogen evolution to $2\frac{1}{2}$ to $3\frac{1}{2}$ times the normal rate. Some investigators noticed a drop off in hydrogen and breakage of filaments after 5 to 7 days of nitrogen starvation, but Benemann, Hallenbeck, et al. (1977b) adjusted N_2 in the feed gas to maintain a high rate of hydrogen evolution for 18 days.

The water fern Azolla is a candidate for nitrogen fixation to replace fertilizer. Nitrogen-fixing blue–green algae live in cavities at the base of the fern. In contrast to the usual algae, which have a ratio of about 1 heterocyst cell: 15 vegetative cells, the algae in symbiosis with Azolla have a ratio of 1 : 1 and thus have a higher specific rate of nitrogen fixation. The fern may supply biochemical energy to the algal heterocysts so that fewer algal cells are needed for this task. Azolla could be grown in irrigation ditches or in the fields to reduce the needs for fertilizer (R. Valentine, personal communication).

Jeffries and Leach (1978), using the hypothesis that diffusion of reducing agent limits hydrogen formation, tried intermittent light with the filamentous blue–green alga *Anabaena cylindrica*. Hydrogen yields per unit of incident light increased by approximately 70 percent when the illumination was intermittent with periodicities of roughly 10 sec. While increased quantum efficiency is important, the main objective is yield of hydrogen. An implication of these studies on intermittent illumination is that thick algal suspensions could be used if cells can cycle between the illuminated zone and the darker zones to be charged with energy in the former so that hydrogen can be generated later.

Stability

The destruction of the photosynthetic apparatus by bright light is termed *photoinhibition.* With unicellular algae, the damage is a function of intensity and duration. Intense illumination with visible light causes an oxygen uptake rate two to three times that of normal dark respiration. This effect, termed *photooxidation,* is stimulated by CO_2 starvation and by high partial pressures of oxygen. A brief period of photooxidation inflicts no permanent damage, but sustained bright light causes irreversible photoinhibition, which may be accompanied by bleaching of the pigments. Loss of photochemical activity precedes massive destruction of pigment molecules. The effect is far more severe *in vitro* than *in vivo,* where cells have repair mechanisms.

Artificial Photosynthesis

The rationale for mimicking photosynthesis with man-made systems is covered nicely by Siebert, Connolly, et al. (1978). Low photosynthetic efficiencies of natural systems can be overcome by modifying the organism, improving its environment, or isolating the photosynthetic apparatus for optimization *in vitro*. Research groups in several countries are laying firm bases for genetic modification of photosynthetic organisms by extending the understanding of hydrogenase and nitrogenase enzymes and the related biochemical pathways. The environment may be improved with gas richer in CO_2, by better mixing, or by better illumination. Immobilization of intact organisms on a carrier can provide a high degree of environmental control. The *in vitro* systems may be dispersed in solution or could be immobilized. Immobilization allows easy recovery for reuse and has the potential advantage of also immobilizing additional pigments, enzymes, or cofactors in close proximity to associated structures.

The light-driven reactions in photosynthesis that carry out charge separation and electron transport have high yields. The dark reactions for CO_2 fixation and normal metabolism are relatively inefficient; thus bypassing them in some way could greatly increase the overall yield for capturing solar energy. Milne, Connolly, et al. (1978) cite efficiencies of the primary photosynthetic events of 31 percent in photosynthetic bacteria and 25 percent in green plants. The cellular components for the light-driven reactions have been isolated and shown to function *in vitro* at efficiencies from 0.5 to 2.5 percent. This provides a basis for inventing nonbiological devices to mimic the desired portions of photosynthesis. The parts of these devices may be derived from either biological or nonbiological sources, but it will require a great deal of ingenuity to invent a solar converter that is better than the chloroplast.

Mimicking of the cell is referred to as "engineered photosynthesis." One arrangement uses a membrane containing a mixture of pigments and enzymes copying features of chloroplasts with an electron donor compound and an electron acceptor on different sides of the membrane. When irradiated, a potential is generated that could do useful work in an external circuit. When some of the components are derived from living cells and others are man-made, the arrangement is termed a "hybrid" system. The U.S. Solar Energy Research Institute has a research program with emphasis on artificial membranes, stabilization of photosynthetic components, production of carbohydrates by artificial photosynthesis, protecting hydrogenases from inactivation by oxygen, and screening organisms to find better hydrogenases and nitrogenases.

Artificial photosynthesis may have short pigment life because of photochemical destruction. The technology for commercial photocatalysis, which mimics natural photosynthesis, will not be developed quickly, and some protection or repair for the harmful effects of too intense light is essential. Unless these catalytic systems are very cheap or have spectacular efficiency, natural photosynthesis will be more cost-effective.

Membranes play important roles in photosynthesis, and stability may again be a problem. Temperature, pH, and ionic strength affect both membrane function and stability. As both proteins and lipids are major constituents of membranes, protease and lipase enzymes may affect activity. Digestion of chloroplasts with trypsin first impairs portions of the electron transport system and photophosphorylation. Any autolysis in an immobilized system or digestion by enzymes from contaminating organisms would be expected to impair performance seriously.

Chloroplast preparations have been stabilized by empirical variations of ions, pH, and biochemicals in the suspending medium. In general, neutral pH is best. However, activity declines even near freezing temperatures in just a few hours. Bovine serum albumin protects well, and reasonable activity is retained for 24 hr at room temperature. Such stability is obviously inadequate for commercial reactors, and the best stabilizers reported so far are very expensive. Furthermore, these are storage stabilities, and activity declines much more rapidly in the presence of light. Immobilization of chloroplasts on a solid support with a fixation agent such as glutaraldehyde leads to increased stability, but at the expense of reaction kinetics. The quantum efficiency of immobilized chloroplasts was about one-fourth that of the original preparation.

Berezin (1977) described experiments utilizing isolated chloroplasts of plants or microscopic algae in the synthesis of strong reductants. Energy of light quanta is transformed into chemical energy of the reduced acceptor and molecular oxygen. The second stage may be dehydrogenation of the reduced carrier that is recycled. The reactions are:

$$H_2O + A \xrightarrow[\text{chloroplasts}]{\text{immobilized}} AH_2 + O_2$$

$$AH_2 \xrightarrow[\text{hydrogenase}]{\text{immobilized}} A + H_2$$

Theoretical efficiency of this process is estimated at 15 percent. The chief difficulty in applying the preceding scheme is the instability of isolated chloroplasts.

Krampitz (1976) has used the photosystems present in the blue–green alga *Anacystis nidulans* because they remain active and reasonably stable. Water splitting is performed by the algae, and a photobacterium, *Rhodospirillum capsulata* evolves hydrogen.

$$\text{Water} + \text{algae} + \text{NADP} + \text{light} \longrightarrow \text{NADPH} + \text{oxygen}$$
$$\text{NADPH} + \text{oxalacetate} + R.\ capsulata \longrightarrow \text{NADP} + \text{malate}$$
$$\text{Malate} + R.\ capsulata + \text{light} \longrightarrow \text{hydrogen} + \text{oxalacetate}$$

The net reaction is

$$\text{Water} + \text{light} \longrightarrow \text{hydrogen} + \text{oxygen}$$

The system works poorly in mixed cultures of intact organisms because NADP is too large a molecule to diffuse readily through cell membranes; thus it has great difficulty in shuttling between the two reactions in which it must participate. When the cells were lyophilized to rupture the cells, hydrogen evolution improved markedly. Weetall and Krampitz (1980, in press) immobilized the algal components and the bacterial components in separate vessels but substituted *R. rubrum* for *R. capsulata*. The NADP was cycled by pumping medium between the illuminated vessels for prolonged evolution of hydrogen.

The present state of hydrogen production with isolated photosynthetic components is not very good. Evolution persists for several hours, but oxygen must be removed by an expensive method. Furthermore, peroxides are formed that are highly toxic; addition of catalase enzyme decomposes hydrogen peroxide but is another expense.

The purple membrane of photosynthetic bacteria is relatively easy to isolate and can be used to develop an electrical potential when illuminated *in vitro*. Hydrogen production has not yet been achieved, but the results with mixed systems of algae and bacteria augers well for future discoveries.

Mutation of photosynthetic organisms to block normal pathways, thus diverting energy to hydrogen production, seems possible, but such cells must be able to grow; otherwise contaminating organisms would overwhelm the desired species. Although algae have no alternate means of coupling photosynthesis to growth, photosynthetic bacteria have several pathways that could support growth while hydrogen production was aided by a mutational block. Unfortunately, the bacteria need added organic substrates.

Problems with Hydrogen

Hydrogen is a diatomic gas with very small molecules of very high diffusivity. It penetrates steel; losses in a common pipeline would be high. A toy balloon that deflates in a few hours with its customary helium holds hydrogen for only a few minutes. If plants evolve hydrogen in place, there would have to be a transparent cover to admit sunlight and to retain hydrogen. There are some overwhelming obstacles to commercializing hydrogen production from algae. The enclosures must be tight to avoid escape of the products and must be spread over vast areas. Clear transparent plastics are quite expensive; thus glass seems the only choice. Cost could be minimized by selecting thin glass, but breakage due to settling of the supports or storm damage from hail or strong winds would be totally unacceptable. Benemann, Hallenbeck, et al. (1977b) propose a target of $5¢/ft^2$ $(50¢/m^2)$, but this seems unrealistic. Dust and dirt would have to be removed periodically to maintain high illumination, and cleaning might break thin material. For a cheap product such as hydrogen, it is very unlikely that glass enclosures could be constructed, sealed, and maintained at costs anywhere near those required for profitable operation.

Should the enclosure problem be solved for hydrogen production, there would still be major engineering challenges for gas separation. Plants in sunlight require carbon dioxide. If supplied as air, there would be extreme dilution of the hydrogen produced. Separation of hydrogen from other gases with molecular sieves or by cryogenic procedures would be very costly. More logical is supplying plants with CO_2 from bioconversion gases or from combustion gases. A counter-current flow scheme might cause most of the CO_2 to be utilized, leaving the hydrogen and oxygen to be separated. Of course, this is an explosive mixture.

Two concepts have been advanced for producing hydrogen within special reactor vessels. Soviet scientists have proposed using photosynthetic algae to fix hydrogen into biochemical acceptor compounds and releasing it later in a closed vessel. There would be problems of removing gases and pumping the algal suspension through reactors with immobilized hydrogenase enzymes. (Berezin, personal communication). Separation of the steps solves the purification and containment problems by creating new problems in immobilizing fragile enzymes, recycling enzyme cofactors, designing reactors, and devising economical arrangements for pumping large volumes over long distances throughout the growing area. Another ingenious plan has been tested by a U.S. Department of Energy contractor (Pelovsky, 1978). Very thick algal suspensions are illuminated by use of fiber optics. Sunlight falling on a large area of lenses focused on fiber tips will be conducted through the fibers to the algal reactor. Fibers inside the reactor would be scratched to allow escape of light through the imperfections to provide diffuse illumination for the algae, which prefer dim light. Evolution of both hydrogen and oxygen again poses a separation problem.

The reaction of oxygen with hydrogen is greatly promoted by just the sort of catalysts that might be used for hydrogen evolution. The very great diffusivity of hydrogen would present engineering difficulties in designing a reactor in which hydrogen and oxygen were kept separate. Oxygen diffusion to the hydrogen catalyst would cause a direct, serious loss.

Hydrogenase enzymes are highly sensitive to oxygen and function well at low redox potential only. Inhibition by oxygen can be either reversible or irreversible. Lien and San Pietro (no date) encourage research for microbial hydrogenases with less sensitivity to oxygen and genetic manipulation, which also aims at more practical enzymes.

If redox dyes or other agents are required to maintain low redox potential, their recycle is essential. All this is made much more difficult by the low value of hydrogen because there is little margin to pay for sophisticated process schemes.

Berezin (1979) has pointed out problems in the storage of hydrogen. In gas cylinders the mass of the system is 99 percent as the container and only 1 percent hydrogen. With metal hydrides for storage, the hydrogen reaches 4 to 5 percent of the mass. By contrast, over 12 percent of methanol is hydrogen. The enzymatic reaction

$$CH_3OH + H_2O \rightleftharpoons CO_2 + 3H_2$$

is readily reversible and has a small free-energy change. However, this reaction cannot be considered seriously at present for hydrogen storage because methanol is the more valuable product. Instead of methanol as a storage system, it makes more sense to forget the hydrogen economy and to transport and use methanol as a fuel or chemical feedstock. There is no great incentive to find a gaseous fuel, particularly when the present gas storage and distribution system is unsuitable for hydrogen.

All things considered, biophotolysis of water to produce hydrogen is very far from commercial fruition. It can only be demonstrated in the laboratory by using expensive reagents to maintain low oxidation–reduction potential. Major biological and engineering breakthroughs are needed for biophotolysis to have any chance. Biophotolysis also fails to meet the criterion of easy acceptance of the product into existing distribution systems. New pipelines or modifications of existing pipelines for transport of hydrogen would take years to construct and would require great amounts of capital.

Fuel Cells

Coupling photosynthesis to the generation of electricity could be mediated through hydrogen fed to a fuel cell. An electrode of polyvinylpyridine and the lithium salt of the aromatic cyano compound

$$\left[\begin{array}{c} NC \\ \diagdown \\ NC \end{array} C \hspace{-2pt}\diagup\hspace{-6pt}\diagdown\hspace{-2pt} C \begin{array}{c} \diagup CN \\ \\ \diagdown CN \end{array} \right] = 2\ L_i^+$$

allows electron transfer from the biological reactions when immobilized hydrogenase enzymes are used. An alternate method is to polymerize dimethyl viologen plus the same lithium salt plus immobilized enzyme. Berezin (1977) has studied the reactions

$$2H_2 \xrightarrow[\text{hydrogenase}]{} 4H^+ + 4e^-$$

$$O_2 + 4H^+ + 4e^- \xrightarrow[\text{laccase}]{} 2H_2O$$

One electrode has immobilized hydrogenase, and the other has immobilized polyphenoloxidase or laccase. The system is stable for several days. The reactions are spontaneous, with a potential of about 1.22 V; efficiency approaches 90 percent. Immobilization of hydrogenase on the electrode material so that there was direct electron transfer from the active center of hydrogenase to the electrode would obviate the necessity for the carrier and would greatly simplify the electrode design. This would be a major theoretical and practical accomplishment.

As substitutes for the hydrogen electrode, other reductants may be used in a biochemical fuel cell. There are many enzymes catalyzing electron transfer between different organic compounds, and many of these processes can be used for electrosynthesis.

Very little has been done thus far in the design of an oxygen electrode. Suitable enzymes do exist; thus there are good departure points for new lines of research.

Generation of Light

The firefly is the best known biological generator of light, but similar reactions occur in microorganisms. A jar full of fireflies or a flask of luminous organisms provides enough dim light to read newsprint. Most organisms emit blue–green light, but Ruby and Nealson (1977) have a strain of *Photobacterium fischeri* that emits yellow light.

Bioluminescence is catalyzed by the enzyme luciferase. A reduced molecule such as flavin mononucleotide is oxidized with production of visible light. Although it is difficult to envision a practical illumination system that uses bioluminescence, there may well be something in the future. There are analytical devices that detect substances based on their abilities to promote or to inhibit bioluminescence.

10

Costs

Economics of Fuels from Biomass

The more exotic schemes for energy from biomass are not ready for meaningful cost analysis. For example, a hydrogen economy is presently poorly defined, and most of this book will be obsolete long before processes for hydrogen could be reduced to practice and a distribution structure built. However, it is appropriate to consider cost of biomass, cost of alternative processing methods, and product values. Inflation is so rapid that costs presented here are certain to be low by publication time—even the relative values may be greatly distorted by the price gyrations of petroleum. Nevertheless, this book has evaluation of options as one of its goals, and economy is the key yardstick.

Cost of Biomass

The U.S. Department of Energy has sponsored several systems studies to survey biomass resources and to project prices. These do not have exactly the same bases, but the general conclusions are shown in Table 10.1. Unless some crop has some unique properties, initial cost and processing cost will dominate in selecting the feedstock. Thermochemical processing is roughly the same for any dry biomass, and the fact that trees are denser than corn stover would seem to offer the best margin for profit. Of course, municipal solid waste would be used when available, but the supply is not sufficient for a major impact with very large conversion plants.

Bioconversion factories can select a cheap feedstock difficult to hydrolyze or a more expensive crop with juices that can go directly to fermentation without hydrolysis. The crops with fermentable juices or those containing starch are well suited to existing processing technology and would be the only reasonable choices if plants must be built now. Crops that require hydrolysis will probably win out in the long run, but there will have to be clear cost advantage for a new technology to displace an established process that has been optimized.

Crop prices will probably fluctuate widely if competing uses create unusual demand. Furthermore, crops with which farmers are relatively inexperienced will not be grown unless the financial incentives are well defined and convincing.

Table 10.1   Comparison of Biomass Feedstocks

Crop	Approximate Cost ($/dry ton)	Ease of Bioconversion
Trees	$18 to $35	Highest subdivision cost, expensive to hydrolyze
Corn stover	$25 to $35	Easier subdivision, slightly easier to hydrolyze
Grasses	$60 to $80	Comparable to corn stover
Corn grain	$80 to $130	Easy extraction of starch that has very easy hydrolysis
Sugarcane	$20 to $40	Easy extraction, no hydrolysis
Sweet sorghum	$30 to $50	Same as sugarcane
Aquatic plants	$100 to $200	No subdivision, probably easy hydrolysis

Existing farms may convert to a new crop if the anticipated financial return is higher than that for their usual crops, but the risks cannot be too great. A relatively high guaranteed price for the first year would encourage planting the new crop, and a negotiated, lower price would be possible in subsequent years when data become available for accurate estimation of production costs. New farms are a different matter. If area acquisition and preparation costs are $2000/acre ($5000/ha), a low estimate, little venture capital from private sources could be attracted to the project unless the annual value of the crop were well in excess of $700/yr·acre. With a yield of 10 tons/acre, this translates to $70/ton. In other words, private firms expect an investment to be repaid in 2 or 3 yr. Government money is different because politics rule over economics. Much of the reasoning for biomass as fuel has been based on utility financing, which uses interest rates in the range of 6 to 12 percent. It must be realized that utilities have had long times to amortize their capital outlays, and there are monopoly structures that create stability and low risk. The fuels and chemicals from biomass will have to compete vigorously; thus risk may be far too high to justify utility interest rates.

Some private conversations with Dr. William A. Sheller of the University of Nebraska provided insights into corn grain as a feedstock for alcohol fermentation. He points out that the corn prices listed on commodity exchanges include transportation and storage. An alcohol plant located in corn country could eliminate most of the transportation costs and could save on storage by using some grain directly and storing the rest in the plant's own facilities. For example, with corn quoted at $2.60 per bushel, the price in the farming areas can be as low as $2.10 per bushel. Furthermore, corn that is damaged by weather, disease, or partial spoilage is available locally at very depressed prices but is usable for making alcohol.

Another powerful argument of Dr. Sheller's concerns pricing. There are many stories and articles in newspapers and magazines, reports by consultants and various organizations, and speeches that state costs, analyze, or compare fuels. Critics of biomass are quick to point out any unfavorable prices based on heating value of biomass products versus those of oil, coal, or natural gas. What is overlooked are the perturbations to a free economy. Fossil fuel prices are low because of such factors as oil-depletion allowances and price controls. Crop prices are high because of parity laws and subsidies to the farmers. There is an excellent chance that biomass will succeed anyway, but free markets and less government intervention might accelerate the advent of very large biomass fuel systems.

Lipinsky (1978b) points out that cost targets for biomass based on price per BTU that is competitive with fossil fuels would equate to a revenue for the farmer of only $160 to $225 per acre ($400 to $560 per hectare). This is insufficient for intensive cultivation; thus it may be advisable to trade off high yield against production cost. The situation is much different when the biomass for energy is a coproduct with biomass for food and fiber because a high-value use carries most of the cost burden. It is suggested that corn-based fuels should integrate use of grain and corn stover by using treated stover both for cattle feed and for making fermentable sugars. Some grain would be fermented to alcohol fuels, particularly after the stover is used up because stover deteriorates rapidly if stored whereas grain can easily be stored for extended periods.

Storage of biomass is estimated to cost from $3 to $5/ton·yr depending on its bulk density when using a shed with sidewalls (Miles, 1977). The cost doubles for totally enclosed storage with a concrete floor. This same reference mentions that rodent infestations can spoil biomass intended for feed. Whereas mice are easily controlled by cats, rats can be a major problem. Rats can nest in sodium hydroxide-treated straw. In one case, rat control by poison and elimination of habitats was found to cost $1 per rat.

Dauve and Flaim (1979) have analyzed the costs for collecting crop residues using corn stover and wheat straw as examples. Only commercially available systems were considered. The dimensions of typical bales or stacks and operating costs are listed in Table 10.2. Loose chop is not a feasible system unless hauling distances are very short because the biomass is light and bulky. The large tight bales offer a very great advantage for storage because the interior of the bale is very well protected from rain; thus no silo, shed, or covering is needed. The collection costs provide a good index to the cost of biomass. Residues have some value to the farmer; $5/ton is a reasonable estimate of their worth for erosion control and for supplying organic matter. Addition of this value and some profit to the collection costs brings residues into the $30/ton range, which is higher than

Table 10.2 Comparison of Residue Collection

System	Dimensions	Total Cost, $/ton
Conventional	Bales 14 in. × 18 in. × 50 in.	24.17
Large, round	7 ft. diameter, 6 ft. long (up to 1.5 tons)	16.93
Big rectangular	4 ft. × 4 ft. × 8 ft.	25.85
Stacking	Up to 15 ft. × 24 ft. (1 to 12 tons)	15.37
Loose chop	Not compacted or packaged	12.63

many of the estimates used for assessing the economics of prospective processes. Whereas an ethanol factory might be able to afford $30 or more per ton of feedstock, an anaerobic digestion process would have little chance for profitability.

Energy Balances

Growing biomass and converting it to fuels by inefficient processes that devour fossil fuels would be self-defeating. Use of a high fraction of the fuels from biomass to power the processes is little better. In fact, there have been rough analyses that indicate that fuel for tractors, energy to produce fertilizer, power for irrigation, and the other energy expenditures for site preparation, sowing, cultivation, and harvest can exceed the fuel value of the crop. The obvious lesson in this is that cheap energy is incompatible with the same intensive agricultural practices used to produce valuable foods. Both production and conversion of biomass must be extremely energy conscious and use every opportunity to conserve and economize.

Energy balances for ethanol from carbohydrate crops in Brazil have been developed by da Silva, Serra, et al. (1978) Table 10.3 shows some of their results. Credits were taken for burning wastes to produce steam, but stillage was considered to be of unknown value. There could be credits for using stillage for feed or fertilizer or debits if it goes to a waste treatment unit. It can be seen that all these crops have a favorable energy production: consumption ratio, and the production of alcohol is more costly by far than the agricultural costs of growing the crop. Sweet sorghum is not a commercial crop in Brazil; thus its costs were based on information from the literature. Although sugarcane has a more favorable energy ratio, high total productivity of sweet sorghum would mean a lower investment. Cassava does not appear attractive, except in unusual circumstances such as growing on poor soils unsuitable for the other crops.

Extrapolation of economic estimates for Brazil to the United States is risky, but there is definitely a good chance for favorable energy balances. Opportunities for major cost reduction in factories producing alcohol could greatly improve these estimates. In the United States, lignocellulosic biomass is being given high

Table 10.3 Energy of Ethanol Production

Crop	Agricultural Yield, tonnes/ha · yr	Alcohol, l/ha · yr	Energy, Mcal/ha · yr			
			Expended		Produced	Ratio
			Agriculture	Factory		
Sugarcane	54	3,564	4,138	10,814	36,297	4.53
Cassava	14.5	2,523	2,573	8,883	13,271	1.71
Sweet sorghum (two crops)	62.5	5,165	8,021	16,500	46,278	3.39

priority for energy research, and credits for by-products seem to be crucial to economic success.

Agricultural Energy Economics

One major cost elimination mentioned earlier is avoidance of yearly tilling and sowing. Perennial plants are more economic than annual plants. Among plants that are harvested after several years, such as trees, coppicing species that grow back from the roots are favored. Of course, a superproductive species that re-quires replanting after each harvest could show better overall economics, but it would take quite a bit of extra biomass to pay the costs. A wild variety of peren-nial corn was recently discovered in Mexico. Breedings with high-yield strains could produce a supercorn for the future.

Water is relatively expensive in many areas of the United States. The lawn around your home provides a good illustration of the need for water. During a wet, hot week, the lawn may need mowing twice; however, without water, mow-ing may be delayed for over 2 weeks. In arrid areas, water is just too valuable for food-crop irrigation to be used for energy crops. The normal vegetation sup-ported by sparse rainfall might be taken for energy purposes in dry areas, but at the risk of creating dust bowls. Biomass does not seem competitive with solar boilers, photovoltaic collectors, or windmills in the regions where there is little water.

The water requirements for various processes for producing fuels or generat-ing energy are discussed by Harte and El-Gasseir (1978). A very disturbing point was raised with regard to oil shales in Colorado because they are found in strata that are an integral part of the ground water system. Mining of these strata could upset the freshwater system and release saline water from artesian deposits with disastrous consequences. The authors present a sorry picture for bioconversion. If sugar beets were the main crop, 8.5 percent of the total land in the 48 cotermi-nous states would be sufficient for supplying all needed energy. The irrigation requirements would exceed *all* current U.S. water consumption by a factor of 3. This indicates that high-irrigation methods should not be used for fuels from bio-mass and that recycle of water is essential.

Povich (1976, 1978) also feels that water limitations seriously prejudice the prospects for fuels from biomass. Furthermore, nutrients containing nitrogen or phosphorus may become overly expensive as mineral sources become depleted.

Thermochemical Conversion

For thermochemical processing, it is particularly important to establish the cor-rect cost per BTU because the key products are intended for open competition with present fuels. Protagonists quote current prices for fossil fuels when trying to show that biomass fuels are uneconomic but fail to point out that regulations and direct or indirect subsidies keep free-market forces from operating. Thermochemical processing of biomass will be analogous to that for coal, except that some of the biomass designs incorporate drying with lower-grade heat. Estimates of the economics of advanced coal liquefaction are presented by Neben (1978). A hypothetical plant located in West Virginia would be fed 30,000 tons of coal per day to produce liquid fuel, hydrocarbons, gas, sulfur, and ammonia. The factory would cover 1 mi^2 of land, but the coal mine spreads beneath 70 mi^2. The liquid fuel output would be 66,000 barrels/day (2.8×10^6 gal), which dwarfs the sizes being mentioned for biomass where outputs would be roughly 75,000 gal/day (280,000 l/day). Large-scale operation seems to be essential to economic operation of coal conversion. The hypothetical factory is estimated to cost $2.1 billion in 1977 dollars. A factory using biomass instead of coal would probably be 30 percent larger and somewhat less efficient because of the moisture and lower density of biomass. The biomass thermochemical conversion technology developed so far is for relatively small scale units that may operate economically for special sources of feedstocks. For thermochemical conversion of biomass, the investments would be very great, major technological problems at high temperatures and pressures remain to be solved, and it will be difficult to beat coal. Gasification can build on established technology and would represent the least of evils if a decision were made to process biomass immediately.

Biomass could be at a severe economic disadvantage to coal if underground gasification is successful. Coal gastification *in situ* is being tested in Wyoming (*Chemical Engineering News,* December 3, 1979, p. 19). Projected costs are about one-third of those for gasification above ground, and biomass has no way for comparable elimination of transportation or handling costs. A material balance for gasification is presented in Table 10.4. The reaction with oxygen supplies the latent heat of evaporation of the moisture in the wood. A cryogenic process would supply the oxygen. The corresponding energy balance is given in Table 10.5.

Much of the char and unreacted carbon are carried in the exit gas and must be removed. Cyclones designed for coal char work poorly with wood char because the particle density is lower; thus development of improved separation methods is needed. The gases must also be cooled if not used nearby, and steam can be generated with them.

Tables 10.6 and 10.7 show costs for a 100-dry-ton/day gasification plant designed by Garrett (1977). The break-even gas price is $2.10/million BTU. There is an ash disposal problem, but no charge was made because the ash may have sufficient value as fertilizer supplement, soil additive, or fill to pay for its disposal.

Methanol

One of the main hopes for a liquid fuel from biomass is methanol from synthesis gas because the technology has been well developed. Biomass moisture reacts in the process; thus there is no need for drying as in other thermochemical proc-

Table 10.4 Wood-Gasification Material Balance[a]

	Tons per Hour
Input	
Wet wood	83.4
Water	293.0
Oxygen	21.0
Dryer air	164.2
Total	561.6
Output	
SNG	11.9
Ash and unburned carbon	1.0
CO_2 to stack	52.0
Dryer stack	180.5
Treated wastewater	126.3
Evaporation losses	189.8
Sulfur	0.1
Total	561.6

Source: Kohan, Barkhordar, et al. (1979).

[a] Basis: 1000 tons/day of wood.

Table 10.5 Wood-Gasification Energy Balance[a]

	Millions of BTU per Hour	Kilocalories per Hour	Percent
Input			
Wet wood	796.7	200.8	100.0
Total	796.7		100.0
Output			
Synthetic natural gas (SNG)	501.6	126.4	63.0
Heat rejected to cooling	250.6	63.2	31.4
Stack	35.1	8.8	4.4
Ash and unconverted carbon	7.1	1.8	0.9
Insulation and miscellaneous losses	2.3	0.6	0.3
Total	796.7	200.8	100.0

Source: Kohan, Barkhordar, et al. (1979).

[a] Basis: 1000 tons/day of wood.

Table 10.6   Capital Investment for a 100-ton/day Plant for Gasification of Manure

Description	Direct Equipment Cost
Concrete storage pad	$ 25,000
Front-end loader	40,000
Ramp with guard walls	15,000
Feed hopper	15,000
Lump breaker	20,000
Rotary valve	10,000
Vacuum dryer–screw conveyor	66,000
Condenser	5,000
Heat exchanger, cooler for quench fluid	4,000
Vacuum pump	5,000
Quench column and settling drum	10,000
Total pyrolysis equipment cost	$ 215,000
Factored costs of installed equipment	462,000
Multiple hearth furnace, 18 ft. (includes installation and all other costs)	960,000
Total capital investment for pyrolysis (gasification) plant	$1,637,000
Gas purification equipment	
Reciprocating compressor, other auxiliary equipment, installation	100,000
Amine glycol gas purification system for H_2S, CO_2, and H_2O removal	150,000
Total	$ 250,000
Total initial investment	$1,890,000

esses. Improvements in catalysts have lowered the pressure required for formation of methanol, and new heat-recovery schemes have increased the efficiency of the process.

A condensed flow sheet for methanol from wood is given in Fig. 10.1. Crude gas from the gasifiers is cooled and scrubbed to remove organics. After compression, CO_2 is removed with alkali or carbonate. Cryogenic separation removes methane and hydrocarbons first and then 99 percent of the nitrogen to leave a mixture of carbon monoxide and hydrogen. This gas is not in the correct proportions for synthesis gas of $H_2/CO_2 = 2$, so a shift conversion at 27 atmospheres reacts part of the CO with water vapor to obtain more hydrogen. The CO_2 also produced in the shift reaction must be removed. At 170 atmospheres a 95 percent yield of methanol is obtained that must be distilled to produce a high-grade.

Both coal and wood contain considerable amounts of partially oxidized molecules, and the result is a lower methanol conversion for either than from natural gas. The process yields are: gas 91 percent, coal, 85 percent, and wood only 51

Table 10.7   Annual Operating Costs for Manure Gasification Plant

Item	Medium-BTU Gas, $/yr	Dollars per Million BTU
Labor		
Eight operators (two per shift),		
24 hrs/day, 365 days/yr, $8/hr	140,000	
Two mechanics at $20,000/yr	40,000	0.802
One superintendent	40,000	
Plant overhead cost, 25 percent of labor		
and supervision	52,000	0.198
Outside maintenance materials,		
4 percent of initial investment	76,000	
Outside maintenance labor,		
1.5 percent of total initial investment	28,000	
Insurance, 2 percent of total initial		
investment	37,000	0.141
Return on investment (10 percent) and		
depreciation (7 percent), 17 percent of total		
initial investment	320,000	1.221
Utilities (other than self-		
generated), power, water, etc.	20,000	0.076
Total if manure is free	$743,000	2.835
Manure @$7/ton (dry basis)		0.801
Total		3.636

percent. Investment costs for plants using either coal or wood are about the same and roughly three times the cost of a plant with natural gas as the feedstock. If wood costs $34/dry ton, the methanol would have to sell for almost $1/gal versus the current price of $0.38/gal.

Methanol from hydrogen and carbon oxides was produced for a number of years using a zinc–chromium catalyst at pressures of about 340 atmospheres and temperatures to 700°F (370°C). Developments by ICI (Imperial Chemical Industries) led to the commercialization of a copper-base catalyst permitting the more economical synthesis conditions of 50 to 100 atmospheres and temperatures below 520°F (270°C).

Lurgi has subsequently commercialized a similar development, also using a copper catalyst. The Lurgi converter is unique in that it is a vertical shell-and-tube heat exchanger with the catalyst inside tubes surrounded by boiling water. The heat of the synthesis reaction is thus recovered by generating medium-pressure (34- to 41-atmosphere) steam. Other licensors are offering low- and medium-pressure processes featuring their own copper catalyst formulations.

Recently, Chem Systems, Inc. reported on a liquid-phase methanol reactor using a three-phase fluidized system with the gases CO, CO_2, and H_2 and liquid passing cocurrently through a catalyst bed. This process provides improved temperature control and greatly increases the conversion per pass. This process has the potential of reducing the cost of methanol by 50¢ to 75¢ per million BTU (MBTU). The mass and energy balances for methanol from wood are shown in Table 10.8.

The methanol costs (Kohan, Barkhordar, et al., 1979) are estimated at $5.18 to

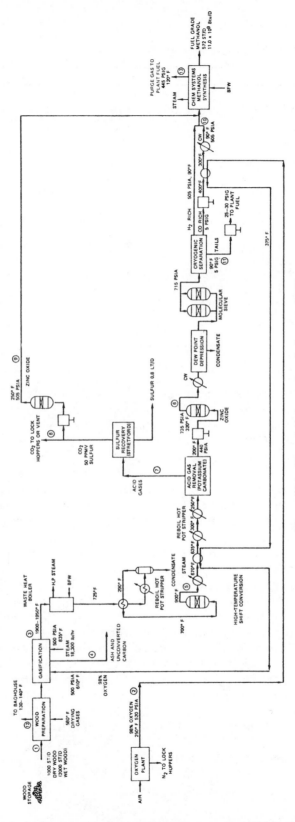

Figure 10.1 Flow sheet for methanol. Courtesy NTIS.

Table 10.8   Material Balance for Methanol from Wood

	Tons per Hour
Input	
Wet wood	83.4
Water	387.8
Oxygen	21.0
Combustion and drier air	165.6
Total	657.8
Output	
Methanol	23.9
Ash and unburned carbon	1.0
Drier stack	186.8
CO_2 to stack	41.7
Treated wastewater	161.6
Evaporation losses	242.7
Sulfur	0.1
Total	657.8

Energy Balance for Methanol from Wood

	Millions of BTU per Hour	Percent
Input		
Wet wood	796.7	99.1
Electricity	7.5	0.9
Total	804.2	100.0
Output		
Methanol	458.8	57.1
Heat rejected to cooling	281.2	35.0
Stack	44.4	5.5
Ash and unconverted carbon	7.1	0.9
Insulation losses	8.0	1.0
Miscellaneous losses[b]	4.7	0.5
Total	804.2	100.0

Source: Kohan, Barkhordar, et al. (1979).
[a]   Basis: 1000 tons of wood per day.
[b]   Because of mechanical inefficiencies.

$6.44 per million BTU, which is higher than the cost from coal. This results from lower production rates with biomass and lower carbon content than that of coal.

As processes go, thermochemical conversion is relatively straightforward, and opportunities for energy savings are fairly obvious. The capital investment may be staggering; thus great attention should be paid to a less costly design. Unfor-

tunately, safety considerations demand very conservative design and extra rein-
forcement, which drive up costs.

Comparsion of fuel alcohol by fermentation with methanol from coal gasification is of dubious value because time frames are different and technology is
advancing. Nevertheless, to produce 2 million tons of methanol per year, each
coal gasification factory will have a capital cost of roughly $2 billion. Completion
could be expected by 1990, and each factory would produce about 0.5 percent of
U.S. need for liquid fuel. Fermentation factories can be built relatively rapidly
and be on stream by 1984. A plant producing 0.1 million tons of ethanol per year
would cost about $50 million and would equate to 0.03 percent of U.S. liquid fuel
needs. This crude analysis shows that fermentation alcohol can be ready faster
than methanol from coal gasification and at a lower investment, but many, many
factories would be distributed across the nation to substitute for gasoline.

Ammonia

Ammonia is widely used as a fertilizer or in the manufacture of synthetic fertilizers and is an important industrial chemical. Prior to World War II, about 90 percent of ammonia was produced from coal. Today the figure is less than 15, as
most of the ammonia is made by steam reforming of natural gas or naphtha. Gasification of biomass followed by reaction of nitrogen with hydrogen to form ammonia is a logical means of converting biomass to a chemical that represents a
large investment of energy. A process flow sheet is given in Fig. 10.2.

The mass and energy balances for ammonia from wood are shown in Tables
10.9 and 10.10.

Anaerobic Digestion

Production of methane from cattle manure has been nicely demonstrated by
units in Colorado and Florida. Methane is used locally or sold, but protein in the
digester solids is used for cattle feed and has twice the value of the methane. The
impact of anaerobic digestion in agricultural practices may be important in relation to manures and to animal husbandry, but there can be only a very minor
contribution to national energy needs. Calculation of methane costs depends
heavily on the credit assigned to cattle feeding; thus market prices of other feeds
affect the estimates. Nevertheless, it is safe to state that anaerobic digestion of
cattle manure shows a favorable return on investment when digester solids are
used to supplement the ration fed to the animals.

The options for a very large methane factory are restricted in comparison to
those for digestion of sewage sludges. The digester residue or sludge must be
dried prior to disposal or sale. The usual method for sludges from municipal
waste treatment plants is placement on sand beds for drainage and air drying.
Several days are required, and there is often no covering to protect from rain.

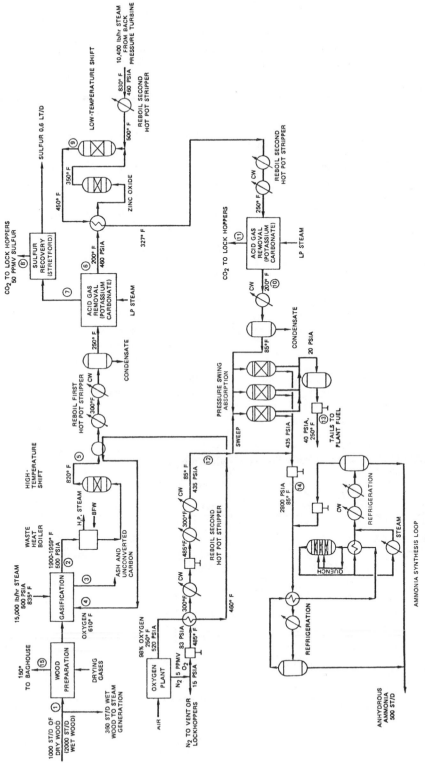

Figure 10.2 Process flow sheet for ammonia. Courtesy NTIS.

Table 10.9 Material Balance for Ammonia from Wood

	Tons per Hour
Input	
Wet wood	83.4
Oxygen	17.2
Combustion air	82.0
Nitrogen	19.4
Water	415.6
Total	617.6
Output	
Ammonia	20.8
Ash and unburned carbon	1.0
Drier stack	115.0
CO_2 to stack	62.4
Treated wastewater	167.9
Evaporation losses	250.4
Sulfur	0.1
Total	617.6

Source: Kohan, Barkhordar, et al. (1979).
[a] Basis: 1000 tons of wood per day.

Table 10.10 Energy Balance for Ammonia from Wood

	Millions of BTU per Hour	Percent
Input		
Wet wood	796.7	100.0
Total	796.7	100.0
Output		
Ammonia	372.9	46.8
Heat rejected to cooling	335.6	42.1
Stacks	59.6	7.5
Ash and unconverted carbon	6.2	0.8
Insulation losses	16.5	2.1
Miscellaneous losses	5.9	0.7
Total	796.7	100.0

Source: Kohan, Barkhordar, et al. (1979).
[a] Basis: 1000 tons of wood per day.

Large areas of beds are required; thus this method is impractical for large factories. Furthermore, labor costs are high for collecting the dried cake. Expensive vacuum filters are used in some waste treatment plants, and there are costs associated with chemicals to condition the sludge for good filtration.

Thermophillic digestion (roughly 60° C) should be best for large factories because shorter detention time equates to less capital investment than for mesophilic digestion (roughly 40° C). Both capital and operating costs depend on the solids concentration of the feed; total cost for processing 20 percent solids is only about $\frac{1}{8}$ of the cost of processing 3 percent solids. Unfortunately, it is very difficult to handle or to mix slurries that approach 10 percent of biomass solids. A reasonable cost for capital (digesters and vacuum filters), mixing, heating, and dewatering is $3/ton of dry solids processed (DeRenzo, 1977). This does not include cost of purifying methane to pipeline specifications. One ton of biomass should yield 6000 ft^3 of methane with heating value 1000 BTU/ft^3 or less than $20 worth at current prices. Only a very cheap feedstock would permit production of methane at a profit.

Anaerobic digestion uses almost primitive vessels—either a simple tank, a floating-roof tank, a lined covered ditch, or the like. Even so, the principal product, methane, is so cheap that equipment costs must be pared to the bone. Very inexpensive digesters are discussed in the section about biomass for methane in foreign nations.

Clausen, Sitton, et al. (1979) have prepared a preliminary design of a factory for producing methane from farm crop residues. The feedstock cost was assumed to be $10/dry ton, which is probably too low. Quotations for corn stover in Indiana seem to be in the range of $25 to $30/ton, and phone inquiries about various residues in Georgia have prices up to $70/ton. Reactor size was based on kinetics determined with oat hulls, refuse, or cornstalks. The bases were 4460 tons of biomass per day to produce 50 million ft^3 (1.4 million m^3) of purified methane with four digesters of 5-million-gallon (20,000-m^3) capacity in series. With better mixing of the digester, residence time can be reduced with a sizable saving in the cost of reactors, but optimum conditions for digestion of these residues has not been determined.

The return on investment for the hypothetical factory would be 11 to 17 percent, but college professors are notorious for overly optimistic costs estimates. For example, this design allows for only one day of feedstock storage in silos, which is little better than omitting storage. A 5-million-gallon tank is enormous, and safety considerations could dictate heavy reinforcing and very thick walls. Based on the economics of digesting cattle manure, which depend on refeed credits to show a margin for profit, digestion of tougher, more expensive crop residues is probably not commercially feasible.

Ashare, Buivid, et al. (1979) have analyzed the costs of methane produced by anaerobic digestion of agricultural crop residues. With no by-product credits from using components of the digester effluent, the economics for methane were poor. Although pretreatment of residues can increase digestibility several fold, the economic benefits were questionable because chemicals for pretreatment add significantly to the operating cost. This report tends to refute the conclusions of Clausen, Sitton, et al. (1979). Furthermore, the energy efficiency of the process ranged from only about 8 percent for small factories to about 30 percent for large factories using the most favorable feedstocks. If methane is desired from biomass, a thermochemical conversion appears better than bioconversion at the present status of technology.

In the author's opinion, proposals to generate methane from kelp grown in the oceans are irresponsible. As discussed in an earlier chapter, kelp yield estimates have been very unrealistically high. Furthermore, if kelp were available at reasonable prices, there are better products than methane. Published cost analyses for methane from kelp show only marginal commercial prospects even if the wild assumptions are accepted (Dynatech, 1978a, b).

Grain to Alcohol (Gasohol)

The case against gasohol is neatly presented by Reilly (1978), who shows that the process is energy inefficient. A bushel of corn has a heat of combustion of 376,000 BTU, whereas 17.1 lb of ethanol from this bushel has 218,000 BTU. Addition of the energy of 17 lb of dry stillage by-product for 130,000 BTU closes the gap, but the steam to prepare the extracts, concentrate, ferment, and distill is equivalent to 340,000 BTU/bushel. Electrical energy is also used. On a thermodynamic basis, the conversion of corn to ethanol appears to be very inefficient. In Reilly's analysis, each 50¢/bushel increase in the price of corn adds 19.2¢/gal to the price of ethanol. He feels that credits for dried stillage may be small because the market would be saturated by a few large ethanol factories. It is difficult to counter the argument that a major program for gasohol from corn would drive up corn prices to make gasohol uncompetitive.

An economic analysis by David, Hammaker, et al. (1978) of making alcohol from grain supported Reilly's conclusions. The product consumes more energy than is produced, and the capital investment would be $1 to $1.50/gal of alcohol depending on the size of the plant. A price of $1.16 to $1.35/gal was used. A good case might be made for government subsidies for alcohol to stimulate the economy and to ease demand for foreign oil, but gasohol is not economical by itself if old-fashioned technology is used to make the ethanol.

Chambers, Herendeen, et al. (1979b) have evaluated the energy balances for ethanol production very critically and concluded that there may be a small net gain depending on the assumptions and on how the system is defined. However, there is absolutely no doubt that fuel alcohol is a complete substitute for petroleum because nearly all the energy inputs can be obtained by burning part of the biomass feedstocks or nonpetroleum fuels such as coal.

Another analysis of net energy considered sugarcane (Hopkinson and Day, 1980). The worst case considered was burning fossil fuel to run the process;

the overall ratio of energy output:inputs was 0.9 when agricultural and industrial expenditures were included. The best case where all bagasse was used to generate steam had a favorable energy ratio of 1:8.

Ethanol from grain in the winter of 1979/80 is selling at over $1.70/gal, and demand outstrips supply by far. There is a premium for alcohol made from renewable sources because of the Federal law remitting the 4¢/gal gasoline tax for alcohol blends, and several states offer additional incentives. Public enthusiasm for gasohol and its ability to substitute for unleaded gasoline have created a large market. The economics are mysterious and could be ruined by a sharp increase in the cost of agricultural commodities.

Other Substrates

It is pertinent to analyze costs of ethanol by various fermentation processes to see what steps are most cost-sensitive and to predict the commercial implications of current research advances.

F. C. Schaffer and Associates, Inc., an engineering firm experienced in construction of sugar mills and ethanol plants, has prepared an analysis of present costs (Lipinsky, Birkett, et al., 1978a). This is based on conventional technology, and it is instructive to analyze where new developments would impact. The basis is 10,000 tons of sugarcane per day to produce 150,000 gal/day (570,000 l/day) of anhydrous ethanol. A commented flow sheet is given in Fig. 10.3.

Note from the comments that new technology such as the Tilby separation process will be used at almost every step. Juice extraction from Tilby pith will be more compact and efficient with less water, which may or may not be warm. Substitution of reverse osmosis (RO) for the evaporation steps will eliminate much of the demand for steam but will add electricity to drive pumps.

Whereas the plant designed by F. C. Schaffer and Associates is energy self-sufficient because of burning of bagasse, the plant with new technology will need to purchase electricity and coal. Steam needs will be much, much smaller because of substitution of RO for evaporation and because of dramatic savings in distillation energy. The savings in electricity by substituting sedimentation for centrifugation will be significant, but pumping for RO will outweigh them. However, high-quality fiber instead of bagasse will add product value that may be considerably greater than the value of the ethanol. By distributing costs over two main products instead of just ethanol, it is likely that a major price reduction for ethanol will be possible.

Table 10.11 gives ethanol yield from the sugars or starches in various materials. When corn sells at $2.70 per bushel, the alcohol raw material cost alone is over $1/gal. Working backward, it appears that a source of sugar at 6 or 7¢/lb (13 to 15¢/kg) would leave sufficient margin for processing cost to provide ethanol at well under $1.50/gal (39¢/l) selling price. The only chances for

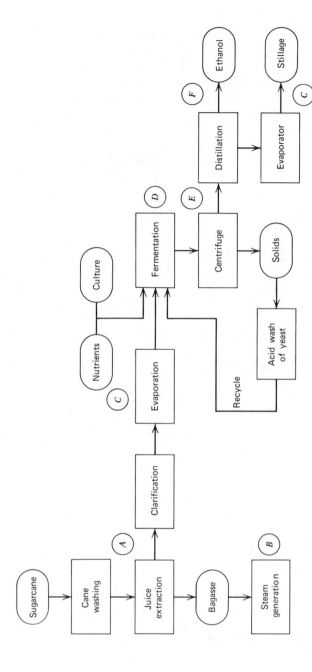

Figure 10.3 Ethanol from sugarcane. Comments: (*A*) mechanical fractionation will simplify or eliminate washing, juice extraction, and clarification; (*B*) no bagasse because high-value fiber instead; (*C*) reverse osmosis should supplant evaporation; (*D*) low-investment fermentation system proposed; (*E*) lamellar settler will supplant centrifugation; (*F*) drying could supplant part of distillation.

Table 10.11   Ethanol Yield

2.6 gal/bushel of corn
28 gal/ton (116 l/ton) of potatoes
0.4 v/v of molasses
120 gal/ton (500 l/ton) of sugar

finding large supplies of cheap sugars are hydrolysis of lignocellulosic materials or a multipurpose crop with valuable constituents to defray the cost of the fermentable fractions.

Ethanol at 60¢/gal from corn in a 20-million-gallon (76-million-l)-per year plant is claimed by Chemapac of Woodbury, N. J. (*Biomass Digest*, April 1979). By-product credits from corn oil, edible meal, and fodder yeast are important, and anaerobic digestion of waste streams will provide methane to power the factory. The distillation step would be energy efficient because of good heat exchange and columns designed for fuel alcohol instead of beverage alcohol. The ACR Process Corporation of Westfield, N. J. also has a highly efficient distillation system for ethanol using novel means for breaking the azeotrope. Vulcan Cincinnati, Inc. uses ethyl ether rather than benzene for the azeotropic distillation, and their recovery system uses 20,000 BTU/gal of absolute ethanol. This figure is typical of really excellent ethanol recovery and is far below the energy for recovering beverage alcohol.

Sheppard (1978) has analyzed the economics of producing ethanol and furfural from corn. The furfural process is shown in Fig. 10.4. Many different feedstocks can be used instead of corn stover, and product selling price is highly dependent on raw material costs. The price from the analysis based on estimated capital and operating costs matched very closely to the actual current price of furfural (about 50¢/lb).

Profitability of ethanol from corn presently hinges on selling the stillage for cattle feed. Large-scale operations for fuel alcohol will overwhelm the market for cattle feed and depress prices. Adverse impacts on soybean prices and the possible protests from farmers have already been mentioned, but the net effect may be a realignment of the systems for animal nutrition. Diversion of major

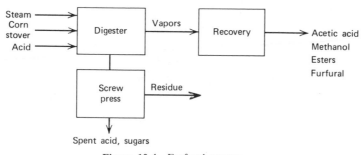

Figure 10.4 Furfural process.

amounts of corn to alcohol should raise corn prices so that less grain would be fed to animals, and dried stillage (distillers dried grains) would have growing markets that would be shared with soybeans and other cheap sources of protein. If there were no credits from the stillage, alcohol from corn would have to increase 30 to 40¢/gal (8 to 11¢/l). The real problem may not be losing the stillage credits because of market saturation, but the complicated matching of supplies to the shifting of a commodity market. With two products, ethanol going to the fuel sector and distillers dried grains to agricultural markets, pricing could be a nightmare. Although corn could be the ultimate feedstock for manufacturing ethanol, gyrations in the price of distillers dried grains are likely to be so troublesome that other feedstocks not dependent on cattle feeding will be much more attractive.

Ethanol from Lignocellulosic Biomass

There is a very interesting history for cost analyses for ethanol from wood or other plant materials. Of course, there were real cost data from a full-scale plant built just after World War II using sugars produced by acid hydrolysis of wood. A few months of testing were sufficient to show that the plant could not compete with alcohol from petroleum. Hokanson and Katzen (1978) have estimated costs for fermentation alcohol when the sugars are obtained by acid hydrolysis of wood. The required selling price of $1.90/gal compares poorly with the current price of alcohol from corn. There have recently been several cost analyses based on enzymatic hydrolysis.

Prior to 1977 there were very rough cost estimates performed by people closely involved in R & D on the conversion processes. In their enthusiasm, these people tended to produce low estimates. However, technological breakthroughs for lowering costs may compensate for their errors. The U.S. Department of Energy contracted for unbiased cost estimates, and one of the first available was by SRI International (1978). This study was based on the Natick process plus the developments from Wilke's laboratory at the University of California, Berkeley. Eight of Wilke's publications and 12 from Natick are cited. The estimated cost of ethanol was $3.34/gal without even including administrative costs, sales cost, or profit. The reaction from the Department of Energy was to consider deemphasis of the fermentation program. However, some reflection showed that the analysis missed some important opportunities for savings, and the process was obsolete in view of newer technology. The key reason for such a high alcohol cost was the charge for raw materials ($2.04/gal ethanol). This comes from using only the 34 percent of hexosans in the feedstock, getting about 50 percent hydrolysis yield and realizing the usual 50 percent weight yield from sugars to ethanol. This means less than 1 part of ethanol from 10 parts of feedstock ($0.37 \times 0.5 \times 0.5$) so that material balance alone shows that 2¢/lb feedstock gives 20¢/lb or $1.32/gal of alcohol. The $2.04 figure includes nutrients, reagents, and feedstock used to produce cellulase enzymes. The proponents of fermentation requested another cost analysis that incorporated the nearly quantitative hydrolysis of cellulose now achieved by several contractors and that allowed by-product credits for hemicellulose and lignin. There were also indications that expensive proteose–peptone used in laboratory fermentation media could be replaced by much cheaper ingredients.

Several RD contractors prepared their own cost estimates and an unbiased analysis by Jenkins, Reddy, et al. (1978) has appeared. This analysis was very conservative with regard to storage of biomass and materials of construction to resist strong acids for pretreatment. Stillage was considered to be of unproven value and required a very expensive waste treatment unit for its disposal. The by-product to be made from hemicellulose was furfural, which was credited on its heating value, not as a valuable chemical. Lumping furfural and ethanol, Jenkins, Reddy, et al. (1978) found the cost to be $15.40 per million BTU, a figure still highly upsetting to the Department of Energy, which sees fossil fuels at $3 to $4/MBTU. Since plant biomass is $2 to $5/MBTU as delivered for conversion, it will be impossible to add much processing cost and produce extremely cheap fuels. When the customer pays $1/gal for gasoline, this is about $8/MBTU, which is not too much different from the projected cost of ethanol. The author feels that correction of Jenkin's figures for cheap storage of feedstock, less expensive acid-resistant equipment, and better by-product credits would show ethanol from hydrolyzed biomass at $1.50 to $2.00/gal.

Credence can be given to a cost estimate by Emert and Katzen (1979) because a factory was to be built. The Gulf process with simultaneous saccharification and fermentation using data from a pilot plant with 1 ton of feedstock per day provided the bases for the estimate. The Gulf Science and Technology pilot plant at Pittsburg, Kansas has increased from a 1 ton/day to a 50-ton/day operation (*Chemical Engineering News*, April 16, 1979, p. 39). The factory could be operational in 1983 with a target selling price of $1.62/gal. Design bases are 1000 dry tons/day of feedstock to yield 75,000 gal/day (280,000 l/day) of 95 percent ethanol. There would also be 267 tons/day of by-product animal feed. The plant is atypical in two respects: the feedstock is wood waste plus municipal solid waste; and utility financing may be acceptable because municipal waste disposal is incorporated. Table 10.12 shows costs. Note that raw materials are a major item, and this plant has cut these to the bone by using municipal solid waste. With tax-free municipal bond financing, a selling price of about $1.02/gal derives acceptable profit whereas $1.62/gal is required with investor financing.

Energy balance shown in Figure 10.5 is instructive in that the energy efficiency is about 50 percent. Fuel oil is purchased, although coal or gas may be more reliable in 1983. Residues are burned for power; finding more valuable uses could make the factory much more profitable. This design uses steam for evaporation; thus there is a strong possibility of savings with RO. It is very encouraging to see private industry moving ahead on making ethanol from lignocellulosic materials. However, the need to use the very cheapest feedstocks of limited availability confirms some of the other opinions that ethanol

Table 10.12 Gulf Oil Chemicals Company: Cellulose Alcohol—Operating Cost and Economics, 25 Million U.S. gallon/yr Production (1983)

	Annual	Dollars per Gallon
Operating cost		
Fixed charges	7.75	0.310
Raw materials, chemicals	10.04	0.402
Utilities	5.30	0.212
Labor	2.50	0.100
Total production cost	25.59	1.024
By-product credit ($120/ton)	(10.48)	(.419)
Sales, freight, G & AO (10¢/gal)	2.50	0.10
Net operating cost	17.62	0.705
Economics—municipal bond financing (80 percent)		
Total fixed investment (required 1980)	70.4	
Company equity (1983)	25.95	
Total plant capital (1983)	83.15	
Total income	25.53	1.021 (selling price)
Interest (municipal & WC)	4.02	0.161
Federal taxes (50 percent)	0	0
Net profit after taxes	3.89	0.156
Percent return on investment (equity) after taxes		15.00
Economics—100 percent investor equity financing		
Total fixed investment (required 1980)	70.4	
Total plant capital	75.92	
Total income	40.40	1.616 (selling price)
Federal taxes (50 percent)	11.39	0.456
Net profit after taxes	11.39	0.456
Percent return on investment after taxes		15.00

from lignocellulosic crops can't quite be profitable with proven technology. The new technology plus some process development has excellent prospects for economic production of alcohol fuels from crops in the near future.

In August 1979 the Gulf process and the associated R & D people were sent to the University of Arkansas, where there may be financial support by the U.S. Department of Energy. Although this move was claimed to be a corporate decision based on business priorities and not lack of confidence in the process, the Gulf process must not have appeared to be an outstanding investment opportunity to their management.

The University of Pennsylvania–General Electric Company team has a preliminary cost analysis for their process (Nolan, unpublished). The feedstock is chipped populus trees; 9 mi^2 (23 km^2) of forest producing 10 tons/acre per year would supply a plant producing 30 million gallons/year (113 million l.) of

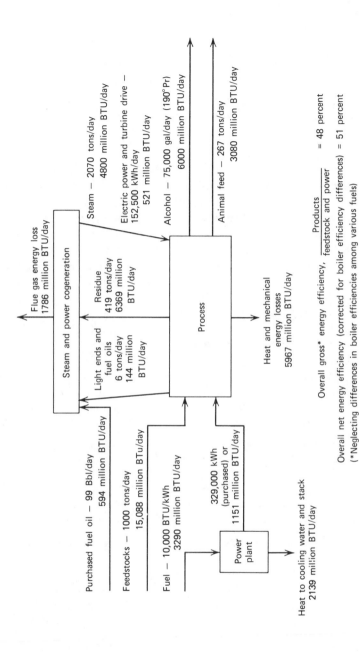

Figure 10.5 Energy balance for ethanol from cellulose. Overall gross (neglecting differences in boiler efficiencies among various fuels) energy efficiency, products/(feedstock and power) = 48 percent. Overall net energy efficiency (corrected for boiler efficiency differences) = 51 percent. Courtesy NTIS.

275

ethanol. A cost of $15/ton of chips has been quoted, but this may be low. Chemical analysis shows

50 to 57 percent	cellulose
16 to 26 percent	hemicellulose
16 to 24 percent	lignin
8 to 12 percent	ash

Lignin is extracted with hot alcohol, and the remaining solids are hydrolyzed with enzyme. A thermophillic organism allows a 60°C continuous fermentation with ethanol held at 4 to 5 weight percent by removal with vacuum. It is claimed that metabolic heat and heat for evaporation are balanced; thus there is an important saving in distillation energy. The cost analysis does not include site preparation, pretreatment, enzyme costs, and sterilization. Equipment costs for fermentation and alcohol recovery were $6.3 million. Production costs are listed in Table 10.13. This production cost led to a selling price of $1.25, but recall that several costs were not included. Although the feedstock costs can be defended by spot quotations, $15/ton is definitely in the low range and may not hold up if a large demand is created. The author feels that the Penn/GE analysis is not inconsistent with lignocellulosic crops to ethanol at $1.50 to $2.00/gal.

One objection to thermophilic operations is the higher partial pressure of ethanol in the vapor. Commercial operatons have been at relatively low temperatures to minimize evaporative losses of ethanol in the large volume of CO_2 that is evolved. It has not been considered cost-effective to increase productivity while adding refrigerated condensers to collect ethanol from the CO_2 stream when the fermentation temperature is elevated.

An obvious question is, "How can there by a high degree of optimism when most analyses point to ethanol at over $1.50/gal?" There are two answers: (1) "the costs should come down with process development and with important breakthroughs in alcohol purification" and (2) "by-product credits should be much greater as better use is made of hemicellulose and lignin." Fermentation

Table 10.13   Costs for Penn/GE Process

	Cost, $/gal Ethanol
Materials	
Poplar wood chips	0.45
Nutrients and chemicals	0.076
Operations	
Labor	0.001
Utilities	0.064
Administration, maintenance, etc.	0.006
Capital charges	0.096
Total	$0.70

products from sugar in hydrolysate of hemicellulose can approach the value of the ethanol from cellulose. This would spread costs over two products and could lower the price of ethanol to perhaps $1.00/gal. Lignin of the very high quality obtained with the mild pretreatments proposed for biomass is worth over 20¢/lb, but markets are small. Expansion of these markets or conversion of lignin to valuable derivatives would further reduce the ethanol price to, say, 80¢/gal. These advances will take a few years, but progress is being made at an impressive rate.

Disposal of aqueous wastes from fermentation plants can be costly if treatment is required. However, sewage effluents and even raw sewage are used for irrigation of crops, and Woodwell (1977) has reviewed the concept of recharging groundwater with sewage effluents that have passed through biological communities that take up their nutrients. With sewage, the main concerns are pathogenic organisms and toxic materials that endanger human health. It is very unlikely that effluents from an alcohol factory would contain pathogens, and toxic materials can be avoided by employing only processes with safe reagents. Therefore, irrigation of crops with factory wastewater should be the most cost-effective option.

Ethanol as a By-product

In a sense, ethanol from corn stover is a by-product to corn grain, but the operations are divorced. The Tilby process or something similar for sugarcane or sweet sorghum can fractionate to give three or four components: waxy epidermis, fiber, juices, and extracted pith. The wax will be valuable but small in amount. The components are shown in Table 10.14. The epidermis is hardly worth considering, except that processing it costs little more than disposal. After recovering the wax, the residue might be mixed with the pith. If the fiber is to be pulped, moisture is no problem. Pith is the main component of sugarcane and major source of sugars. Extracted pith has short fibers of some industrial use as filler. If the price were low, pith would be an excellent feedstock for the lignocellulosic route to ethanol. The net mix of products is shown in Table 10.15.

Wet cane costs about $15/ton. The very rough value of the product mix from one ton is 81 lb of long fiber ($40.50), 520 lb of pith fiber ($2.60), and 180 lb of sugar ($7.20). Obviously, the long fiber can carry the whole process, and the other fractions are relatively unimportant. The 180 lb of sugar would give 26 gal of alcohol that at $1/gal would be an excellent credit. The situation should be much the same for sweet sorghum.

J. E. Atchison Consultants, Inc. (1977) under contract to U.S. Department of Energy analyzed the commercial prospects of the Tilby separator. Prototype units have been operated at 5-, 20-, and 40-ton/day scales with excellent results. Sugarcane has had most of the emphasis, but sweet sorghum also processes very well. There has been mention of obtaining the fibrous fraction dry, but microscopic examination of sugarcane shows some channels containing juice in the rind layer. In fact, the sugar content of rind is too high for some of the proposed

Table 10.14   Sugarcane Fractions from Tilby Process

Component	Percent of cane	Fiber	Moisture	Sugar	Wax
Pith	79	33 %	59 %	8.0	
Rind	19	12.8	72.1	15.1	
Epidermis	2	32.2	55.6	8.7	2.5

Source: Lipinsky, Birkett, et al. (1978a).

uses. Fortunately, after the waxy epidermis is removed by the Tilby separator, it is relatively easy to extract juices from the rind. Rind fiber has excellent strength and has potential for binding with resins to produce a product that could substitute for plywood or for hardwood flooring. The short fibers and pith could compete with wood flour as a filler for phenolic plastics. Only burning is proposed now for getting some value from the epidermis, but it is rich in waxes and oil that would command good prices if an extraction process were developed.

Pith and short fiber might be cheap enough to be considered as feedstocks for fermentation to alcohol fuels, but long fiber will be too attractive for better uses. The main implication of the Tilby process is not for cheap cellulosic materials, but for cheap sugar juices. By upgrading the value by the by-products and producing clean juices that are easy to process, the net cost assigned to fermentable sugars might be as little as 4¢/lb.

Valuable by-products can be obtained from tree bark. Stewart (1976) has described a process in which Douglas fir bark is extracted to obtain wax. The spent bark can be used as an entender in plastics, or a complex mechanical screening operation can yield cork and fiber fractions.

As an aid to program planning, the U.S. Department of Energy contracted for a preliminary economic analysis of the MIT direct fermentation of biomass, even though the research is still at a laboratory stage (Jenkins and Reddy, 1979). Using the same bases as with other analyses by the same group, a very favorable economic prospect was reported for the MIT process that appears to have high potential for profit and for considerably less capital investment than other processes for producing fuel alcohol from lignocellulosic biomass. A conceptual plant for processing corn stover was designed; a portion of the plan is shown in Fig. 10.6. Distillation operations are not shown because these can be about the same for any ethanol factory. Of course, the newest and most efficient distillation

Table 10.15   Products from Tilby Process

	Percent of Wet Cane	Percent of Dry Cane	Value, ¢/lb
Long fiber	2.43	6	50
Pith fiber	26.1	67	5
Sugars	9.18	24	4
Wax	0.05	0.1	50

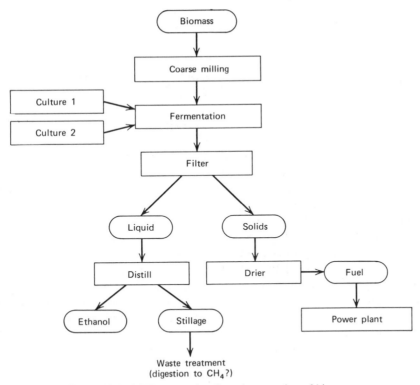

Figure 10.6 MIT process for direct fermentation of biomass.

schemes should be used. An anaerobic digestion system for methane generation from various plant wastes is not shown either; such designs are fairly standardized.

Two cases were considered. Case A is based on current experimental results in which by-products such as acetic acid are produced at too low a concentration to justify recovery. Such by-products are converted to methane in the waste treatment section of the plant. Case B assumes that successful research on increasing the ratio of ethanol to by-products will achieve almost complete supression of by-products while increasing ethanol concentration proportionately. With a feed of 1500 tons/day of corn stover, case A results in 27 million gal/yr (102 million l) of ethanol compared to 41 million gallons per year (155 million l) for case B.

The features of the MIT process that provide the most significant cost advantages are the simple preparation of the feedstock and the easy processing of waste fractions. Unreacted feedstock is burned to provide steam and electricity. The power exceeds process needs by a factor of about 2, thus electricity may be sold. Liquid wastes are anaerobically digested to methane to produce supplementary fuel and to reduce the concentration of potential pollutants. Each of these operations is carried out with relatively inexpensive equipment. Compared to acid hydrolysis or enzyme hydrolysis and the respective methods of pretreatment and required subsidiary steps, the MIT process is far less complicated and needs no costly materials to resist corrosion.

The groups at MIT and Purdue have grasped a concept that has important implications to processing of biomass. Basically, they propose taking the easily removable and fermentable materials from lignocellulosic biomass and diverting the residue to combustion fuel or to animal feed. Although the opposite approach of deriving valuable products from all the biomass fractions has been promoted in this chapter, it makes sense to start with the easy processes now instead of delaying until more sophisticated schemes are practical. One Purdue idea is to use dilute acid to obtain cheap sugars from the hemicellulose in biomass. Hardly any development work is needed, and the sugars are estimated to cost about $100/ton. The residue is still worth about $20/ton for burning; the materials costs are shown in Table 10.16. This simplistic analysis merely indicates that crediting the hemicellulose sugars at a value that is attractive for fermentation feedstocks still leaves a margin for paying the processing cost.

As better uses are found for the residue, the factories can be modified to realize new opportunities. For example, removal of hemicellulose sugars is partial pretreatment for cellulose hydrolysis, and the residue is a more valuable feedstock because of its enrichment in glucose. An implication of this approach is that the factories should be designed with sufficient flexibility and ease of expansion for ready incorporation of advances in technology and additions of new steps.

Fermentation Practice

Conventional fermentation has dealt with either high-value products such as antibiotics, enzymes, and vitamins or intermediate-value products such as citric acid. In the days of cheap energy, it was not cost-effective to seek minor process economies by reducing needs for steam or for cooling water. Lower-value products such as ethanol, acetone, or butanol have been made more cheaply from fossil fuels. Now ethanol from sugars again is economically feasible, but the context is not the same as it was several decades ago. The constraints for penicillin fermentation are inappropriate for producing cheap ethanol. Relatively crude equipment but with good engineering may reduce costs drastically. Although this approach has been dubbed "low technology," very sophisticated biochemical engineering may be required to devise a process with a favorable return on investment.

Table 10.16 Materials Cost Balance for Hemicellulose Sugars

Basis:	100 tons (20 percent hemicellulose) @ $20/ton	$2000
Products:	20 tons of sugars @ $100	2000
	80 tons of residue @ $20	1600
		$3600

Fermentation Configuration

Beer has been made in open vats for centuries; only recently has closed equipment been used to achieve better process control. Evolution of CO_2 in the vat sweeps out air, and foam on the surface offers some protection for anaerobiosis. Alcohol is produced at a declining rate because the product is inhibitory to the yeast at concentrations above 6 to 12 percent depending on the strain.

The justification of continuous fermentation is maintaining the culture at very high productivity. Since alcohol production and growth are not coupled, a single-stage continuous fermentation is unlikely to optimize both. A multistage system could obtain growth in some units and have sufficient residence time in later stages to reach high alcohol concentration. Plug flow, especially with yeast recycle, could be highly effective in attaining rapid growth in the leading sections and long detention times for the rest of the unit.

Productivity versus Capital Cost

It is dogma to engineers that high production rates are desirable. Usually, the costs of equipment, labor, and utilities are such major factors that it is obvious that higher productivity will save money. Production of cheap alcohol may be an exception to this casual thinking. As the saying goes, we must look at the bottom line on the balance sheet. Fermentation is but one step in the process, and distillation costs are also important. For distillation, a higher alcohol concentration is less costly to process. By this reasoning, a trade-off should be made between higher alcohol concentration and lower production rate caused by inhibition. The alcohol fermentation also has great potential for saving in that crude fermenters could have very low capital cost, labor charges could be low for large, slow fermenters, and there may be very low utility costs because mechanical agitation is probably not required. In other words, a low production rate to reach a high final ethanol concentration is quite acceptable if capital expenditures for equipment and operating costs are low.

Low Investment Technology

Ethanol was produced by fermentation for centuries before the discovery of microorganisms. Primitive practice sometimes employed cooking of the medium, but fruits and juices usually give a good yield of ethanol if untreated. This is possible because yeasts are hardy and grow rapidly while the production of alcohol and lowering of pH retard other organisms. Bacterial "diseases" of wine were common until Louis Pasteur discovered their causes and developed protective measures. For very large-scale fuel fermentations, it is likely that strict asepsis with rigorous sterilization could be avoided; the savings in capital for fermenters would be a major factor in commercialization. Some aids to avoiding serious contamination are (1) continuous fermentation with lowered pH and a protectively high ethanol concentration, (2) high dilution rate to wash out contaminants before they can become established, and (3) separate feed streams of the various nutrients so that none has a complete nutrient balance for contaminants. The main need is a reorientation of thinking from high technology of elegant processes based on cheap energy and special materials of construction to low technology that tolerates some irregularities in output but is expedient and economical (Bungay, 1980).

In the United States, low technology has a different connotation than in other nations where labor is relatively cheap. For example, composting of organic wastes to produce fertilizer requires occasional turning over of the solid mass to assist aeration. Placement of a pile of organic matter in a shed and turning it every few hours seems to be low technology, and composting is successful in various parts of the world. Engineers in the United States can't automate the process economically, and composting based on using tractors or back hoes has failed to be profitable.

When the product is more valuable, low technology may work well, even in the United States. Mushrooms grown in the dark on solid beds of organic matter are quite profitable. However, high technology using aerated, aseptic, deep-culture fermentations can produce a mycelial mass of mushrooms at a lower cost and with excellent flavor. Were it not for the preference of consumers for solid chunks of mushroom over the soft soup from stirred fermentations, low-technology mushrooms could not compete.

Cellulose degradation is particularly amenable to low technology. Fungi are effective in attacking cellulose and related molecules because their filaments can invade a solid structure (J. E. Smith, 1976). Extracellular enzymes digest passages ahead of the hyphae for further penetration. The tip extension of filamentous fungi provides a great advantage in colonizing a substrate. Furthermore, translocation of nutrients along filaments can maintain better balance at growth sites.

The potential of low technology has been discussed by Poole and Smith (1976). For single-cell protein, it is noted that production from straw is seasonal. To maintain production through the year, straw would have to be stockpiled during the harvesting season and drawn on continually. Straw is bulky; therefore, its transport and storage are costly. Low technology could make use of slow-rate processes during the storage period by inoculating straw with microorganisms and by adding moisture and nutrients. After a period of time, there would be microbial protein plus partially digested straw. This mixture is good for cattle feed, or the microorganisms could be washed off and harvested, leaving the straw ready for further processing with less need for costly pretreatment and subdivision. Such low technology requires little capital and makes good use of space and existing facilities.

A low-technology ethanol fermentation might use pits scooped from the earth and lined with cheap material. Yeast creates some natural mixing by evolution of carbon dioxide, but some investment in mixing equipment could be wise. Covering is necessary to capture the carbon dioxide, which might, in turn, be used to nourish algae or other plants that constitute the biomass feedstock for the factory. Separation of yeast from alcohol solutions is highly desirable so that cells can be reused many times. Yeasts, however, differ from bacteria in that one cell cannot divide forever. A bud on a yeast leaves a scar, and no new bud can form there. When there are 10 to 20 bud scars, a yeast cell stops reproducing. For this reason, 5 to 10 percent of the yeast mass should be harvested. There is a ready market for animal feed, or spent yeast could be hydrolyzed and returned to the fermenter as nutrients.

Climatic changes may make temperature control costly. Enormous volumes in the fermenters will resist temperature fluctuations, but it may be advisable to switch cultures on a seasonal basis to use strains that thrive at prevailing temperatures.

Low-technology fermenters may substitute for storage containers. Seasonal biomass would require storage until needed, and silos would be expensive. If stored in piles, the feedstock for a large biomass conversion plant would cover scores of acres to a height of 10 to 20 ft (3 to 6 m). Any rotting, insect or rodent damage, spontaneous combustion, or other losses of stored material could be disastrous. It seems sensible, therefore, to process the biomass to a fermentable form on harvesting, and then to inoculate in very inexpensive vessels for a long fermentation. Alcohol concentrations of over 20 percent are attainable at prolonged times, and distillation would be facilitated. It would be crucial to operate with very concentrated media to avoid absurdly large vessels. For example, if 10 cylindrical vessels of 12-ft (4-m) diameter were to hold all the biomass sugars from the crop for a plant with a capacity of 50 million gal (190 million l) per year, assuming 20 percent alcohol v/v in the fermentation broth, each vessel would be 2.36×10^3 ft or 0.447 mi (0.72 km) long. Lined and covered ditches of this length in a spiral or hairpin configuration would be no more expensive than several very large conventional fermenters and much cheaper than fermenters plus silos.

Low-investment fermenters might be constructed with plastic pipe. Previously the largest polyolefin pipe was 48 in in diameter, but Spiral Engineered Systems has announced a 144-in size (*Chemical Engineering News,* April 9, 1979, p. 15). A fermenter of one million gallon capacity would have 1182 ft (0.36 km) of this 12-ft (4-m) diameter pipe and would cost roughly $400,000.

If placed above ground, pipe fermenters would be subject to ambient temperature variations. Sunk into the ground, their low surface:volume ratio and the poor heat transfer through dirt would simplify temperature control. Heating, which is seldom needed because considerable metabolic heat is evolved, would be easy to accomplish by feeding hot, pure water from the bottom of the ethanol rectification column. No coils or jackets would be required for heating if hot water were added directly. Cooling could be achieved in similar fashion by direct addition of cold water, but dilution would be excessive. It might be less costly to select thermophilic yeast strains and process conditions for which cooling is never needed because metabolic heat is matched to losses. The soil above the fermenters could be used for agriculture or other purposes, thus lowering costs and greatly reducing the unsightliness of the factory.

Recycle

Reuse of yeast depends on its collection prior to distillation. Simple filtration performs poorly, centrifugation is costly, and ultrafiltration needs more development. Sedimentation is usually a very inexpensive step. Yeast cells have low settling rates and settle only after CO_2 evolution has ceased. With long residence times to maximize ethanol concentration, mixing by evolution of CO_2 should subside and not disturb sedimentation.

Deep vessels for sedimentation require long detention, thus plate settlers seem much more appropriate. Parallel plate, inclined settlers should work well with yeast (Walsh and Bungay, 1979). The settling path is short; thus compact units are possible even though a settling rate of only 1 cm/hr was observed for *S. cerevisiae*. A collection unit is shown in Fig. 10.7. If CO_2 evolution does prove to be a problem, the suspensions can be heated to a temperature that stops respiration but does not kill the cells. Small energy costs are involved because the product stream will be heated for distillation anyway, and the return of warm, settled yeast to the fermenter is not an appreciable heat load.

Complete recycle of yeast is not desirable because unknown inhibitory factors may accumulate (Cysewski and Wilke, 1977). With roughly 80 to 90 percent recycle of dense yeast, light bacterial contaminants should pass through the settlers with a very low collection efficiency, thus aiding control of the yeast culture.

Nitrogen, phosphorus, and trace elements must be provided to the fermentation. These will be present in nonvolatile compounds that will reach the still bottoms. Along with cell debris, these still bottoms may have dissolved and particulate materials with nutrient value. Recycle of some proportion directly to the fermenter is possible, and the remainder may be sold for animal feed or sprayed on nearby agricultural land. Ultimately, both recycle and cattle feeding may

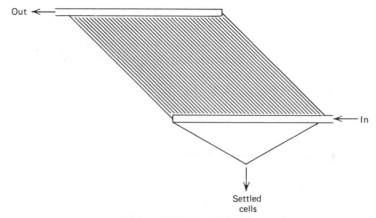

Figure 10.7 Lamellar settler.

benefit from fractionating and purifying the still bottoms. There might be valuable by-products present.

Pasteurization

Low pH and high ethanol concentration protect the fermentation from contamination. It is possible that the feed streams to the fermenter could be kept separate so that none offers a proper nutrient mix for undesired organisms. Some streams may require heating to kill contaminants and should be kept concentrated to reduce costs. Filtration sterilization might be practiced on streams free of appreciable suspended solids.

The Brazilians have been pasteurizing fermentation media and have also had good results with no heating at all. However, pasteurization costs very little if performed with efficient countercurrent heat exchange. The distillation unit can furnish large amounts of hot, distilled water and might be able to heat fermentation nutrient streams if these are used as condenser cooling water. The point is that energy needs for fermentation may be nil if fermentation fluids serve as a sink for distillation heat.

Distillation

Distillation is the separation of volatile substances by making use of differences in composition between boiling liquid and vapor. In contrast, evaporation usually deals with but one volatile component; thus its removal achieves an increase in concentration of nonvolatile components of the liquid. Several of the prime candidates for fuels from biomass, for example, methanol, ethanol, acetone and butanol, occur with water in the various processes. Recovery of products by distillation is a likely step, and its costs can be crucial to overall economics.

Consider the properties of two volatile liquids with different boiling points. If heated at atmospheric pressure, a boiling point will be reached that depends on the concentration of the particular sample. Typical plots for the liquid and vapor in equilibrium are shown in Fig. 10.8. The first sketch is for an ideal case in which the two liquids form no strong associations with each other. Where there is interaction, the curves may depart markedly from ideal behavior. Case B shows a low-boiling azeotrope, and case C shows a high-boiling azeotrope. At the azeotrope, the liquid and vapor compositions are identical, and distillation at these points accomplishes no enrichment. This is developed further with regard to ethanol.

A large difference between the liquid curve and vapor curve at a given temperature means that distillation will achieve considerable separation. Note, however, that the vapor will condense to a liquid that is *not* free of the less volatile component. In other words, simply boiling a mixture and cooling the vapor yields condensate enriched in the more volatile components whereas the original mixture becomes enriched in the less volatile component.

Boiling of the condensate gives vapor even more enriched in the more volatile

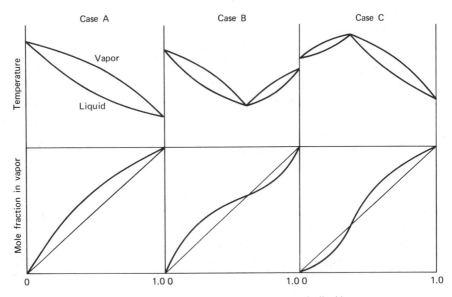

Figure 10.8 Liquid–vapor equilibrium curves.

component; thus repeated boiling and condensing can approach complete separation. As the liquid and vapor curves come closer together, less is accomplished by a distillation step so that absolute separation is impossible, except with an infinite number of steps. Furthermore, a full step will result only if equilibrium is attained between liquid and vapor. In practice, columns are used to establish contact between liquid and vapor. The columns may be packed with material that wets to permit intimate contact between liquid and vapor, or bubble-cap columns may be used. Bubble-cap columns are segmented into sections called "plates." Each plate has a layer of liquid and means of dispersing vapor into the liquid. A diagram of one of the many possible designs of a bubble-cap column is shown in Fig. 10.9. With provision for rising vapors and downflowing liquid, a bubble-cap column has more reliable operation than a packed column that may flood with liquid at conditions far from those of optimum vapor–liquid contact.

Fractional distillation requires liquid in the column, and this is accomplished by returning some of the condensate. Such return is called *reflux,* and the reflux ratio is the amount returned over the total amount. Most complete fractionation is obtained with total reflux, but, of course, no product is collected. The reflux ratio affects the separation attained in a step; there are a number of methods for determining the number of steps required at a given reflux ratio to reach a desired concentration.

With batch distillation, the concentration of less volatile component in the boiling pot keeps increasing as a product rich in the more volatile component is taken off. Thus the separation becomes more difficult, and conditions in the fractionation column change during the run. In large-scale operations it is much

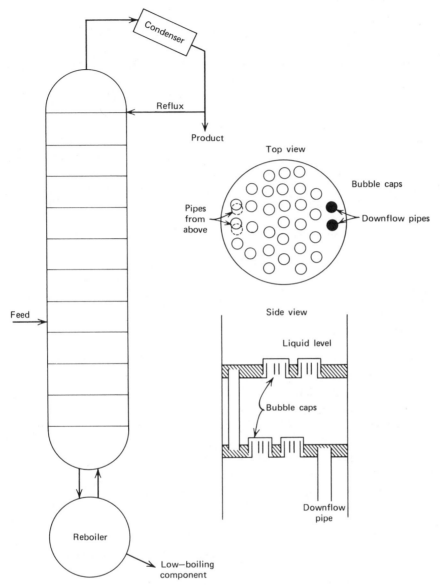

Figure 10.9 Bubble-cap column.

more common to run continuous distillation with the feed going into the column at a plate where the composition is about the same as that of the feed. Two product streams are collected, nearly pure volatile component from the top of the column and concentrated less-volatile component from the bottom. Another boiling pot is needed for the liquid leaving the bottom of the column to provide rising vapor.

Ethanol has a peculiar vapor–liquid diagram as shown in Fig. 10.10. At one point the lines come together; this means that liquid and vapor in equilibrium have the same composition, and distillation cannot achieve separation. Such a point determines an azeotropic mixture. The concentration at which an azeotrope occurs usually depends on pressure. In the case of ethanol, the azeotrope has shifted off the graph at 70 mm of mercury. Industrial use of low pressure to avoid the ethanol–water azeotrope seems to have had limited success. The wide distance between the liquid and vapor curves for aqueous ethanol means that it takes relatively few steps for separation to 85 percent ethanol and many steps to 95 percent because the curves pinch. For example, six or eight plates in a distillation column are sufficient to obtain 85 percent ethanol, whereas 40 to 50 plates are needed to approach the azeotropic concentration. However, the product from the top of the column is only about 95 percent ethanol. This may be acceptable for fuel or chemical uses, but absolute (completely dry) alcohol would be better. Multicomponent mixtures have different azeotropes or no azeotropes, and several methods are known for recovering absolute alcohol from various mixtures. Another 40 to 50 plates are needed in this distillation column. Some additives such as benzene can allow recovery of absolute alcohol while at the same time denaturing (adulterating) the alcohol so that it is unfit for human consumption and can be taxed at a low rate.

Also shown in Fig. 10.10 are operating lines marked a and b. The reflux ratio determines the slope of the operating line, but for ethanol there is an additional restraint because the operating line must not intersect the vapor composition line. Line a shows the operating line tangent to the vapor composition line. Although the lines are widely spaced at low ethanol concentration and good enrichment would be possible in the first steps of distillation, there would be no further enrichment when the composition corresponded to the touching lines. Line b represents a reflux ratio that could give product approaching the azeotrope, but the steps would be very small in the pinched region.

The liquid–vapor diagram for n-butanol is of the type of case b in Fig. 10.8. However, the azeotrope occurs at a point where n-butanol and water form a two-phase liquid mixture. Combination of two distillation steps with separation of liquid phases leads to recovery of absolute n-butanol.

Scheller (1978) has pointed out that the distillation system for beverage alcohol is not required for fuel alcohol. In a liquor distillery, fermented mash of about 10 percent alcohol is fed continuously to the beer still through a heat exchanger to recover heat from the beer still vapors. Direct steam strips the alcohol from the mash. Stillage from the bottom of the unit can be concentrated for use as animal feed and also has value as a nutrient stream for other fermentations. The condensed vapor gives about 50 percent alcohol. After mixing with a recycle stream, it is fed to the "selective still," where again heating is accomplished with live steam. This step fractionates the aldehydes, esters, and other organic components termed "fusel oil" from the ethanol. While these are undesirable in beverage ethanol, they are perfectly acceptable in fuel alcohol.

The addition of live steam and recycle keeps the bottoms of the selective still

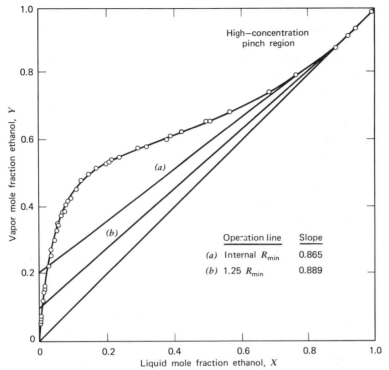

Figure 10.10   Liquid–vapor curve for ethanol. *Source*: Wilke and Blanche (1979b).

diluted to about 20 percent ethanol. Another distillation column is required to produce 96 percent ethanol, which is drawn off a few plates down from the top of the column. The uppermost plates have liquid richer in more volatile components; thus it is recycled to the selective still. In similar fashion, the column that further purifies the top fractions from the selective still has bottoms rich in ethanol that are worth recycling. The fusel oils are less soluble in water; thus decantation can be used to effect separation.

Scheller proposes that fuel alcohol processing would use the beer still but then depart radically from beverage alcohol technology. The selective still and its energy-inefficient dilution would be eliminated. A single rectifying column would deliver alcohol approaching the azeotropic composition, and other organics would carry along in the product. Denaturation by addition of adulterants might not be necessary because fuel alcohol would have the unpleasant tastes and odors of the fusel oil and other organics. Instead of roughly 75.2 lb of steam consumed per gallon of 98 percent ethanol, the simplified distillation scheme would require about 44.4 lb/gal. This saving in energy and the reduced costs of equipment are very significant.

Although fusel oils are acceptable in mixtures of gasoline and alcohol, it may be necessary to remove them during alcohol distillation. At some stage of distillation there can be a mixture of alcohol, water, and organic compounds that separate into two phases. Action of two phases at a plate in a distillation column is unpredictable in terms of heat transfer and circulation of the liquid. The two-phase mixture can be separated, and the fusel oils can be recovered easily.

Danziger (1979) has presented a strong case for distillation with vapor recompression. A problem with conventional distillation is that the vapors from the top of the column contain the most volatile materials and must thus condense at a lower temperature than that of the feed. Heat from the vapors could be used to raise the temperature of cold feed on its way to distillation, but not for the more energy intensive step of vaporizing the feed. However, compressing this vapor increases the partial pressures of its components so that they condense at a higher temperature; thus there would be a thermal driving force for boiling part of the feed stream. In other words, the heat in the vapors represents energy that can be useful if the temperature of condensation is raised by adding a relatively small amount of mechanical energy for compression.

Energy in the compressed vapors can be reused in the same distillation step or elsewhere in the factory. However, it is unlikely that the cost of process piping to carry the vapors very far can be justified. Commonly, the compressed vapors are used in one of the distillation operations very near where they are generated.

Mechanical compression of vapors can be avoided by operating the distillation under pressure. The entire column is at a higher temperature; thus the vapors cannot exchange heat with the boiler in the same step. The higher pressure vapors are well suited, however, for boiling the liquids in another, lower-pressure distillation step. When there are several distillation steps at various pressures or at partial vacuum, it can be very cost effective to exchange heat between process streams. Control of heat-integrated distillation columns is discussed by Tyreus and Luyben (1976).

A very exciting development in ethanol purification may save 50 percent of the distillation energy (Ladisch and Dyck, 1979). Whereas about 20 to 25,000 BTU/gal are required to power an efficient distillation system for absolute ethanol, only 9000 BTU/gal is sufficient for 85 percent alcohol if condensate heat warms the feed stream. In addition to the small steps for each equilibrium contact mentioned previously for the liquid and vapor, there is a matter of reflux ratio, the fraction of condensate returned to the column to provide liquid for contacting the vapor. A large reflux ratio is required in the pinched portions of the equilibrium curve; thus energy efficiency is reduced by the extra needs for cooling water for condensing and steam for boiling. For advancing beyond 85 percent ethanol, drying is proposed.

It may be possibly to dry the liquid, but an unexpected result is intense drying in the vapor state. Vapors with various percentages of ethanol were dried with different agents as shown in Table 10.17. The heats of adsorption of water vapor on the drying agents tend to be small; thus regeneration will require little energy. The agents that are feedstocks to the conversion process need not be regenerated. There is little information yet on the capacities of the agents for water, but this approach deserves highest priority for cost reduction. Other means for recovering or using ethanol from 85 percent solution should be investigated.

Table 10.17 Drying of Ethanol Vapors

Agent	Inlet Vapor, Percent EtOH	Outlet Vapor, Percent EtOH
Cornstarch	73	99
Sucrose	72.5	90.7
Corn (cracked)	77	97.7
Cellulose	84.8	99.8
Corn residue	85.2	92
Sodium hydroxide	80.7	97.6
Calcium sulfate	90.1	98

Recovery of Several Products

The components of fusel oil can have sufficient value as solvents and chemical intermediates to justify their separation from ethanol. In batch distillation with a good fractionation column, products can be collected sequentially in order of their volatilities. Continuous distillation has to have each component going somewhere, and a fractionating column could deliver them from the top or the bottom. Often the minor components can be removed in a bleed from an intermediate tray of the column. This bleed is fractionated in another column that yields purified minor component plus material to be fed back to the original column. A column is needed for each of the minor components (Abrams, 1975).

Evaporation

In large-scale processes for fuels it is highly likely that recycle of aqueous streams will be practiced. As with any recycle system, some wastage is essential to prevent accumulation of toxic substances. This wastage will probably consist of a fraction of the still bottoms; it may have value for feeding animals, or there may be vitamins or other biochemicals that can be recovered economically. At some point in the scheme for handling wastes, concentration may be needed. If no other waste treatment is practical, wastes can be evaporated to dryness and the solids disposed of by burial or incineration. The large volumes in a fuels-from-biomass factory make total evaporation unlikely. Purification with ion exchange or carbon adsorption to remove toxic substances has a good chance of bringing spent streams to a quality fit for irrigation of crops.

Evaporation, like distillation, loses a little heat to the surroundings, but the main loss is degradation of heat by accepting steam or high-temperature process streams and producing lower-temperature spent streams. Economics depends on the ability to use further the heat in these spent streams, but driving forces are low; thus equipment becomes prohibitively large. Exact calculations of costs requires analysis of the entire system. In general, evaporation is expensive and dilute solutions should be avoided.

Membrane Concentration

Reverse osmosis uses pressure to force solvent molecules through a membrane while solute molecules are retained. The physical pressure must be greater than the osmotic pressure of the solution. As the solution becomes more concentrated, its osmotic pressure increases, and the pressure required to drive the process may become impractical as pumping and pressurized systems become too costly. Nevertheless, RO has a great potential advantage over evaporation because the energy needed to cause a phase change from liquid to vapor is not required. Candy companies have changed from evaporation to RO for concentration of sugar solutions; thus fermentation factories will probably also find RO superior for achieving concentration. To remove 1000 gal (3780 l) of water, evaporation costs about $15, and a membrane process can cost under $2.

Alternatives to Distillation

Gaudy, Goswami, et al. (1977) have studied stripping of volatile organic compounds from biological waste treatment fractions. At an aeration rate of 4 volumes per volume per minute, less than 3 percent of the initial chemical oxygen demand (COD) in a fraction was found in cold traps for the exiting gas. An aerobic process should not have very high concentrations of volatile organic compounds, and the reactor COD was probably carbohydrate, proteins, and other nonvolatile materials. It would be of interest to circulate carbon dioxide through anaerobic waste treatment media where volatile organics may be abundant. Nevertheless, organic acids that make up most of the volatile matter have low vapor pressures. To obtain free acids in the first place, acidification with mineral acid may be necessary. Reagent costs and costs of chilling the carrier gas make it very unlikely that stripping could be practical.

Extraction of fermentation products with water-immiscible solvents is discussed in Chapter 8 as a means for continuous removal to avoid feedback inhibition of production. In that application, the solvents must not be toxic to the fermentation organisms. When solvent extraction is not integrated into the fermentation process, toxicity does not matter; thus many more solvents can be considered. Commercial applications of solvent extraction are very common, and this approach to recovery of various fermentation products deserves high priority. However, ethanol can probably be purified more economically by other means.

An extraction process using CO_2 liquified near its critical point is being developed at Arthur D. Little, Inc. (*Inside R&D*, May 21, 1980). Ethanol couples with the CO_2 and separates from the fermentation broth. Shifting the conditions slightly to get gaseous CO_2 yields ethanol with about 5 percent moisture. This process is of great interest because the claimed energy consumption is less than half of that for a highly efficient distillation system.

Small-Scale Bioconversion

Large-scale processes make a great deal of sense when the feedstock is localized. In other words, direct connection to a pipeline or proximity to a depot for crude oil allows a refinery easy access to relatively cheap raw material, and valuable products are shipped to markets that may be quite distant. Economics favor constructing a very large refinery to reduce proportionate costs of labor, management, and auxiliary equipment and services.

Biomass is by its nature distributed over wide areas; thus it would be costly to transport to a few central plants. A smaller factory would have much more attractive collection costs. Compare, for example, one factory in the center of a large square with factories in each of nine equally small squares comprising the large square. The ratio of hauling distances would be 354/118. For a given tract, the optimum number and location of factories can be estimated from hauling costs, factory construction costs, operating costs, and other restraints such as geographic barriers.

Transportation costs can add significantly to a system for biomass fuels. Adler, Blakey, et al. (1978) have analyzed direct and indirect costs for transporting wood ships to a large wood-fired power plant. The State of Vermont was selected as an example because wood is already an important fuel there because coal is scarce. Some interesting comments are:

- Federal gross weight limit for a five-axle trailer is 73,280 lb. Some states have different limits; a special permit may be sold to allow a higher limit.
- At 1977 prices, new tractors cost about $40,000, and trailers cost $11,000; used equipment is a questionable bargain as higher maintenance costs are encountered.
- Based on the experiences of paper companies, pulp mills, and trucking firms, the cost of hauling per ton-mile is 4 to 6¢.
- Rail transportation can be comparable to trucking when negotiated for large contracts, but rates for occasional jobs can be three times higher than trucking. Rail costs are comparatively better for longer hauls.
- Rail spurs are expensive and inflexible as harvesting shifts to a different site. Trucks would probably bring chips to the railroad.

A 50-MW power plant collecting wood within a 30-mi (48-km) radius would require 105 truckloads each day, each with a 30-ton payload. On the basis of $3\frac{1}{3}$ trips per day, 32 trucks would be required plus three more to substitute during maintenance periods. The logistics of $\frac{1}{3}$ of a trip were not clear. In any event, hauling and storing biomass can be significantly costly and can pose serious engineering problems.

Of course, hauling is not a simple function of area and distance. Topography, roads, and fuel costs are important, but labor may be the dominant cost. Use of trucks should be coordinated with reasonable work shifts. With the single factory and long hauling distances, it may be difficult to get trips to fit into the work shift in even multiples. If the crews are at their home base with insufficient time to make a round trip before the end of the shift, there will be idle time if they wait for the quitting whistle or overtime if they make another trip. Small plants with short runs will have easier scheduling. In fact, small plants may have good success with drag lines, conveyor belts, or mixed means of material handling that would be impractical for a large central plant because of large investment and geographic barriers.

Let us consider what are the bad features of decentralized biomass factories. The economics of scale level off with increasing size, and a small plant that is not too small may not be at a significant disadvantage in operating cost. Except for very small plants, shipment of products should cost just about the same for any plant that has tank truck or railroad tank car quantities for sale. Labor costs could be high for a small plant, because it doesn't require any more people to operate a large fermenter than a small fermenter. However, it is quite likely that the factories will use fairly standard equipment so that a large plant has much the same fermenters, tanks, filters, and the like as a small plant, but more of each. In this case the labor cost per unit output will not be strongly dependent on size.

Small plants could be at a severe disadvantage for sales and marketing, except that there isn't much to advertise about alcohol or other basic chemicals. A few dependable customers such as a trucking company, a bus fleet, or a chain of filling stations could obviate the need for more than simple bookkeeping by a small company. It is very difficult to find any serious flaws in the concept of small factories, especially if the processes don't require highly skilled labor and much laboratory support. Brazil is having good success with small alcohol factories, and the United States may do best with a small-business approach. The first investments we see may be small plants for alcohol springing up in the agricultural areas where residual biomass from corn or other crops is readily available and local demand for fuel is sizable and dependable.

Additional Reading

For additional reading see Ashare and Wilson (1979); David, Hammaker, et al. (1978a); Mix, Dweck, et al. (1978); Paul (1979); and Report of the Alcohol Fuels Policy Review (1979).

11

Environmental
Impacts

Introduction

Any activity undertaken on the scale necessary to contribute significantly to U.S.
energy needs has the potential for profound upsets of climates, ecological balances, and other factors in the environment. There would be major relocations of
water for irrigation of biomass grown on land, or redistribution of subsurface
waters for upwelling to feed marine plants. These shifts would influence evaporation, transpiration from plants, cloud formation, fogs, and rainfall. Detailed discussion of such impacts is beyond the scope of this book, and no means are yet
available for accurate prediction, anyway. Nevertheless, it seems appropriate to
alert people to the potential hazards and to encourage addressing impacts early
in R&D programs.

Ocean Systems

Massive support structures and the associated plant growth for a species such as
kelp would create drag to impede ocean currents. Waves generated by wind
would work against biomass, thus producing damping. The world's climates are
profoundly affected by the Gulf Stream, the Japan current, and many other cir-
culating elements. With impediments to flow in both horizontal and vertical di-
rections plus surface temperature alterations from absorption of sunlight, metabolic activities, and upwelling of cold, deep water, it is reasonable to expect
climatic changes. If moist air passes over cold upwelled waters, fog formation
could impair illumination of the plants, thus destroying productivity.

Hruby (Dynatech 1978a) mentions release of heavy metal ions as support
structures corrode and formation of toxic organic molecules from polymers used
in cables and nets. Although there are unknown dangers of this type, most people
would assess the risk as being very slight. Toxins produced by the biomass crop
or its contaminants or epiphytes are potentially more serious, but it is unlikely
that concentrations would be high or that chemical stability would allow persist-

ence while the toxins were transported from distant growth areas to human populations.

A serious objection to oceanic biomass farms is possible interference with ships. In an extreme case, a ship caught in a storm might lose directional control because it was fouled by a biomass net. Sinking of an oil tanker and release of its cargo would not only ruin the biomass, but would present a hazard to marine life and to beaches. The cost of ensuring against or controlling such accidents could add very significantly to initial investment and operating costs.

Impacts of biomass farms on fishing could be negative, but it is much more likely that fishing would benefit. The presence of biomass, upwelled nutrients, and extra plankton should provide food and refuge for many species. Unless these species eat the main crop, mixing and excretion by these communities should be a positive effect, whereas obstruction of penetration of sunlight is a drawback.

Hruby discusses publications of others on the effects of cloud banks. These would reflect the sun's rays and would thus lower the earth's temperature. Although the effect would be small, it would be cumulative from year to year.

Animal biomass would be brought to the surface with upwelled water and would probably die. There is the possibility of population shifts and ecological upsets because of this constant killing of species from a given depth.

Land-Based Aquatic Farms

Competition of biomass farms with other requirements for land, water, and materials will cause serious physical and economic upsets. If a high level of recycle of water and nutrients can be achieved, the biomass farms may not be serious polluters. However, the farms will use such enormous amounts of water and conversion factories will be so large that even 99 percent recycle will leave 1 percent waste streams with very large amounts of organic matter or other potential pollutants. A goal of 100 percent recycle of water might seem reasonable for a crop grown on land, but an aquatic crop cannot be maintained on a pond or impoundment without a bleed stream or a purification step. Considering evaporation alone, salt would eventually build up to intolerable levels.

Ponds

Consider the environmental impacts of enormous algae ponds. Whether of fresh or salt water, these impoundments can provide sanctuary for fantastic population of birds. Fish might not be present because of low dissolved oxygen at night, but algae and epiphites would be eaten by some avian species. In turn, bird excreta would benefit the ponds by supplementing nitrogen. Fish grasped by birds in other waters could be dropped into ponds; thus organisms attached to or handled by birds would be inoculated. Special structures for growing aquatic plants above the pond surface might also be very attractive roosts for birds. Hordes of birds could be a nuisance to people living near energy farms.

Carbon Dioxide and Climate

The combustion of fossil fuels pours carbon dioxide into the atmosphere at an alarming rate. Although volcanoes, forest fires, controlled burning of wood, and industrial processing of mineral carbonates make a contribution, these sources are dwarfed by the CO_2 produced from petroleum, coal, and natural gas. Within a few centuries, the concentrated organic carbon stored in sedimentary rock throughout the ages is being converted to carbon dioxide. Baes, Goeller, et al. (1977) have expounded on the problems and the possible climatic effects. Since the start of the industrial revolution, the increase in atmospheric CO_2 has been about 12 percent, but the rate has been logarithmic. Before it peaks in the next century, the CO_2 concentration may double. While CO_2 is transparent to the wavelengths of incoming solar radiation, it absorbs part of the IR emissions sent by earth to space. This "greenhouse effect" should increase our average temperature, but the magnitude is highly uncertain.

Atmospheric concentrations of CO_2 are seasonal. Release is greatest when the weather is cold, and this coincides with slowest uptake by photosynthesis. A current average is 700 Gton (Gton represents a gigaton, or 10^9 tons) total carbon in the atmosphere. Estimates of total biomass on land are about 1800 Gtons of carbon, and this is much less than the carbon stored in the deep oceans. Substantial portions from each of these pools are circulated annually in the carbon cycle, and the exchange rate of photosynthesis is affected by atmospheric concentration. The increase in biomass due to CO_2 stimulation seems to be small, however.

Baes, Goeller, et al. (1977) claim that if humans could increase the amount of living biomass by 1 percent per year, the rate of carbon dioxide released by combustion would be more than counterbalanced. Unfortunately, human activities reduce biomass by cutting forests faster than they can regrow; by paving over land to create highways; and by building cities, homes, and factories.

Oceans are great reservoirs of CO_2 and carbonates at a total concentration of about 0.002 moles/l to give 39,000 Gtons of carbon. Dead organic matter in the oceans is about 1650 Gtons of carbon. Only about 1 Gton of living carbon is found in the oceans, mostly as plankton near the surface. Solid mineral particulate carbon, predominantly precipitated calcium carbonate, circulates well enough to be considered available; it amounts to about 400 Gtons of carbon.

The ultimate capacity of the oceans to deal with excess CO_2 in the atmosphere is far more than adequate. The reaction

$$CaCO_3 + CO_2 + H_2O = Ca^{+2} + 2\ HCO_3^-$$

could capture CO_2 until its partial pressure was very small. This mechanism may be much too sluggish to cope with rapid burning of fossil fuels, which may be 5600 Gtons in recoverable reserves. If oil shales become a commercial source,

7300 Gtons of more carbon are added. It seems highly likely that CO_2 will be a factor in climatic change because the oceans, our main hope for absorbing excess CO_2, cannot keep up.

The "greenhouse effect" may be coming upon us at a particularly bad time. It is 10,000 yr since the last glacial period, and a general warming trend is continuing. If natural warming is compounded by human activities that release CO_2, there may be pronounced melting of snow and ice. This would mean less reflection of solar rays and even more warming. Ocean levels could rise, with disastrous consequences to coastal cities. The effects of warming on cloud cover and on respiration of land or aquatic plants are not predictable. A warmer climate might be beneficial with longer growing seasons and lower fuel demands during mild winters. Rapid changes in climate could be very upsetting to plant species that have become adapted over many years. Even a temporary decline in fitness or productivity while plants are adapting could be serious. Water balances could be upset by greater evaporation due to increased temperature. If the levels of major freshwater bodies such as the Great Lakes are lowered, navigation, sewage disposal, and water supply would be adversely impacted.

Broecker, Takahashi, et al. (1979) have analyzed claims that cutting and burning of forests is a major source of carbon dioxide in the atmosphere. They rejected the hypothesis because CO_2 uptake by regrowth could greatly reduce the net production, and, furthermore, combustion of fossil fuels appears to be much more significant. The effects of oceanic uptake and models for CO_2 transport and reactions with minerals were reviewed.

Garrels, Lerman, et al. (1976) have modeled the cycles in nature of O_2 and CO_2 with emphasis on reactions with sedimentary rocks. Their results are in general agreement with various statements about rise in CO_2 and decline of oxygen concentrations in the atmosphere, but their time scales are greatly extended. For example, if human activities should cause a sudden and permanent cessation of life in the oceans, millions of years would pass before oxygen concentrations in the atmosphere would not support human life.

The U.S. Department of Energy maintains an advisory committee on the effects of CO_2 in the atmosphere. Lepkowski (1977) has reported the opinions of various individuals and committee members; there is much concern that the data are unreliable. The National Oceanic and Atmospheric Administration found a 5 percent increase in CO_2 from 314 to 330 ppm (parts per million) between 1958 and 1976 in Hawaii and similar trends at four other sites. A complicating factor is a warming trend in the oceans that may be consistent with past cycles for our planet. This warming releases CO_2 thus it is difficult to decide how much comes from the oceans and how much is produced by combustion.

A mysterious oscillation of temperatures in the southern Pacific Ocean affects CO_2 concentrations. Hanson (1977) reports a 3 to 4° C rise extending over thousands of miles and lasting for 6 or 7 years. Great amounts of CO_2 are released to the atmosphere.

Switching from fossil fuels to biomass would emphasize the carbon cycle at the photosynthesis stage more than at combustion. To produce the same number of BTUs, more CO_2 might be produced from biomass because its heating value is lowered by water content. Greater productivity of biomass is the cornerstone of the fuels-from-biomass program, and at some point there would be a balance between production and consumption. Other solar energy technologies will avoid the CO_2 problem entirely by converting radiation directly to heat, electricity, or steam. Nevertheless, some convenient fuels are essential to transportation and other special applications.

In summary, biomass should impact favorably on the CO_2 problem and buy time for equilibration with the oceans to lower CO_2 in the atmosphere. Controlled recycle of CO_2 from bioconversion to growing areas may halt atmospheric buildup.

Legal Considerations

The laws that apply to biomass or to products from biomass present a very complicated tangle of interdependencies and conflicts. Schwab (1979) has briefly summarized the present situation. Authorization for R&D comes from the Nonnuclear Energy Research and Development Act of 1974 and the Solar Energy Research, Development, and Demonstration Act of 1974. The U.S. Department of Energy administers these programs. The mandate for participation by the U.S. Department of Agriculture is part of the Agricultural Research, Extension, and Teaching Act of 1977. The agencies have been instructed to cooperate, but some confusion exists.

Congressional apprehension that new technology might reach fruition slowly had led to legislation for loan guarantees and for tax incentives that encourage use of biomass fuels. There is great activity in the private sector as investors plan ways to exploit these new opportunities. Several states have provided additional incentives for commercialization of fuels from biomass. It is highly likely that new ventures will become sufficiently profitable that governmental incentives can be phased out after a few years.

The federal government owns about 20 percent of the commercial forests in the United States. Except for some portions under control of the Bureau of Land Management, most of this resource is managed by the U.S. Forest Service. There are rules and regulations that must be reconciled with biomass programs. Clear-cutting, the cutting of all woody material, is a feature of most schemes for biomass fuels. Although the U.S. Forest Service allows clear-cutting, the courts have decided in favor of environmental groups opposed to this practice. The basis was a provision in the Forest Service Organic Act of 1897. It appears that opponents of biomass fuels could create long delays by bringing suits.

Converting animal manures to useful fuels would seem to be desirable from any point of view. However, protein from the digester is used as cattle feed which is more valuable than the methane. Legal considerations for animal feeding relate to land use, regulation of water quality, and nuisance actions. Feedlot operations must consider the Federal Water Pollution Control Act Amendment of

1972 as well as various state regulations. The burden might be on the biomass people to prove that their actions are beneficial and cause no harm to a concerned party. It is conceivable that manure piles would be acceptable to a regulatory agency, but wastes from digesting this same manure could be disposed of only by a costly method.

Municipal solid wastes are the subject of many federal and state laws. The Resource Recovery Act of 1970 encourages demonstration, construction, and application of solid waste management. Recovery of materials and energy are stated goals. No major legal hurdles to use of municipal wastes are evident. The supply of waste could be uncertain. Very often an urban area has city, town, and county jurisdictions, and some areas overlap state boundaries. A biomass–energy facility might not be able to contract with enough of the jurisdictions to ensure sufficient supplies of wastes. Some states prohibit importation of wastes, and this could hamper efforts to establish regional facilities for biomass fuels. Processed biomass such as densified waste could be reclassified to bypass regulations against importing wastes.

Gas from biomass is already admitted to the pipeline system on a very small scale. The Natural Gas Pipeline Company petitioned the Federal Power Commission to exempt gas from biomass from the price regulations for natural gas. The request was denied. The gases comingle in the pipeline and storage systems and are transplanted interstate. It is unlikely that new laws will distinguish between gases because of their sources.

Damage to Drinking Water

Soil erosion was discussed in Chapter 2 with regard to growing of biomass. Increases in erosion deprive future generations of topsoil reserves and create economic hardships. However, there are no serious hazards to health because the runoff from agricultural lands to water supplies has been dealt with in the design of treatment plants for drinking water. Pesticides and herbicides that are dangerous to humans have been controlled or banned, and biomass for energy should just be an extension of present agricultural practices.

The main water-pollution potential of refining of biomass to fuels is waste disposal. Concentrated organic wastes cannot be allowed in lakes or streams because microbial decay consumes oxygen and may result in septic, anaerobic conditions that create bad tastes and odors and suffocate fish and other desirable species. The means of treating such organic wastes to reduce their concentration are well known, but the capital and operating costs can be incompatible with economic goals for cheap fuels. In other words, treatment of organic wastes is likely to be too expensive if by conventional practice.

It is unfair to consider biomass wastes as equivalent to municipal wastes that contain pathogenic organisms that cause human diseases. Wastes from biomass processing could have toxic materials, but pathogenic organisms will be absent except when manures are the feedstock. The thermochemical processes and most of the bioconversion processes establish conditions where pathogenic organisms are killed. This means that waste treatment need deal only with problems of toxicity and disposal; contagious diseases can be disregarded. Inorganic chemicals coming from the factory are likely to be simple acids, bases, or salts from neutralization. These degrade the quality of drinking water by contributing to hardness but are not health hazards at the low concentrations expected.

When biomass factories are located near growing areas, it makes a great deal of sense to consider waste waters for irrigation. As sanitary wastes are sometimes placed on agricultural lands, there is little danger in applying biomass wastes. The additional expense of using ammonia, nitric acid, or phosphoric acid instead of cheaper reagents in the factory would be repaid partially by their fertilizer value. Organics in the waste would contribute nutrients to the plants. Toxic materials would tend to be adsorbed on soil. Water entering the groundwater table would be relatively pure, and much could be recycled to the factory by a system of wells. Transport to lakes and streams could be negligible. Of course, it would be poor practice to irrigate steeply sloping fields adjacent to a water course. It seems that good design can ensure safe disposal of aqueous wastes.

Air Pollution

As with water pollution, there is a well-established technology for air-pollution control. Again, the costs may not be compatible with cheap fuels from biomass. The biomass problems tend to be less than those for coal because of lower sulfur content, but biomass tends to have two or three times the amount of low-density particulate emissions when burned. Bioconversion will have almost no air-pollution problems, except for those from the power plant; thermochemical conversion will have pollution similar to that for synfuels from coal. Regulations and specifications of government agencies must be met. There will be no health hazards, but the price tag may be excessive.

Other Factors

Noise should be no problem because the factories are relatively quiet compared to other industries. Biomass refining will usually be quite remote from population centers, anyway.

Workers will not face any unusual dangers. Empty fermentation tanks may be unsafe because carbon dioxide is the main gas present with insufficient oxygen for life. Thermochemical conversion will have heat and pressure that have potential for explosion, but no more than occur in chemical factories. A good program for safety should reduce risks to a fully acceptable level.

Conclusion

Energy from biomass has generally favorable or acceptable environmental impacts. However, on a global scale the effects on climate are unpredictable and have potential for harm. The economic consequences of new fuels, competition of fuels and chemicals with food and fiber, new jobs as old jobs disappear, and shifting life-styles will benefit some people and devastate others. Free-market capitalism with the government enforcing fair practices and keeping the losers from starving can lead to speedy transition. If legislators think that a major industrial change can be managed by government agencies instead of by free-market forces, the damage to our economy can be grave.

Additional Reading

For additional reading see DiNovo, Ballantyne, et al. (1978); and Dunwoody, Takach, et al. (1980).

12

Conclusions

Perspective

Each chapter has presented pieces of the biomass puzzle, and several patterns have emerged. The strongest message is that fuels and chemicals can be made from biomass right now by a wide variety of methods, some of which are practical. Biomass technology has progressed rapidly, even in the brief period during which this book was written. After many years when fossil fuels were so cheap that no incentive existed for developing biomass processes, there are critical shortages compounded by price manipulations such that new energy sources can and must be found.

It is fair to say that the oil, gas, and petrochemical industries are mature, with established, efficient processes. Certainly there can be improvements, new processes, and new products, but the margin for major breakthroughs is small. In contrast, biomass processing is in its early infancy. Practically every R&D group embarking on biomass research makes a significant improvement. As progress is made it becomes more difficult to sustain the rate, but biomass processing is far from the time of diminishing returns on investment in R&D. Furthermore, biological improvement works differently from chemical and engineering improvement. A chemical reaction with 80 percent yield can be improved, but not much. When recovery of a product reaches a good yield, further process improvement is less cost-effective. On the other hand, the limit for biological reactions is not known. Witness the improvements from a fraction of a milligram per liter to 30 or so grams per liter for commercial penicillin fermentation, a stepwise increase of about 10,000 fold over 35 years. More recently we see the 10-fold increase in oleoresins in pine trees by injection with paraquat. Who can predict how much oleoresins can be produced in trees or in tissue culture when the mechanisms for formation are better understood? Biological rates also offer great potential for advances. Some industrial fermentations take several days, but microorganisms in the laboratory can be induced to double their numbers in as little as 6 min. We may find that the limitations of biological processes will not be in the organisms themselves, but in our abilities to supply gases, nutrients, light, or mixing. The point is that pouring funds into biomass R&D is already paying dividends. Continuing success will lead to sophisticated processes certain to reshape the energy picture.

Crop Selection

The first biomass factories can and should use available feedstocks such as agricultural residues, wood wastes, and municipal solid wastes. As demand outstrips these supplies, there will have to be energy crops. There can be an intermediate phase where marginal lands or set-aside acreage are used for energy crops. A very rough timetable might show 5 to 15 yr of factories based on available residues and wastes followed by 10 to 30 yr of bringing in new lands before there is outright competition between energy crops, food crops, and fiber crops.

An alternative scenario uses dual-purpose crops. It is reasonable to seek an energy crop that has edible components or high-quality fiber in addition to portions to be converted to fuels. Dual-purpose crops can be developed in parallel to the factories using available wastes, and marginal lands will come into play differently. Instead of seeking cheap new lands because the products are cheap fuels with too narrow profit margins, the dual-purpose crops may have a valuable product to carry the costs for the cheap fuel product. Farmers may find these crops more profitable than conventional crops; thus there may be major impacts on food production. Hopefully, the dual-purpose crop could be something like a corn variant that sacrificed a little grain yield to get a lot of biomass for energy. However, there is a very likely chance that the dual crop could have a food protein component with which the public is unfamiliar. Competition for good land could drive up prices of such commodities as grain to where a revolution occurred in eating habits because beef or other conventional foods were unaffordable.

Energy feedstocks that seem most important are shown in Table 12.1. Each of these feedstocks has been discussed in detail in earlier chapters. Points that bear repeating are:

1.   Single-purpose crops must operate on very tight energy budgets because energy for planting, harvesting, and converting can be a high percentage of the energy in the products.

2.   Multipurpose crops can have a product such as food or fiber for which we expect to spend energy. By apportioning a fair share of costs to these high-value products, the energy products can have much more attractive energy balances and costs.

3.   Although the author has been highly critical of kelp and current work with small algae, aquatic plants are probably going to become the ultimate energy crops because productivity can be very high and because the oceans will be relatively unexploited when all the good croplands are in use.

As energy crops become widespread, there will be marked changes in agricultural practices. Farmers may grow trees as field crops and harvest them every 2 or 3 years. Crops will be tailored to the geography, climate, and soil; Minnesota is not likely to grow the crops favored by Louisiana. There will be new machinery and

Table 12.1   Promising Feedstocks

	Comments
Single-purpose materials	
Municipal solid waste	Best for immediate commercialization, but too small a resource
Trees	Good productivity, widely available, but difficult to convert
Corn stover	Available for immediate commercialization, insufficient for very large scale
Dual-purpose materials	
Sugarcane	Food sugar plus stalks for energy; very limited geographic range.
Sweet sorghum	Fermentable juices plus excellent fiber; wide geographic range.
Pine trees	Oleoresins plus wood; do not coppice.
Ill-defined materials	
Algae	High productivity; could supply food and energy
Marine plants	Main hope when all fertile land is in use

an emphasis on sheer density of growth because of the bulk of biomass required and the need to achieve high productivity to minimize the needed acreage. There may also be devices for fractionating biomass as it is harvested, and some processing may be done on the farm. New businesses and new jobs will be created.

Biochemistry and Biology

The United States is the world leader in medical research and in many areas of pure biochemistry and biology. Agricultural research is also very strong. An area of relative weakness is applied microbiology, where Japanese investigators outnumbers those in the United States by as much as 10:1. However, the United States has more than an adequate base for the R&D needed for commercialization of biomass. A problem exists because some of our brightest people have a bias against industry and shun practical projects. There have been many complaints that the Department of Energy has a very lopsided program for support of process development and construction with grossly inadequate funding of basic research. The kelp project would support this contention in that there is an expensive effort to develop support structures and upwelling pumps in the oceans with only a small project to learn how to grow kelp. In most cases, however, the investigators who propose basic research seek support money on faith that something good will result. Their arguments have been unconvincing because of the nebulous connection between their research and practical applications. We need better communication between the applied and the basic people to develop an organized, focused approach. The financial needs for laboratory research are small compared to those for pilot plants and demonstration factories, so basic research can expect generous support when their goals are better defined and explained well to funding agencies.

In addition to very practical work on plant breeding, microbial genetics, and agricultural practices, there can be wide scope for molecular genetics and genetic engineering. There is excellent research under way on cellulase enzymes, but this can be expanded to other organisms and to mixtures of enzymes from various sources. Biochemical research is needed on fermentation pathways, chemical inhibition or genetic blocking of undesired steps, and process optimization. New products should be studied by both biochemists and microbiologists. Obviously, there should be expanded roles for bioscientists, and this group is presently underutilized in the United States because the supply of trained graduates has exceeded the number of good jobs available.

Thermochemical Conversion

It offends one trained in biochemistry to see a wealth of interesting biochemicals lumped into an ill-defined category called "biomass" and treated together by a thermochemical process. Intuitively, there must be a more selective way to fractionate biomass and to convert it economically to valuable products. However, pyrolysis, liquefaction, and gasification are closer to commercialization than are the better bioconversion processes. Dry biomass such as solid wastes and wood residues can be handled quite well by thermochemical technology. Even simple combustion can lead to immediate contributions to U.S. energy needs. Nevertheless, coal seems superior to biomass for very large central stations for producing fuels thermochemically. Continued R&D on small units should be cost-effective, but large units using coal can lead the way if biomass can eventually follow. The author feels that bioconversion will soon overtake thermochemical conversion and will have better products from factories that require relatively small capital investment.

Bioconversion

Anaerobic digestion is the most attractive technology in terms of low investment, but the product is methane, a very simple and cheap product. It is apparent that methane will be made economically only from very special feedstocks such as manure. The kelp advocates are pushing methane, but kelp probably has much better uses for protein and chemicals. Support for research on methane can soon be phased out because the main research objectives have been nearly reached.

Anaerobic digestion to organic acids rather than methane has merit, and research should be continued. Economical recovery of the acids is very difficult, but membrane processes and solvent extraction are showing some success. Digester residue contains microorganisms with protein value and sludge rich in lignin of little value. Enormous quantities of sludge could be generated, and a better use than soil conditioning should be found.

Fermentation of biomass may be on the verge of becoming a great new industry. Much will hinge on making good use of the hemicellulose and lignin fractions. There is potential for using very low cost equipment, but development of special cultures and very clever engineering are needed.

Photochemical Reactions

Use of sunlight with either chemical or biological catalysts to split water presents a particularly intriguing R&D challenge. However, hydrogen is not valuable enough to allow for very much investment. There is very likely to be a long interval between demonstration of a feasible process and development of a practical process. In the opinion of many scientists, the time scale for commercialization of a process for biological production of hydrogen is many decades.

Environmental Impacts

Except for the competition for limited resources, the impacts of fuels from biomass should be highly positive. This does imply that making energy production more profitable than growing food could drive up prices of both. If the number of customers stays constant, the United States may be able to produce food and fiber and energy with available land, water, and fertilizer, but as world population surges upward, there may be severe shortages.

A fuels-from-biomass factory should not be an environmental hazard. Fundamental to pollution is the concept that concentrated point sources are more dangerous than distributed sources. Biomass factories will be large, but pollution will be relatively small and very widely dispersed. A well-designed bioconversion facility will create very little thermal pollution, almost no air pollution, and negligible water pollution if efficient recycle can be implemented. Thermochemical conversion will have some pollution problems, but not as severe as those for coal containing sulfur.

Commercial Acceptance of Biomass Processes

Even as fuels and chemicals from biomass have favorable economic estimates and have been demonstrated in pilot plants (probably at government expense), there may be prolonged delays before the industrial sector invests in this technology. There are lessons to be learned from coal-based chemicals in that some processes that seem attractive are not being adopted commercially. Escalating prices and the uncertainty of supplies of imported petroleum have shifted attention to coal as a source of chemicals. Frank and Leonard (1977) have pointed out that ammonia and methanol can be produced from coal at competitive prices, but plants are not being built. The ammonia technology is 20 years old, and coal gasifiers are already operating in several foreign countries. Even with coal resources located near large corn belt ammonia markets to provide excellent logistics, coal does not seem to be the choice feedstock for making ammonia. New plants are being planned—all based on natural gas.

Although coal-derived ammonia can compete economically with natural gas, the differences are not great. The deciding factor may well be the very high capital investment needed for coal gasification. The argument has been advanced that selecting natural gas over coal for an ammonia plant means that the investor considers the risks of feedstock supply of natural gas to be less than the capital risk associated with capital-intensive and less commercially proven coal-based technology. Other factors are reluctance to build the initial, very expensive plant, and the rumors that there may be financial assistance from the government for prototype plants for coal conversion. No company wants to make an early commitment if a new national energy policy changes the rules in the near future.

Oil companies may diversify into biomass and coal processes or may function as competitors. Domestic prices operate fairly rationally, but imported oil is valued by different criteria, including politics, national pride, and attitudes. The expansion of oil companies into the organic intermediates business has been selective and in areas where they have a good technological position such as with ethylene glycol, styrene, and terephthalic acid (study of First Boston Corporation of New York City as reported by *Chemical Engineering News,* November 14, 1977). The oil companies could price their chemical products from petrochemical raw material grossly differently from prices of chemicals based on other feedstocks.

McKelvey (1979) points out that only six feedstocks are used to make 90 percent of the organic chemicals in the United States. These are shown in Table 12.2. Gross sales of the products were $50 billion, and 400,000 workers were employed.

About a dozen chemicals are vulnerable to competitive pricing by oil compa-

Table 12.2   Organic Chemical Feedstocks in 1977

Compound	Amount consumed, Thousands of Tons
Synthesis gas	35
Ethylene	12.5
Propylene	6.5
Butadiene	1.5
Benzene	5.5
p-Xylene	1.5

nies that could transfer feedstocks from the refinery to the chemical plant at cost. The advantage could be from 10 to 50 percent. The OPEC nations could create havoc by temporarily lowering prices for oil.

Basic petrochemical production in OPEC countries will impact heavily on U.S. prices. Shipping of final products is no more difficult than shipping of raw materials, and profit margins can be much larger. One of the first units will be a 2000-ton/day plant for methanol being built in Saudi Arabia (*Chemical Engineering News,* January 9, 1978, p. 6). Most of the plants will be joint ventures. United States companies will be well represented and can provide sensible advice on pricing. The Saudi Arabian venture will become a large petrochemical complex; initial capitalization is $3 billion.

New plants are not really needed now because petrochemical companies had expanded just before the recent business slump. There is overcapacity at present, and a steady but not spectacular market growth should utilize this capacity while creating some demand for new construction. Income from basic petrochemicals is not keeping pace with productions costs (*Chemical Engineering News,* November 7, 1977, p. 8).

One view of the preceding picture is that a new technology with competitive pricing of its products and with low capital investment could be commercially successful in today's economy. Biomass processes that mimic those for conversion of coal will be capital intensive; thus private investment will very likely be as slow in coming as that for the coal processes. Bioconversion could be the winner if carried out with "low technology" and its characteristic low capital requirements. If located near biomass supplies and potential markets, bioconversion factories could be very attractive to groups of small investors.

Government control of patents is another sensitive issue. It seems unjust to the taxpayers to pay a government contractor to do research and then to give up patent rights. Nevertheless, a liberal patent policy is very much in the national interest because there must be incentives for private investment. If the government owns a patent and offers licenses to many parties, no group may wish to go first. Development of the markets and starting up the first plant for a new process are risky and expensive. If some companies can wait while others take the risks but can share easily in the profits, it may be difficult to find a pioneer. Chance for profit is the engine that drives the capitalistic system, and private enterprise is almost always far more efficient than government. The group whose efforts lead to a patent are likely to have the skills and enthusiasm necessary to achieve commercial fruition, so the federal agency should carefully consider granting exclusive rights. The taxpayers will often be well repaid from the jobs and revenues generated by the new ventures.

The regulatory environment can influence the decision to invest in new technologies. State Public Utility Commissions rule on applications for price increases. Usually, capital investment by a utility is good justification for a price increase, but there are recent cases where such requests were denied. There are some precedents for rulings prior to the capital investment, and these guarantees expedite spending on improvements and new ideas. Local pollution-control boards can also affect prices by demanding expensive control devices. A group of investors for a biomass conversion factory would be encouraged by knowing the ground rules for price regulations and pollution control early in their planning.

Economic and political considerations in some countries lead to successful commercial ventures that would founder elsewhere. In India molasses is used to make polyethylene on a scale of 15,000 tons/year (Davies, 1974). Ethanol factories are important as providers of employment, and small distilleries are scattered over the nation. Although alcohol sells for roughly 60 percent of the price of petroleum naphtha, it is difficult to buy and to transport all the alcohol required for a large polyethylene plant. The alcohol plants are owned by agricultural cooperatives subsidized by the government; thus the true production cost is not known. Because of the balance-of-payment problem, India has a strong incentive to substitute fermentation ethanol for petroleum whenever possible.

Global Economics

Petroleum prices do not obey the simple laws of supply and demand because domestic oil is regulated by the government and world prices are set by OPEC. Although there is some dissention within OPEC, the trend has been steep escalation of prices. As with any wise seller, OPEC tries to establish a price that will maximize profits. Unfortunately for the United States and other importers, lack of competition in OPEC allows them to set a high price.

If biomass becomes an economically viable feedstock for energy, there will be serious competition with petroleum. Instead of continued escalation of petroleum prices, OPEC could lower prices for a few years in hopes that biomass industries would founder. Perhaps present trends are keys to the future. Several of the OPEC nations with greatest petroleum reserves were until recently very underdeveloped in terms of the wealth, employment, prospects, and aspirations of their average citizens. Most of OPEC have an autocratic strong central government. With their new riches, these nations are showing concern for their inhabitants by building schools, hospitals, and factories. Some can well be accused of spending money indiscriminately. It is very likely that R&D on biomass processes can lower costs more rapidly than the cartels can respond, and rather than new biomass industries being bankrupted, there may be oil-producing nations thrown into depression. There are several examples of nations with rich mineral resources riding high for many years and being plunged into the economic doldrums when world demand for their ores slackened.

Another scenario would be blackmail by OPEC before biomass can become a significant factor in the marketplace. In desperation, the United States might sign long-term contracts for purchase of petroleum. The biomass factories would come on stream with clean fuels at attractive prices, but prospective buyers would be committed to petroleum for a period of years.

Although speculation is a poor basis for decisions, several reasonable recommendations can be made in view of possible international responses to the threat of competition by biomass:

1. Petroleum should be stockpiled to allow for time in which to react to sudden cessation of oil from OPEC. With domestic production and these reserves, the U.S. economy could survive by drastic rationing.

2. Biomass will have to be developed with federal funds or guarantees to be somewhat immune from destructive pricing practices of OPEC.

3. A fuel system based on biomass should be established quickly before OPEC has a financial stranglehold on the world and before they can plan and implement a counteroffensive.

An Action Plan for Biomass

Background

In 1972 some research activity on energy from biomass was supported by NSF through its RANN Program (research applied to national needs). A very modest program was maintained for several years, mostly on anaerobic digestion to produce methane. These projects and some of the NSF staff were transferred to the new ERDA (Energy Research and Development Administration; successor to Atomic Energy Commission) in 1975 and given a budget of about $2.5 million (ERDA became part of the Department of Energy in 1977. Support level growth is shown in Table 12.3. Academic research programs for the science and technology to support commercialization of biomass could thrive on $6 to $8 million/yr; thus support above this level will be used for growing crops, building experimental facilities, and demonstration plants. In addition to the biomass–energy systems program, the Department of Energy also has a Division of Conservation that supports biomass research. Special crops and residues are the mission of biomass–energy systems, and wastes are assigned to the Division of Conservation.

The enabling legislation for the Department of Energy does not require formal review of proposals, but some of the Department of Energy staff members prefer peer review and/or panels for rating proposals. Unfortunately, the total Department of Energy budget, in the billions of dollars, leads to "pork barrel" politics. Research facilities and demonstration plants bring construction money

Table 12.3   Department of
Energy Funding for Biomass
R&D

Year	Millions of Dollars
1975	0.6
1976	7.1
1977	12.7
1978	20.2
1979	52.0
1980	56.0

into a state and create new jobs. Any conscientious senator or congressperson wants these benefits for his or her state and does not hesitate to use his or her political clout. With the staff who handle the proposals trying to judge by merit and attempting to maintain their integrity and self-respect and, politicians trying to serve their constituents, Department of Energy administrators are caught in the middle. No one is really good or bad or completely right or wrong, but common, everyday politics seem awfully rough to scientists and engineers employed by the Department of Energy. Almost all of the really good proposals are approved, as are a high percentage of the pretty good proposals. Regrettably, some very bad proposals are funded because politicians can influence administrators.

The Solar Energy Research Institute (SERI) was established at Golden, Colorado with the mandates of managing portions of solar energy development, promoting commercialization, and becoming an outstanding laboratory for solar energy research. There were delays, and adequate laboratory space did not become available until 1979. The biomass group has some very talented people, and photosynthesis is a prime topic for research. As mentioned elsewhere, this area will not move from the laboratory to commercial fruition for many decades, and the research on biomass at SERI may not impress the general public. Much of the routine work of judging proposals, making awards, and managing contracts has been delegated to SERI by the Department of Energy.

Pilot Plants

Private industry has become more and more reluctant to build pilot plants because they are very expensive. Engineering advances have led to good confidence for moving directly from the laboratory bench to full production. Some mistakes are made in scaling up, but correcting them is usually less costly than would be building and operating a pilot plant.

Government agencies have a different philosophy than does private industry. The tendency is to move from the laboratory bench to a small experimental facility with pumps, pipes, and controls. There may be two more stages of scaleup to a demonstration plant that is roughly 10 to 25 percent of full size. As the Department of Energy adopts more of a NASA-crash effort approach, scaleup is attempted before economic answers have been reached on some of the intermediate steps. This is referred to in poker parlance as "betting on the come." Just as the poker player hopes for a miracle card or two, the Department of Energy contractor needs technological breakthroughs to occur during construction of the plant. Perhaps in the American compulsion for results, deliberate and logical progress is not an acceptable alternative to an inefficient but exciting (and expensive) crash effort.

Energy technologies are not as well developed as those of the petroleum or chemical process industries, and pilot plants have their place. However, judgment must be exercised in devoting resources to crucial problems and in making prudent use of good engineering when direct scaleup from the laboratory bench is possible.

Research Needs

It is an easy task to analyze costs of producing fuels from biomass using existing technology. For example, ethanol from a cheap sucrose source such as molasses has well-known economics. Thermochemical conversion has not been run commercially since the early 1930s, and economic estimates are much less exact. Certain steps, especially those thermochemical operations that require flow and metering of irregular suspensions of biomass at high temperature and pressure, definitely need testing on a larger scale before feasibility costs can be determined. There is a spectrum of relative priorities and of depth of knowledge required for various steps in biomass processing, and it would be fiscally irresponsible to give each project equal attention.

Just as the first step in preparing chicken soup is to find a chicken, the primary biomass task is to procure abundant biomass. If this initial feedstock is not cheap, fuels produced from it will not be economically attractive. After years of research that have led to the world's greatest agricultural system, it is hard to believe that the United States is not ready to grow biomass. In fact, much of our elegant, expensive agriculture is poorly suited to really cheap biomass. Economies in water, fertilizer, labor, and equipment are urgently required to produce biomass for energy.

Growing of abundant biomass will need very complicated research. There are many variables, such as plant strains, soil types, seed preparation, spacing, irrigation practices, weather, climate, and control of diseases or pests. Only one or two runs can be made per year because an entire growing season is needed. With trees, a run may take several years. Trying every permutation of the variables would take years and would be prohibitively expensive, but good statistical design allows high confidence levels in assessing effects. The judgments of experts can be invaluable in selecting the most promising avenues of research and in minimizing the number of uninspired experiments. The proverbial "green thumb" could be worth many millions of dollars if it means rapid identification of good ways of obtaining high biomass productivity.

Multipurpose crops further confuse the research picture. An edible component or valuable fiber should probably be the prime target whereas residue for conversion to fuel is subsidiary. However, market size could dictate agricultural goals if there were an enormous demand for the residue and less demand for the high-value component. It is not likely that some novel food will be developed quickly. Sweet sorghum as a fiber crop might be adopted rapidly, but residues from present-day crops such as corn are the most likely energy feedstocks.

Many aspects of biomass processing are good enough now because the cost sensitivity is minor. There may be worthwhile improvements in chipping or subdividing, in storage methods, and in hauling, but present methods are reasonably effective. Slurry pipelines could convey biomass inexpensively, but no large-scale research is needed, especially as it is not yet known what are the best feedstocks or their likely state of subdivision.

A high proportion of research expenditures should go to conversion. Direct burning and anaerobic digestion to methane deserve relatively low priorities because the former is well understood and the latter has a product that is too cheap. Both thermochemical conversion and fermentation have high priorities as very promising alternatives.

Costs

Priority ratings should and must change. For example, anaerobic digestion of cattle wastes has had a higher priority because the biomass was available in some locations with essentially zero cost. As feedlot demonstration plants are being built, diminishing returns can be expected from supporting research. Some improvements in process performance should be forthcoming, but more payoff should result from research in other areas. A wise manager must know when to cease work on yield improvement and shift his or her resources to projects with more potential.

Research costs can be discussed in abstract terms, but results depend on people. Genius, creativity, and industriousness lead to progress. Just as some authors turn out one great literary work after another while countless others produce very little of merit, some researchers string successes together while others labor in vain. Money is a poor index of potential for progress; unproductive people are a poor bargain at any price. Research must be distinguished from routine testing. For example, determination of the optimum spacing of a plant for biomass productivity is mostly testing, whereas economical nutrition of the plant is sophisticated research involving agronomy, biology, biochemistry, engineering, and other disciplines.

The biomass research areas have some general features that determine costs. Projects concerned with growing biomass have land-acquisition costs, purchase of expensive equipment, intensive labor for crop management and data collection, and a long time frame because several years of production are required to confirm the data. Transportation projects are relatively short term but may necessitate purchase of unusual, specialized large equipment. Conversion research at the bench is short term on some steps but prolonged on others. Scaleup of thermochemical steps is very expensive in terms of equipment, whereas bioconversion scaleup is relatively inexpensive because high pressure and temperature are not required. It would cost very little to rent a large fermenter for a few months to examine some problems.

Market analyses and impact assessment are pencil-and-paper exercises that are relatively inexpensive. Such work should continue so that new data and perspectives can be used to refine estimates. Although the best processes and crops have not been identified, it is important to prepare for a major national change such as energy farming.

Periods of Change

There is much to be learned from an analysis by Berg (1978) of the history of transition from wood to coal to oil as major fuel. He points out that price and availability were less important factors than technological innovation. Furthermore, large bursts in R&D efforts came in use of the new fuel while innovation withered for the old fuel. Coal first was better than wood in blast furnaces and later enabled the Bessemer steel processes. The higher temperatures possible with coal allowed Portland cement to be produced at a large scale. The price of coal itself was relatively unimportant to a new, continuous process for making glass. Coal can be pulverized for better control of combustion. Coal was not prized so much for its BTUs and easier handling, but for new and more productive processes. Conversion to coal enlisted the excitement and talents of some of the best scientists and engineers in the world, whereas practically no one attempted to find new processes using wood.

Oil also caught the imagination of talented people. The refining of petroleum to gasoline, illuminating oil, lubricants, and petrochemicals attracted so many good scientists and engineers that coal science entered a prolonged decline. Now that a return to coal is being forced on us, there are few academic scientists who have long experience with coal.

Use of coal for generation of electricity kept usage up despite the technological challenges from petroleum. However, this blunted the need for new uses and prolonged the decline in coal research.

Berg feels that we may be misled by thinking only in terms of dollars per million BTUs and should integrate the fuel with its use. New methods of fabrication, economies in operating furnaces, efficient use of lasers, and other investions may change the basic patterns of energy use. This is completely in accord with the hopes of a fuels from biomass program realizing that other energy sources are more suitable for power generation. Biomass needs new processes and new products that represent net energy savings far in excess of the heat produced by burning these products.

Initial Adventures

If biomass fuels were to take off immediately, there are only a few reliable and proven technologies. The most direct of these is simple combustion, and wood or other relatively dry biomass could be burned in conventional equipment if modifications were made. Only minor changes need be made if the fuel is coal with biomass added. Other thermochemical conversion processes are in the developmental stage and not ready for commercialization. A possible exception is gasification, because this is a well-known reaction between a carbonaceous material and water to produce synthesis gas, a mixture of CO and H_2. It is reasonable that biomass could substitute for coke or coal and thus copy existing designs.

Two types of bioconversion are ready for commercialization: anaerobic digestion and ethanol fermentation. Digestion works best with wet, disintegrated biomass; thus certain feedstocks are much more suitable than others. Manure, aquatic growth, or a biomass that is easily shredded and dispersed in water are good for anaerobic digestion. Ethanol fermentation as presently practiced uses sugar as the main substrate. This means that only a portion of very special biomass such as sugarcane can be used. Starchy plants must have the starch extracted and hydrolyzed to glucose that is a good ethanol substrate.

The point of this discussion is that the immediate opportunities for fuels from biomass are with selected biomass. Later developments such as inexpensive hydrolysis of cellulose to glucose will broaden the options and permit many different biomass feedstocks to be employed.

Closing Comments

No alternate energy source is as cost-effective as conservation. There must be an end to our energy gluttony, or both old and new energy sources will soon be overwhelmed by the increasing demand of a growing population. We must recognize the merits of arguments of Lovins (1977) and others that end uses be addressed, instead of charging ahead with gigantic power stations. Electricity and fuels are not goals, but means to desired ends. Better insulation for homes, more efficient appliances, energy-conscious heating and cooling systems, lighter automobiles, and a myriad of other conservation measures can greatly reduce our energy needs. When energy costs are set at replacement values, coal and oil and gas will seem very expensive, and renewable sources such as biomass will become bargains.

Conservation can be stimulated in many ways. Rationing of fuels or very high prices can force reduced consumption, but high gasoline prices do not seem to have cut its usage very much in the United States or other nations. Power rate structures could affect conservation. Usually, cost per unit of electricity goes down as the total units increase. However, some states have mandated low rates for certain small users. For example, electricity for irrigation in Texas costs less per unit than that sold to large factories. A constant rate or a rate that increased with use would put great economic pressure on large users, and this is where significant improvement can be made.

Conservation on a per capita basis will be futile if population continues to increase logarithmically. New energy sources can only buy time to stabilize the population growth. These solutions are far beyond the scope of this book, but we may note that temporary answers such as feeding of starving people today means more of their children to starve in the future. Whether we do anything or not, population will ultimately stabilize as further growth is limited by food supplies. Severe competition for land and water resources can favor fuels from biomass that use nonedible portions of food plants. In other words, the multiple-product concept becomes more important when demand drives costs so high that valuable uses must be found for each component of the crop. Logic may not triumph. Wars, desolation, the strong taking from the weak, dictatorships, and turmoil are mankind's past and future. Our generation and our children may prosper, but extrapolation of present trends into the future looks grim.

In the next few decades, biomass fuels and chemicals can supplant fossil fuels and petrochemicals. The costs are already competitive, and R&D have much promise for improvement. There will be new companies with new opportunities for investment. New jobs will be created, and suppliers to these companies will have major new customers. In keeping with U.S. traditions, the capitalistic system may reach its finest hours by using competition to implement new technology to satisfy an urgent national need. The biomass options will help to avoid the bleakest aspects of overpopulation and will become a major pillar in the foundations of our future civilization.

Appendix A

Some Definitions and Conversion Factors

Acre	Multiply by 0.405 to get hectares.
Aerobic	Life process conducted in the presence of oxygen.
Anaerobic	Life process conducted at very low oxygen concentration.
Aseptic	Free of contaminating organisms.
Autolysis	Breakdown of all or part of a cellular system by self-produced enzymes.
Bagasse	Residue of sugarcane after the juice has been extracted.
BTU	British thermal unit: heat to raise 1 lb of water by 1°F.
Buffer	A solution that resists changes in pH; salts of weak acids or weak bases react with additives to cushion shocks to the acid-base equilibrium.
Bushel	Multiply by 35.24 to get liters. (One bushel of corn weighs 56 lb.)
Coniferous	Relating to evergreen trees or shrubs. These plants produce cones, for example, pine trees.
Coppice	Growth of plants from stumps or shoots.
Corn stover	Residue after corn grain is collected.
Endothermic	Process that consumes heat energy.
Epiphyte	A plant that infests another plant.
Exothermic	Process that evolves heat energy.
Fluidized bed	Reactive media are suspended in the flow of a gas or liquid.
Gallon	Multiply by 0.1337 to get cubic feet; by 3.785 to get liters.
Grams per square centimeter	Multiply by 44.6 to get tons/acre; by 16,377 to get kg/ha.

Hectare	Multiply by 2.471 to get acres.
Methanogenic	Organisms that tend to produce methane.
Mutant	Organism produced by mutation; a genetic variant of the parent.
Oleoresins	Turpentine and resinous material produced by coniferous plants.
Oligosaccharide	An organic molecule composed of two or more sugar units.
Quad	Quadrillion BTU.
Ratoon	Usually refers to sugarcane or sweet sorghum or other plant that grows back from basal buds on the stem after cutting.
Saccharification	Process for obtaining sugars by hydrolysis of a polymer.
Single-cell protein	Microbial cells grown as food for humans or animals.
Symbiosis	The living together of two dissimilar organisms in close association.
Tons per acre	Multiply by 2.241 to get tons (metric) per hectare.

Appendix B

Summary of U.S. Department of Energy Fermentation Program

The U.S. Department of Energy program for fermentation of biomass adopted several working assumptions:

1. Cellulosic biomass is the ultimate substrate because of its low cost and wide availability.
2. Technology for fermentation of sugars to alcohol is well known and needs little R&D.
3. Hydrolysis of cellulose is the main bottleneck.

As research efforts mature and results are analyzed, there are some preeminent conclusions:

1. Concentration on cellulose and ignoring hemicellulose and lignin is uneconomic. All the biomass must be converted to products.
2. Acid hydrolysis must not be ruled out. Although enzymatic hydrolysis has many desirable features, acid or acid followed by enzymes may be more cost-effective. The Department of Energy program has been top-heavy on enzymatic research.
3. Hydrolysate may be more difficult to ferment than is a solution of the same sugars. Removal of inhibitors may be required.
4. Cellulose is more valuable for pulp, paper, or plywood uses so that the cost of by-product sugars from waste hemicellulose or excess cellulose would be very low for ethanol fermentation. See section on sweet sorghum, p. 44.
5. Ethanol recovery is expensive in terms of energy input. Major improvements are expected by using techniques presently under investigation (see Chapter 8).

320

A newsletter is issued periodically for the U.S. Department of Energy efforts on fermentation of biomass, and requests for copies should be sent to Branch Chief, Biomass Refining, SERI, Golden, Colorado.

Natick Laboratories

The U.S. Army has been interested in degradation of cellulose for many years because of problems with military clothing, tents, and the like. Reese, Mandels, and their co-workers found cultures that attack cellulose and isolated cellulase enzymes from many of them. Almost all the pioneering work on hydrolysis of cellulose was done at Natick, and a process was proposed for ethanol production of from cellulose. Current research aims at better titers of enzyme and reduced costs by continuous fermentation. There is a pilot plant for production of ethanol or cellulases, and computer control of the fermenters is under way. Earlier expensive grinding or milling of biomass to facilitate hydrolysis has been superseded by two-roll wet milling, which is less costly. The Natick program attacks the overall process with steps for pretreatment, enzyme production, saccharification, and fermentation.

University of California, Berkeley

Under Wilke's leadership there has been a comprehensive study of the engineering problems of biological conversion of biomass to fuels and chemicals. All the individual steps have been operated at bench scale, and the product of one step has been the feed for the next. A wide variety of biomass feedstocks has been analyzed and evaluated. Innovations such as vacuum fermentation for ethanol have been developed to the point where reliable economic estimates can be performed. New pretreatments and fermentations using sugars from hemicellulose have been devised. Strains of microorganisms from the Wilke group's work and those from other investigators are given careful process development to optimize titers of cellulase.

Massachusetts Institute of Technology

There have been many interesting innovations by this group. Remarkable progress has been made in developing a fermentation for acrylic acid. Initially the MIT group had a questionable spot that could be acrylic acid on a chromatogram. The process now gives yields of a few grams per liter with the potential for improvement to a commercially important concentration. Another significant development is direct production of ethanol from cellulose without the addition of enzymes for saccharification. Although ethanol concentration at a practical level has not yet been demonstrated, there is good alcohol tolerance by the organism, *Clostridium thermocellum*, and a very favorable ratio of ethanol to acetic acid. A second organism, *C. thermosaccharyliticum,* is added to ferment hemi-cellulose sugars to ethanol.

Other projects at MIT are development of the acetone–butanol fermentation and fermentation of lignin. Results with acetone–butanol are as good as the best reported in the old literature for this fermentation, but further progress is hindered by the strong inhibition by butanol. Mutation to find resistant strains has not been successful; perhaps butanol interferes with membrane action in a fundamental way not easily corrected by mutation. Acetone is relatively nontoxic. Direct fermentation of lignin is a new project based on organisms that break lignin down to small molecules that are the substrates for other members of the culture mix. In other words, no single organism seems capable of producing a valuable chemical directly from lignin, but mixed cultures may do so.

Rutgers University

By using clever means for accentuating detection on petri dishes, this group very quickly found superior strains for producing cellulases. They have also obtained good titers on media made from cheap ingredients. The result has been at least a 10-fold reduction in enzyme costs that previously were the major cost item in the overall process, and enzymes less repressed by accumulation of products have been developed.

Purdue University

The leadership of Tsao has produced several exciting advances. Demonstration of nearly complete hydrolysis of cellulose after solvent pretreatment was a great stimulus to all the contractors working on saccharification. The Purdue process is a contender for the best pretreatment, and any better pretreatment will probably trace its origin to the concepts promulgated by the Purdue group. Previous yields were only slightly over 50 percent of the theoretical amount of reducing sugar from cellulose; thus quantitative yield after pretreatment could cut product cost almost in half. An alternate product, 2,3-butanediol, can be produced from both five- and six-carbon sugars. Yields close to theoretical have been demonstrated.

The Purdue group also has projects on ethanol recovery, a packed fermenter with *Rhizopus* that performs similarly to tower fermentation, and work not sponsored by the Department of Energy. The State of Indiana has supplied funds considerably in excess of those from the federal government because Indiana plans to become the leader in commercializing fuels from biomass.

University of Pennsylvania–General Electric Company

There has been excellent research on the biochemistry of cellulase action. This complements the work of other investigators because different organisms have been used. Most emphasis in process development is given to using poplar wood chips because these trees have a wide geographic range and do not need prime soils. A different pretreatment, removal of lignin with alcoholic solvents, leads to excellent yield for cellulose hydrolysis. The sugars have been used for producing ethanol or acetone–butanol. By using a thermophilic organism at a higher fermentation temperature, there is a better chance of vacuum fermentation becoming practical.

General Electric Company

A novel idea for microbial utilization of lignin as a pretreatment did not work out because substances in wood were inhibitory to growth. Current research is on chemically assisted pretreatment by steam explosion and on direct fermentation of cellulose to ethanol.

University of Connecticut

There is a small project on using immobilized β-glucosidase to assist saccharification. Supplemental β-glucosidase prevents accumulation of cellobiose and speeds the rate of glucose formation.

Dynatech Corporation

This project is greatly different from the others in that the feedstocks are aquatic plants and the desired products are organic acids. Anaerobic digestion is performed with high loadings and conditions that suppress methane formation to favor acid production. The acids are partially purified and subjected to Kolbe electrolysis to form hydrocarbon fuels. Another novel feature is the use of a packed-bed microbial reactor to retain a very high density of active organisms.

Dartmouth University

A flow reactor with rapid heating and cooling allows precise determination of the reaction kinetics for acid hydrolysis. It appears that the upper limit for acid hydrolysis of untreated biomass is about 55 percent of the theoretical amount of reducing sugars. However, mild acid treatment can remove the hemicellulose fraction and pretreat the cellulose such that subsequent enzymatic hydrolysis gives yields of about 90 percent of theoretical.

Auburn University

Batch studies of acid hydrolysis complement the research at Dartmouth with a continuous reactor. Hydrolysis of hemicellulose is fairly selective at 150°C with 0.2 percent sulfuric acid for 90 min. The mixed sugar yield was 83 percent of the hemicellulose whereas glucose yield was 17 percent of the cellulose, and furfural from pentose decomposition was 8 percent. Solids after reaction are shown in Table B.1.

A very significant difference has been noted in that several cultures that fer-

**Table B.1   Major Products of Hydrolysis of
Southern Red Oak Hemicellulose**

Component	Percentage of Solids
Xylose	50
Glucose	26
Polysaccharide	10
Acetic acid	10
Furfural	3
Methanol	1

mented xylose nicely were unable to ferment wood hydrolysate where xylose was the predominant sugar. Fortunately, a few strains were found with good growth on wood hydrolysate.

Iotech Corporation

This Canadian company has a pilot plant for steam explosion. Characteristics of various feedstocks have been determined, and samples of treated materials are supplied to other investigators. The economics should benefit greatly with the use of lignin from the process as a high-value binder for plywood or chipwood boards.

Battelle–Columbus Laboratories

Whereas all the other work focuses on lignocellulosic biomass, this group is exploring sources of sugar juices. Sugarcane is a good feedstock, but it grows in few places in the United States. Corn provides starch, which can be an excellent feedstock, but corn grain has high value for food. If a new major use were developed for corn, demand could cause prices to skyrocket beyond the point where fuel fermentations were economic. The best choice at this point seems to be sweet sorghum, which grows well throughout almost all the United States. Processing to by-products based on the strong stalk fibers and fermentation of the juices are being investigated.

References

Abrams, H. J., *Octagon Papers No. 2,* University of Manchester, U.K., (1975), pp. 49–67.

Adams, J. E., Influence of mulches on runoff, erosion, and soil moisture depletion, *Soil Sci. Soc. Am. Proc.,* **30,** 110 (1966).

Adler, T. J., M. Blakey, and T. Meyer, The direct and indirect costs of transporting wood chips to supply a wood-fired power plant, NTIS, TID-28737 (1978).

Alexander, A. G., and C. Gonzales Molina, Production of sugarcane and tropical grasses as renewable energy sources, NTIS, CRO-5422 (1978).

Alich, J. A., F. Schooley, R. Ernest, R. Hamilton, B. Louks, K. Miller, T. Veblen, and J. Witwer, An evaluation of the use of agricultural residues as an energy feedstock—a ten site survey, Vol. 1, NTIS, TID-27904/1 (1977).

Anderson, E. V., Signs point to erosion of chemical trade, *Chem. Eng. News* (October 3, 1977), p. 11.

Anderson, E. V., Gasohol: energy mountain or molehill, *Chem. Eng. News* (July 31, 1978), pp. 8–15.

Anderson, L. L., and D. A. Tillman, *Fuels from Waste,* Academic, New York (1977).

Ashare, E., D. Augenstein, J. Young, R. Hossan, and G. Duret, Evaluation of systems for purification of fuel gas from anaerobic digestion, NTIS, COO-2991-44 (1978).

Ashare, E., M. Buivid, and E. Wilson, Feasibility study for anaerobic digestion of agricultural crop residues, NTIS, SERI/TR-8157-1 (1979).

Ashare, E. and E. H. Wilson, Analysis of digester design concepts, NTIS, COO-2991-42 (1979).

Atchison, J. E., Consultants, Inc., Preliminary evaluation of new process for separation of components of sugar cane, sweet sorghum and other plant stalks relative to its possible contribution toward increased fuel and energy production from these materials, NTIS, TID-28734 (1977).

Augenstein, D. C., D. Wise, and C. Cooney, Packed bed digestion of solid wastes, *Resour. Recovery Conserv.,* **2,** 257–262 (1976/77).

Baes, C. F., Jr., H. Goeller, J. Olson, and R. Rotty, Carbon dioxide and climate: the uncontrolled experiment, *Am. Sci.,* **65,** 310 (1977).

Bailey, M., T. M. Enari, and M. Linko, Eds., *Symposium on Enzymatic Hydrolysis of Cellulose,* Aulanko, Finnish National Fund for Research and Development (1975).

Balch, W. E., G. Fox, L. Magrum, C. Woese, and R. Wolfe, Methanogens: reevaluation of a unique biological group, *Microbiol. Rev.,* **43,** 260–296 (1979).

Ban, S., M. Glanser-Soljan, and M. Smailagié, Rapid biodegradation of calcium lignosulfonate by means of a mixed culture of microorganisms, *Biotechnol. Bioeng.,* **21,** 1917–1928 (1979).

Bassham, J. A., Increasing crop production through more controlled photosynthesis, *Science,* **197,** 630–636 (1977).

Bechtel National, Inc., Liquefaction project at Albany, Oregon, NTIS, SAN-1338-T1 (1979).

Beck, S.R., Fluidized bed gasification of cattle feedlot manure, NTIS, CONF-7806107-Pl, 269–286 (1978).

Beck, S. R., J. Halligan, C. Lin, R. Bartsch, W. Huffman, B. Landeene, R. Ravi, and P. Hillman,

Partial oxidation–pyrolysis of cattle feedlot manure in the SGFM reactor, NTIS, ALO-3379-1 (1977).

Bellamy, W. D., Role of thermophiles in cellulose recycling, *ASM News,* **45,** 326–331 (1979).

Benedict, H. M., and B. Inman, A review of current research on hydrocarbon production by plants, NTIS, SERI/TR33-129 (1979).

Benemann, J. R., B. Koopman, M. Murry, J. Weissman, D. Eisenberg, and W. Oswald, Species control in large-scale algal biomass production, U.S. Department of Energy, SAN7405-77/1 (1977a).

Benemann, J. R., P. Hallenbeck, J. Weissman, L. Kochian, P. Kostel, and W. Oswald, Solar energy conversion with hydrogen producing algae, UCB-SERL Report 78-2 (1977b).

Benemann, J. R., J. Weissman, B. Koopman, D. Eisenberg, R. Goebel, and W. Oswald, Large-scale freshwater microalgal biomass production for fuel and fertilizer, NTIS, SAN-0034-1 (1978a).

Benemann, J. R., J. Weissman, D. Eisenberg, B. Koopman, R. Goebel, P. Caskey, R. Thomson, and W. Oswald, An integrated system for the conversion of solar energy with sewage grown algae, NTIS, SAN-0034-T2 (1978b).

Benson, W. R., A. Allen, R. Athey, A. McElroy, M. Davis, and M. Bennett, Systems study of fuels from grains and grasses, Phase 1, Final report, NTIS, ALO/3729-1 (1978).

Berezin, I. V., Advances in the development of biocatalysis based on immobilized enzymes, *Proceedings of the Third Joint U.S.—U.S.S.R. Enzyme Engineering Seminar,* H. Bungay, Ed., NTIS PB-283328-T (1978), pp. 7–27.

Berezin, I. V., Informal seminar at Moscow State University (1979).

Berezin, I. V., Bioelectrocatalysis as a new phenomenon, *Proceedings of the Fourth Joint U.S.—U.S.S.R. Conference on Microbial Enzyme Reactions,* G. G. Guilbault, Ed., NTIS PB80-132913 (1979), pp. 369–384.

Berg, C. A., Process innovation and changes in industrial use, *Science,* **199,** 608–614 (1978).

Bernhardt, W., Ed., Proceedings: International Symposium on Alcohol Fuel Technology Methanol and Ethanol, NTIS, CONF-771175 (1977a).

Bernhardt, W., Methanol and ethanol for combustion engines and alternative power units, paper presented at Symposium on Alternate Fuels and Energy Sources for Road Transport in South Africa, Port Elizabeth, South Africa (1977b).

Berry, W. L., Jr., Operation of the biomass liquefaction facility, Albany, Oregon, NTIS, SERI/TP-33-285, (1979), pp. 105–108.

Binder, A. and T. Ghose, Adsorption of cellulose by *Trichoderma viride,* *Biotechnol. Bioeng.,* **20,** 1187–1199 (1978).

Bisaria, V. S. and T. Ghose, Sorption characteristics of cellulases on cellulosic substances, in T. K. Ghose, Ed., *Bioconversion of Cellulosic Substances into Energy Chemicals and Microbial Protein,* Indian Institute of Technology (IIT), New Delhi (1978), pp. 155–164.

Black, W., Marketing biomass, *CHEMTECH,* **8,** 690–698 (1978).

Blanken, J. G., The tower fermenter for lager production, *Brewers' Guardian* (October 1974), pp. 35–39.

Bliss, C. and D. Blake, Silviculture biomass farms V. Conversion processes and costs, NTIS, MTR-7347-5 (1977).

Blotkamp, P. J., M. Takagi, M. Pemberton, and G. Emert, Enzymatic hydrolysis of cellulose and simultaneous fermentation to alcohol, paper presented at AIChE National Meeting, Atlanta, Ga. (1978).

Brake, J., R. Townsend, and H. Silverman, Electrical energy from microorganisms, *Chem. Eng. Progr.,* **61,** 65–68 (1965).

Brenner, W., report in *Chem. Eng. News* (October 8, 1979), p. 19.

Brink, R. A., J. Densmore, and G. Hill, Soil deterioration and the growing world demand for food, *Science,* **197,** 625–630 (1977).

Broecker, W. S., T. Takahashi, H. Simpson, and T. H. Peng, Fate of fossil fuel carbon dioxide and the global carbon budget, *Science,* **206,** 409–418 (1979).

Brooks, R. E., W. Bellamy, and T. Su, Bioconversion of plant biomass to ethanol, NTIS, COO/4147-4 (1978).

Brown, J. C., report in *Chem. Eng. News* (September 12, 1977), p. 50.

Bryant, M. P., Hydrogen as regulator of anaerobic degradation, transcript of presentation at Hastings, Neb., October 9, 1977, in NTIS, COO-2991-28 (1978), pp. 85–105.

Bu'Lock, J. D., Acetone/butanol, butanediol, and other fermentations, *Octagon Papers No. 2,* University of Manchester, U.K. (1975), pp. 5–16.

Bungay, H. R., A low investment approach to alcohol fermentation, *Bioresourc. Dig.,* **2,** 27–31 (1980).

Burton, R. S., and R. Bailie, Fluidized bed pyrolysis of solid waste materials, *Combustion* (February 1974), pp. 13–18.

Calvin, M., Photosynthesis as a resource for energy and materials, *Am. Sci.,* **64,** 270–278 (1976).

Calvin, M., Green factories, Priestly Medal Address, *Chem. Eng. News* (March 20, 1978), pp. 30–36.

Campos-López, E., E. Neavez-Camacho, M. Ponce-Vélez, and J. Angulo-Sanchez, The rubber shrub, *CHEMTECH,* **9,** 50–57 (1979).

Carioca, J. O. B., J. Scares, and W. Thieman, Production of ethyl alcohol from babasu, *Biotechnol. Bioeng.,* **20,** 443–445 (1978).

Casida, L. E., *Industrial Microbiology,* Wiley, New York (1968).

Chambers, R. P., Y. Lee, T. McCaskey, T. Placek, B. Fitch, and L. Mathias, Liquid fuel and chemical production from cellulosic biomass—hemicellulose recovery and pentose utilization in a biomass processing complex, Quarterly progress report to U.S. Department of Energy (June 1979).

Chambers, R. S., R. Herendeen, J. Joyce, and P. Penner, Gasohol: does it or doesn't it produce positive net energy, *Science,* **206,** 789–795 (1979).

Chang, M. and G. Tsao, The effect of structure on reactivity of cellulose, paper presented at AIChE (American Institute of Chemical Engineers) national meeting, San Francisco (1979).

Chibata, I., paper presented at Enzyme Engineering Conference, Henniker, N. H. (1979).

Christian, S. D. and J. J. Zuckerman, Eds., *Energy and the Chemical Sciences,* Pergamon, Elmsford, N. Y. (1978).

Clausen, E. C., O. Sitton, and J. Gaddy, Biological production of methane from energy crops, *Biotechnol. Bioeng.,* **21,** 1209–1219 (1979).

Cochran, B. J. and R. Ricaud, The potential of producing and harvesting sugarcane and sweet sorghum as a renewable biomass energy resource, NTIS, ORO-5373-TI (1979).

Coffman, J. Q., Catalyzed steam gasification of biomass, NTIS, CONF-7806107-P1, (1978), pp. 287–300.

Cooney, C. L., and M. Makiguchi, An assessment of single cell protein from methanol-grown yeast, *Biotechnol. Bioeng. Symp.,* **7,** 65–76 (1977).

CSO International, Inc., Cost analysis of microalgae biomass systems, Final report: Design assumptions and engineering plans for 100-square mile system, NTIS, HCP/T1605-01, UC-61 (1978).

Cysewski, G. R. and C. Wilke, Rapid ethanol fermentations using vacuum and cell recycle, *Biotechnol. Bioeng.,* **19,** 1125–1143 (1977).

Danziger, R., Distillation columns with vapor recompression, *Chem. Eng. Progr.,* **75,** 58–64 (1979).

da Silva, J. G., G. Serra, J. Moreira, J. Conclaves, and J. Goldemberg, Energy balances for ethyl alcohol production from crops, *Science,* **201,** 903–906 (1978).

Dauve, J., and S. Flaim, Agricultural crop residue collection costs, NTIS, SERI/RR-353-345 (1979).

David, M. L., G. Hammaker, R. Buzenberg, and J. Wagner, Gasohol: economic feasibility study, report prepared for Energy Research and Development Center, Lincoln, Neb. (July 1978a).

David, M. L., G. S. Hammaker, R. J. Buzenberg, and J. P. Wagner, Gasohol: Economic feasibility study, NTIS, SAN-1681-T1 (1978b).

Davies, D., Production of ethylene from molasses, *Octagon Papers No. 1,* University of Manchester, U.K. (1974), p. 63.

DeBell, D. S., and J. C. Harms, Identification of cost factors associated with intensive culture of forest crops, *Iowa State Journ. of Res.,* **50,** 295–300 (1976).

Del Rosario, E. J., K. Lee, and P. Rogers, Kinetics of alcohol fermentation at high yeast levels, *Biotechnol. Bioeng.,* **21,** 1477–1482 (1979).

De Menezes, T. J. B., T. Arakaki, R. DeLamo, and A. Sales, Fungal cellulases as an aid for the saccharification of cassava, *Biotechnol. Bioeng.,* **20,** 555–565 (1978).

De Renzo, D. J., Energy from bioconversion of waste materials, Noyes Data Corporation, Park Ridge, N. J., 1977.

Desrosiers, R. E., Process designs and cost estimates for a medium BTU gasification plant using a wood feedstock, NTIS, SERI/TR-33-151 (1979).

Diebold, J. P. and G. Smith, Thermochemical conversion of biomass to gasoline, NTIS, SERI/TP-35-285, (1979), pp. 139–146.

DiNovo, S. T., W. Ballantyne, L. Curran, W. Baytes, K. Duke, B. Bornaby, M. Matthews, R. Ewing, and B. Vigan, Preliminary environmental assessment of biomass conversion to synthetic fuels, NTIS EPA-600/7-78-204 (1978).

Drew, S. W., Chemical feedstocks and fuels from lignin, paper presented at AIChE national meeting, Atlanta, Ga. (1978).

Dubinsky, Z., S. Aaronson, and T. Berner, Potential of large-scale biomass and liquid production in arid lands, *Biotechnol. Bioengr. Symp. 8,* Biotechnology in Energy Production and Conservation, 51–68 (1978).

Dunning, J. W., E. Lathrop, *Ind. Eng. Chem.,* **37,** 24 (1945).

Dunwoody, J. E., H. Takach, C. S. Kelly, R. Opalanko, C. High, and A. Fege, Wood combustion systems: Status of environmental concerns, NTIS, DOE/EV-0064 (1980).

Dynatech R/D Company, Cost analysis of aquatic biomass systems, NTIS, HCP/ET-4000-78/1 (1978a).

Dynatech R/D Company, Reviewers comments on cost analysis of aquatic biomass systems, NTIS, HCP/ET-4000-78/2 (1978b).

Eimers, K. L., Short rotation forestry, *CHEMTECH,* 212–215 (April 1978).

Elliot, D. C. and G. Giacoletto, Bench scale research in biomass liquefaction in support of the Albany, Oregon experimental facility, NTIS, SERI/TP-33-285, (1979), pp. 123–129.

Elliot, D. C. and P. Walkup, Bench scale research in thermochemical conversion of biomass to liquids in support of the Albany, Oregon experimental facility, NTIS, TID-28415 (1977).

Elmore, C. L., New low cost pulping technology, *Chem. Eng. Progr.,* **74,** 75–82 (1978).

Emert, G. H. and R. Katzen, Chemicals from biomass by improved enzyme technology, paper presented at ACS/CJS Joint Chemical Congress, Honolulu, Hawaii (1979).

Enari, T.-M., and P. Markkanen, Production of cellulolytic enzymes by fungi, *Adv. Biochem. Eng.,* **5,** 3–24 (1977).

Engelbart, W. and H. Dellweg, Basic data on continuous alcoholic fermentation of sugar solutions and of mashes from starch containing materials, NTIS, CONF-771175 (1978).

Epstein, E. and J. Norlyn, Seawater-based crop production: a feasibility study, *Science,* **197,** 249–251 (1977).

Ergun, S., An overview of biomass liquefaction, NTIS, SERI/TP-33-285, (1979), pp. 103–104.

Ericksson, K.-E., Enzyme mechanisms involved in cellulose hydrolysis by the rot fungus *Sporotrichum pulverulentum, Biotechnol. Bioeng.,* **20,** 317–332 (1978).

Ernest, R. K., R. H. Hamilton, N. S. Borgeson, F. A. Schooley, and R. L. Dickenson, Mission analysis for the federal fuels from biomass program, Vol. 3: Feedstock availability, NTIS, SAN-0115-T1 (1979).

Espinosa, R., V. Cojulun, and F. Marroguin, Alternatives for energy savings at plant level for the production of alcohol and its use as automotive fuel, *Biotechnol. Bioeng. Symp.*, **8**, 69–74 (1978).

Evans, H. J., Barber, L., Biological nitrogen fixation for food and fiber production, *Science,* **197**, 322–339 (1977).

Eveleigh, D. E. and Montenecourt, B., Review of fungal cellulases, ERDA *Fuels from Biomass Newsletter,* RPI, Troy, N.Y. (August 1977).

Feldman, H., Conversion of forest residue to a methane-rich gas, NTIS, CONF-7806107-P1, (1978), pp. 245–251.

Feldman, H., S. Chauhan, K. Liu, B. Kim, P. Choi, and H. Conkle, Conversion of forest residue to a methane-rich gas, 3rd annual Biomass Energy Systems Conference, NTIS, SERI/TP-33-285 (1979), pp. 439–444.

Figueroa, C. and S. Ergun, Direct liquefaction of biomass—correlative assessment of process development, NTIS, SERI/TP-33-285, (1979), pp. 109–110.

Finn, R. K. and A. Ramalingam, The vacuferm process: a new approach to fermentation alcohol, *Biotechnol. Bioeng.,* **19**, 583–589 (1977).

Flickinger, M. C., and G. T. Tsao, Fermentation substrates from cellulosic materials: fermentation products from cellulosic materials, *Annual Reports on Fermentation Processes,* Vol. 2, D. Perlman, Ed., Academic (1978), pp. 23–42.

Flowers, A., and A. Bryce, Energy conversion from marine biomass, report prepared by the American Gas Association and General Electric Company (1977).

Frank, M. E. and J. Leonard, *Chem. Eng. News* (November 21, 1977), pp. 17–18.

Gaddy, J. L., paper presented at joint CIC-ACS meeting, Montreal (1977).

Gaden, E. L., M. H. Mandels, and L. A. Spano, Eds., *Enzymatic Conversion of Cellulosic Materials: Technology and Applications,* Biotech. Bioeng. Symposium Series No. 6, Interscience, New York (1976).

Gardner, R. J., R. Crane, and J. Hannan, Hollow fiber permeation for separating gasses, *Chem. Eng. Progr.,* **73**, 76–78 (1977).

Garrels, R. M., A. Lerman, and F. Mackenzie, Control of atmospheric O_2 and CO_2: past, present, and future, *Am. Sci.,* **64**, 306–315 (1976).

Garrett, D. E., Conversion of biomass materials into gaseous products, NTIS, SAN/1241-77/1 (1977).

Garrett, D. E., Investigation of woody biomass species for fuel production in warm climate, non-agricultural land irrigated by brackish or saline water, NTIS, DSE-3007-T1 (1979).

Gaudy, A. F., Jr., S. Goswami, T. Manickam, and E. Gaudy, Formation of strippable metabolic products in biological waste treatment, *Biotechnol. Bioeng.,* **19**, 1239–1244 (1977).

Geisser, H. R., and J. Pfeffer, Biological conversion of biomass to methane. The effect of reactor design on kinetics, U.S. Department of Energy Report COO/2917-8 (1977).

Ghose, T. K., Ed., *Bioconversion of Cellulosic Substances into Energy Chemicals and Microbial Protein,* IIT, New Delhi, India (1978).

Ghose, T. K., and V. Sahai, Production of cellulases by *T. reesei* in fed-batch and continuous-flow culture with cell recycle, *Biotechnol. Bioeng.,* **21**, 283–296 (1979).

Ghose, T. K., and R. Tyagi, Rapid ethanol fermentation of cellulose hydrolysate, I. Batch versus continuous systems, *Biotechnol. Bioeng.,* **21**, 1387–1400 (1979a).

Ghose, T. K., and R. Tyagi, Rapid ethanol fermentation of cellulose hydrolysate, II. Product and substrate inhibition and optimization of fermenter design, *Biotechnol., Bioeng.,* **21**, 1401–1420 (1979b).

Ghosh, S., D. Klass, J. Conrad, M. Henry, K. Griswold, and M. Sedzielarz, A comprehensive gasification process for energy recovery from cellulosic wastes, in T. K. Ghose, *Bioconversion of Cellulosic Substances* into *Energy Chemicals and Microbial Protein,* IIT, New Delhi (1978), pp. 479–533.

Goldemberg, J., Brazil: energy options and current outlook, *Science,* **200,** 159–164 (1978).

Goldman, J. C., Fuels from solar energy: photosynthetic systems—state of the art and potential for energy production, NTIS, COO-4151-2 (1978).

Goldman, J. C., J. Ryther, R. Waaland, and E. Wilson, Topical report on sources and systems for aquatic plant biomass as an energy source, NTIS, OST-4000-77/1 (1977).

Goldstein, I. S., Potential for converting wood into plastics, *Science,* **189,** 847–852 (1975).

Grant, G. A., A. Anderson, and Y. Han, Preliminary cost estimates of straw as animal feed, *Biotechnol. Bioeng.,* **19,** 1817–1830 (1977).

Graves, D. and J. Stramondo, Photosynthesis as a model for hydrogen generation, *AIChE Symp. Ser. 158,* **72,** 43–46 (1976).

Greenshields, R. N., *Octagon Papers No. 2,* University of Manchester, U.K., (1975), pp. 20–27.

Greenshields, R. N. and E. Smith, Tower fermentation systems and their application, *The Chemical Engineer,* 182–190 (May 1971).

Gregor, H. P., Development of membrane processes for the recovery of feed values from the anaerobic fermentation of livestock residues, Progress report No. 2 to U.S. Department of Energy (July 1978).

Gregor, H. P. and R. Cardenas, Algal concentration by ultrafiltration, NTIS, COO/4076-3 (1976).

Gregor, H, P. and R. Cardenas, Algal concentration by ultrafiltration, NTIS, COO/4076/2 (1977).

Gregor, H. P. and C. Gregor, Synthetic membrane technology, *Sci. Am.,* **239,** 88–101 (1978).

Gregor, H. P. and T. Jeffries, Ethanolic fuels from renewable sources in the solar age, in W. R. Vieth, and K. Venkatasubramian, Eds., *Biochemical Engineering,* New York Academy of Sciences, New York (1979).

Grove, S. N. and C. E. Bracker, Protoplasmic organization of hyphal tips among fungi: vesicles and spitzenkörper, *J. Bacteriol.,* **104,** 989–1009 (1970).

Gupta, K. G., N. Yadav, and S. Dhawan, Laboratory scale production of acetoin plus diacetyl by *Enterobacter cloacae* ATCC 27613, *Biotechnol. Bioeng.,* **20,** 1895–1901 (1978).

Hägerdal, B. G., J. Ferchak, and E. Pye, Cellulolytic enzyme system of *Thermoactinomyces* sp. grown on microcrystalline cellulose, *Appl. Environ. Microbiol.,* **36,** 606–612 (1978).

Hägerdal, B. G., H. Harris, and E. Pye. Association of beta-glucosidase with intact cells of *Thermoactinomyces, Biotechnol. Bioeng.,* **21,** 345–355 (1979).

Hammond, O. H. and R. Barron, Synthetic fuels: prices, prospects, and prior art, *Am. Sci.,* **64,** 407–417 (1976).

Hamrick, J. T., Development of wood as an alternative fuel for large power generating systems, NTIS, ORO-5682-8 (1978).

Hanson, K., report in *Chem. Eng. News* (October 10, 1977), p. 30.

Hardin, G., The tragedy of the commons, *Science,* **162,** 1243–1248 (1968).

Harte, J. and M. El-Gasseir, Energy and water, *Science,* **199,** 623–634 (1978).

Heitland, H., H. Czaschke, and N. Pinto, The use of ethanol from biomass as an alternative fuel in Brazil, translation of conference, NTIS, CONF-771175 (1977).

Helsel, R. W., Removing carboxylic acids from aqueous wastes, *Chem. Eng. Progr.,* 55–59 (May 1977).

Hellwig, K. C., S. Alpert, C. Johnson, and S. Schuman, Production of phenols from lignin, *TAPPI* (Journal of the American Pulp and Paper Institute) (February 18, 1969).

Heperer, L., Feasibility of producing basic chemicals by fermentation, in H. G. Schlegel and J. Barnes, Eds., *Microbial Energy Conversion,* Oxford Pergamon Press (1977), p. 550.

Hertzmark, D. I., A preliminary report on the agricultural sector impacts of obtaining ethanol from grain, NTIS, SERI/RR-51-292 (1979).

Hertzmark, D. and B. Gould, The market for ethanol feed joint products, NTIS, SERI/RR-352-357 (1978).

Heylin, M., South Africa commits to oil-from-coal process, *Chem. Eng. News.* (September 17, 1979), pp. 13–16.

Hillman, W. S. and D. Culley, Jr., The uses of duckweed, *Am. Sci.,* **66,** 442–451 (1978).

Hoffman, R. T., A. Strickland, and P. Harvey, Ocean food and energy project, subtask 4: preliminary design of substrate and upwelling systems, NTIS, ERDA-USN-1027-3/2 (1976).

Hokanson, A. E. and R. Katzen, Chemicals from wood wastes, *Chem. Eng. Progr.,* **74,** 67–71 (1978).

Hooverman, R. H., Catalyzed steam gasification of biomass, phase II, NTIS, COO-4736-12 (1979).

Hopkinson, C. S. and J. Day, Net energy analysis of alcohol production from sugarcane, *Science,* **207,** 302–303 (1980).

Hosaka, H. and H. Suzuki, paper presented at 2nd meeting of working party on wood hydrolysis, FAO Technical Panel on Wood Chemistry, Tokyo (1960).

Hospodka, J., Industrial applications of continuous fermentation, in I. Malek and Z. Fencl, Eds., *Theoretical and Methodological Basis of Continuous Culture of Microorganisms,* Academic, New York (1966), pp. 495–645.

Howell, J. A. and M. Mangat, Enzyme deactivation during cellulose hydrolysis, *Biotechnol. Bioeng.,* **20,** 847–863 (1978).

Howlett, K. and A. Gamache, *Silvicultural Biomass Farms,* Vol. 2, The biomass potential of short rotation farms, NTIS, MTR-7347-2 (1977a).

Howlett, K. and A. Gamache, *Silvicultural Biomass Farms,* Vol. 6, Forest and mill residues as potential sources of biomass, NTIS, MTR-7347-6 (1977b).

Huang, S. Y., K. Won, H. Wang, and C. Lin, Optimum design of an anaerobic digester for swine waste, paper presented at U.S.–Republic of China Symposium, Philadelphia (1978).

'uffman, W. J., J. Halligan, R. Peterson, and E. de la Garza, Ammonia synthesis gas and petrochemicals from cattle feedlot manure, paper presented at symposium on Clean Fuels from Biomass, Orlando, Fla. (1977).

Humphrey, A. E., A. Moreira, W. Armiger, and D. Zabriskie, Production of single-cell protein from cellulose wastes, *Biotechnol. Bioeng., Symp.,* **7,** 45–64 (1977).

Inman, R. E., D. Salo, and B. McGurk, *Silvicultural Biomass Farms,* Vol. 4, Site-specific production studies and cost analyses, NTIS, MTR-7347-4 (1977).

Intertechnology/Solar Corp., The photosynthesis energy factory: analysis, synthesis, and demonstration, NTIS, HCP-T3548-01 (1978).

Isaacs, S. H. and C. Wilke, Utilization of immobilized β-glucosidase in the enzymatic hydrolysis of cellulose, Lawrence Berkeley Laboratory Report LBL-7857 (1978).

Jahn, E. C., Ed., *Proceedings of the Seventh Cellulose Conference,* Interscience, New York (1971).

James, A. L. and D. Murphy, *Preliminary Design Studies of Substrate and Upwelling Systems,* Vol. 1, Kelp support substrate structures for use in the OFEF project, NTIS, ERDA-USN-1027-3/1 (1976).

Jeffries, T. W., D. Olmstead, R. Cardenas, and H. Gregor, Membrane controlled digestion: effect of ultrafiltration on anaerobic digestion of glucose, *Biotechnol. Bioeng., Symp.,* **8,** 37–49 (1978).

Jeffries, T. W. and K. Leach, Intermittent illumination increases biophotolytic hydrogen yield by *Anabaena cylindrica, Appl. Environ, Microbiol.,* **35,** 1228–1230 (1978).

Jenkins, D. M., T. Reddy, and J. Harrington, Economics of manufacturing liquid fuels from corn stover, report to U.S. Department of Energy from Battelle/Columbus Laboratories (October 20, 1978).

Jenkins, D. M. and T. Reddy, Economic evaluation of the MIT process for manufacture of ethanol, report to U.S. Department of Energy from Battelle/Columbus Laboratories (June 28, 1979).

Jensen, N. F., Limits to growth in world food production, *Science,* **201,** 317–320 (1978).

Jewell, W. J., Ed., Anaerobic fermentation of agricultural residue: Potential for improvement and implementation, NTIS HCP/T2981-07 (February 1978a).

Jewell, W. J., Small farm methane generation potential, NTIS, CONF-7806107-P2 (1978b), pp. 701–736.

Jones, J. L., Evaluation of the proposed "bio-gas" project at Lamar, Colorado, NTIS, TID-29306 (1978).

Jones, J. L., W. Fong, F. Schooley, and R. Dickenson, *Mission Analysis for the Federal Fuels from Biomass Program,* Vol. V, Biochemical conversion of biomass to fuels and chemicals, NTIS, TID-29093 (1978).

Jones, J. L., and S. B. Radding, Eds., Solid wastes and residues, conversion by advanced thermal processing, American Chemical Society Symposium Series No. 76 American Chemical Society, Washington, D.C. (1978).

Jones, J. L., S. Kohan, K. Semrau, F. Schooley, and R. Dickenson, *Mission Analysis for the Federal Fuels from Biomass Program,* Vol. VI, Mission addendum, NTIS, SAN-0115-T4 (1979).

Karube, I., S. Tanaka, T. Shirai, and S. Suzuki, Hydrolysis of cellulose in a cellulase bead fluidized bed reactor, *Biotechnol. Bioeng.,* **19,** 1183–1191 (1977).

Kaspar, H. F., K. Wuhrmann, Kinetic parameters and relative turnovers of some important catabolic reactions in digesting sludge, *Appl. Environ, Microbiol.,* **36,** 1–7 (1978).

Katzen, R., R. Frederickson, and B. Bush, The alcohol pulping and recovery process, *Chem. Eng. Progr.,* **76,** 62–67 (1980).

Kemp, C. C. and G. Szego, The energy plantation, *AIChE Symp. Ser. 158,* **72,** 1–5 (1976).

Kleinpeter, J. A., The outlook for clean liquid fuels from coal in the U.S., paper presented at Am. Chem. Soc. national meeting, Miami, Fla. (1978).

Knappert, D., H. Grethlein, and A. Converse, Partial hydrolysis of cellulosic materials as a pretreatment for enzymatic hydrolysis, in G. Guilbault, Ed., *Proceedings of Fourth Annual Joint US/USSR Conference on Microbial Enzyme Reactions,* NTIS PB80-132913 (1979a), pp. 403–419.

Knappert, D., H. Grethlein, A. Converse, Potential acid hydrolysis of cellulosic materials as a pretreatment for enzymatic hydrolysis, paper presented at AIChE national meeting, San Francisco (1979b).

Koch, P. and T. Nicholson, Harvesting residual biomass and swathe-felling with a mobile chipper, NTIS, CONF-7806107-P1, (1978), pp. 227–243.

Kohan, S. M., P. Barkhordar, F. Schooley, and R. Dickenson, *Mission Analysis for the Federal Fuels from Biomass Program,* Vol. IV, Thermochemical conversion of biomass to fuels and chemicals, NTIS, SAN-0115-T3 (1979).

Kollman, V. H. and D. Kollman, Feasibility of algae production, phase II, final report of Four Corners Regional Commission Demonstration Project, FCRC No. 672-366-022 (undated a).

Kollman, V. H. and D. Kollman, Feasibility of algae production, Final report prepared by KLA Labs, Inc., Four Corners Regional Commission Demonstration Project, FCRC No. 362-332-051 (undated b).

Krampitz, L. O., Biophotolysis of water to hydrogen and oxygen, *AIChE Symp. Ser. 158,* **72,** 47–48 (1976).

Kringstad, K., The challenge of lignin, paper presented at International Symposium on Feedstocks of the Future, Toronto (1978).

Ladisch, M. R., paper presented at Indiana Bioconversion Conference, Indianapolis (1979).

Ladisch, M. R., C. Ladisch, and G. Tsao, Cellulose to sugars: new path gives quantitative yield, *Science,* **201,** 743–745 (1978).

Ladisch, M. R. and K. Dyck, Dehydration of ethanol: new approach gives positive energy balance, *Science,* **205,** 898–900 (1979).

Lai, M. N. and I. Wang, A rapid process for manufacturing vinegar, paper presented at U.S.–Republic of China Seminar, Philadelphia (1978).

Landau, R., The chemical industry 2000 A.D., *Chem. Eng. Progr.,* **74,** 27–31 (1978).

Lanyi, J. K., Light energy conversion in *Halobacterium halobrium, Microbiol. Rev.,* **42,** 682–706 (1978).

Laskin, A. I., Ethanol as a substrate for single cell protein production, *Biotechnol. Bioeng. Symp.,* **7,** 91–103 (1977).

Lee, Y. Y., T. Yue, and A. Tarrar, Acid hydrolysis of oak sawdust, paper presented at AIChE national meeting, Kansas City (1976).

Lee, Y. Y., C. Lin, T. Johnson and R. Chambers, Selective hydrolysis of hardwood hemicellulose by acids, *Biotechnol. Bioeng. Symp.,* **8,** 75–88 (1979).

Leisola, M. and V. Kauppinen, Automatic assay of cellulase activity during fermentation, *Biotechnol. Bioeng.,* **20,** 837–846 (1978).

Lepkowski, W., Carbon dioxide: a problem of producing usable data, *Chem, Eng. News* (October 17, 1977), pp. 26–30.

Lien, S., and A. San Pietro, An inquiry into biophotolysis of water to produce hydrogen, report to National Science Foundation (no date).

Lindeman, L. R. and C. Rocchiccioli, Ethanol in Brazil: Brief summary of the state of the industry in 1977, *Biotechnol. Bioeng.,* **21,** 1197–1119 (1979).

Linden, J. V. and V. Murphy, Improved enzymatic conversion of cellulose by ethylene treatment, NTIS, COO-4546-7 (1978).

Lindenfelser, L. A., R. Detroy, J. Ramstack, and K. Worden, Biological modification of the lignin and cellulose components of wheat straw by *Pleurotus ostreatus,* in L. A. Underkofler, Ed., *Dev. in Ind. Microbiol,* vol. 20 *(1979), pp. 541–551.*

Lindsey, R. R. and C. Wilke, Process design and optimization of cellulose hydrolysis, Lawrence Berkekey Laboratory Report LBL-7864 (1978).

Linko, M., An evaluation of enzymatic hydrolysis of cellulosic materials, *Adv. Biochem. Eng.,* **5,** 27–48 (1977).

Lipinsky, E. S., Systems study of fuels from sugarcane, sweet sorghum, sugar beets, and corn, Vols. 1–4, NTIS, BMI-1957 (1977).

Lipinsky, E. S., Fuels from biomass; integration with food and materials systems, *Science,* **199,** 644–651 (1978).

Lipinsky, E., Carbohydrate crops as sources of fuels. NTIS, SERI/TP-35-285. (1979), pp. 224–236.

Lipinsky, E. S., H. Birkett, J. Polack, J. Atchison, S. Kresovich, T. McClure, and W. Lawhon, Sugar crops as a source of fuels, Vol. 2, NTIS, TID-29400/2 (1978).

Lipinsky, E. S., S. Kresovich, T. McClure, E. Helper, and W. Lawhon, Fuels from sugar crops, NTIS, TID-27834 (1977).

Lipinsky, E. S., S. Kresovich, T. McClure, D. Jackson, W. Lawhon, A. Lalyoncu, and E. Daniels, Sugar crops as sources of fuels, Vol. 1, Agricultural research, NTIS, TID-29400/1 (1978).

Lipinsky, E. S., W. Sheppard, J. Otis, E. Helper, T. McClure, and D. Scantland, Systems study of fuels from sugarcane, sweet sorghum, sugar beets, and corn, Vol. 5, NTIS, BMI-1957A-5 (1977).

Lipinsky, E. S., D. Scantland, and T. McClure, Systems study of the potential integration of U.S. corn production and cattle feeding with manufacture of fuels via fermentation, Vol. 1, NTIS, BMI-2033 (1979).

Litchfield, J. H., Microbial cells on your menu, *CHEMTECH,* 218–223 (April 1978).

Lizdas, D. J., Methane production by beef cattle feedlots, NTIS, CONF-7806107-P2, (1978), pp. 737–757.

Lizdas, D. J. and W. Coe, Experimental anaerobic fermentation facility, NTIS, SERI/TP-35-285, (1979), pp. 291–295.

Lockeretz, W., The lessons of the dust bowl, *Am. Sci.,* **66,** 560–590 (1978).

Loomis, R. S., Agriculture, paper presented at CHEMRAWN Symposium on Organic Raw Materials, Toronto (1978).

Lora, J. H. and M. Wayman, Delignification of hardwoods by autohydrolysis and extraction, *TAPPI,* **61,** 47–50 (1978).

Lovins, A. B., *Soft Energy Paths. Toward a Durable Peace,* Friends of the Earth International, Ballinger, Cambridge, Mass. (1977).

Lurie, J., Octagon Papers No. 2, University of Manchester, U.K. (1975), pp. 18–19 and 70–73.

Macdonald, D. G. and J. Mathews, Effects of steam treatment on the hydrolysis of aspen by commercial enzymes, *Biotechnol. Bioeng.,* **21,** 1091–1096 (1979).

Mariani, E. O., W. Wood, P. Kouchoukos, and M. Minton, The eucalyptus energy farm: Feasibility and demonstration, Phase 1, Site and species selection, NTIS, HCP/T2557-01-UC-61 (1978).

Märkl, H., CO_2 transport and photosynthetic productivity of a continuous culture of algae, *Biotechnol. Bioeng.,* **19,** 1851–1862 (1977).

Martin, J. P. and K. Haider, Biodegradation of ^{14}C-labeled model and cornstalk lignins, model phenolase humic polymers, and fungal melanins as influenced by a readily available carbon source and soil, *Appl. Environ. Microbiol.,* **38,** 283–289 (1979).

Marzola, D. L. and D. Bartholomew, Photosynthetic pathway and biomass energy production, *Science,* **205,** 555–559 (1979).

McCann, D. J. and R. Prince, Agro-industrial systems for ethanol production, *Proceedings: Alcohol Fuels Conference,* August 9–11, Sydney, Australia (1978).

McCarty, P. L., L. Young, J. Healy, W. Owen, and D. Stuckey, Thermochemical treatment of lignocellulosic and nitrogenous residuals for increasing anaerobic biodegradability, NTIS, CONF-7806107-P1, (1978), pp. 787–822.

McKelvey, K. N., Alternative raw materials for organic chemical feedstocks, *Chem. Eng. Progr.,* **75,** 45–48 (1979).

Mears, K., quoted in *Science,* **205,** 564 (1979).

Mendel, D. A., A production cost analysis of *Euphorbia lathyris,* NTIS, DSE-3891-T1 (1979).

Meyers, S. G., Ethanolic fermentation during enzymatic hydrolysis of cellulose, *AIChE Symp. Ser. 181,* **74,** 79–84 (1978).

Miles, T. R., Logistics of energy resources and residues, in D. Tillman, K. Sarkanen, L. Anderson, Eds., *Fuels and Energy from Renewable Resources,* Academic, New York, 1977, pp. 225–248.

Milne, T. A., J. Connolly, R. Inman, T. Reed, and M. Seibert, Research overview of biological and chemical conversion methods and identification of key research areas for SERI, NTIS, SERI/tr-33-067 (1978).

Mix, T. J., D. Dweck, M. Weinberg, and R. Armstrong, Energy conservation in distillation, *Chem. Eng. Progr.,* **74,** 49–55 (1978).

Miyamoto, K., P. Hallenbeck, and J. Benemann, Hydrogen production by thermophilic alga *Mastigocladus laminosus:* Effects of nitrogen, temperature, and inhibition of photosynthesis, *Appl. Environ. Microbiol.,* **38,** 440–446 (1979).

Montenecourt, B. S., D. Eveleigh, G. Elmund, and J. Parcells, Antibiotic discs—an improvement in the filter paper assay for cellulase, *Biotechnol. Bioeng.,* **20,** 297–300 (1978).

Mooney, The cellulose project, *The Orange Disc,* **22** (1977).

Moraine, R., G. Shelef, A. Meydan, and A. Levi, Algal single cell protein from wastewater treatment and renovation process, *Biotechnol. Bioeng.,* **21,** 1191–1207 (1979).

Morris, R. O., The Purdue process applied to paraquat treated loblolly pine chips, M. S. thesis, Rensselaer Polytechnic Institute (1978).

Mueller Associates, Inc., Whey and dry milk products as feedstocks for ethanol production, NTIS, DOE/CS/2098-03 (1979).

Nathan, R. A., Fuels from sugar crops, NTIS, TID-22781 (1978).

National Academy of Science, *Methane generation from human, animal and agricultural wastes*, Report of an ad hoc panel of the advisory committee on technology inovation, Washington, D.C. (1977).

Neben, E. W., The economics of advanced coal liquefaction, *Chem. Engr. Progr.*, **74**, 43–48 (1978).

Neish, I. C., Role of mariculture in the Canadian seaweed industry, *J. Fish. Res. Board Can.*, **33**, 1007–1014 (1976).

Nielsen, P. E., N. Nishimura, J. Otvos, and M. Calvin, Plant crops as a source of hydrocarbon-like materials, *Science*, **198**, 942–944 (1977).

North, W. J., The giant kelp Macrocystis: a potential producer of marine biomass for energy, *Bioresour. Dig.*, **1**, 96–102 (1979).

Nyberg, D. W., Mechanical dewatering increases fuel value of residue, in J. Mater and M. Mater, Eds., *Technology of Utilizing Bark and Residues as an Energy and Chemical Resource,* Forest Prod. Res. Soc., Madison, Wis. (1976), pp. 73–78.

O'Hara, J. B., E. Becker, N. Jentz, and T. Harding, Petrochemical feedstocks from coal, *Chem. Eng. Progr.*, **73**, 64–72 (1977).

Oliver, D. J. and I. Zelitch, Increasing photosynthesis by inhibiting photorespiration with glyoxylate, *Science*, **196**, 1450–1451 (1977).

O'Neil, D. J., An integrated chemical system for whole-tree utilization of complete, paraquat-treated pines, paper presented at annual meeting Lightwood Research Coordinating Council, Jacksonville, Fla. (1978).

O'Neil, D. J., A. Colcord, M. Bery, S. Day, R. Roberts, I. El-Barbary, S. Havlicek, M. Anders, and D. Sondhi, Design, fabrication and operation of a biomass fermentation facility, First quarterly report to U.S. Department of Energy (1978).

Páca, J. and V. Grégr, Effect of PO_2 on growth and physiological characteristics of *Candida utilis* in a multistage tower fermentation, *Biotechnol. Bioeng.*, **21**, 1827–1843 (1979).

Paturau, J. M., *By-products of the Cane Sugar Industry,* Elsevier, Amsterdam (1969), pp. 95–115.

Paul, J. K., *Ethyl Alcohol Production and Use as a Motor Fuel,* Noyes Data Corp., Park Ridge, N. J. (1979).

Pearson, Process for preparing hydrocarbons from methanol and phosphorous pentoxide, U.S. Patent 4133838 (1979).

Peitersen, N. and E. Ross, Jr., Mathematical model for enzymatic hydrolysis and fermentation of cellulose by *Trichoderma, Biotechnol. Bioeng.*, **21**, 997–1017 (1979).

Pelovsky A., Personal communication and short notes in various publications (1978).

Pfeffer, J. T., Biological conversion of biomass to methane, quarterly progress report, NTIS., COO-2917-10 (1978).

Pirt, S. J., Algal culture in solar panels: A primary source of fuel, food, and carbon feedstocks, paper presented at International Chemical Engineering Meeting, London (1979).

Polack, J. A. and M. West, Bioconversions of sugar cane products, paper presented at U.S.-Republic of China Symposium, Philadelphia, Pa. (1978).

Pollard, W. G., The long range prospects for solar energy, *Am. Sci.,* **64**, 424–430 (1976a).

Pollard, W. G., The long range prospects for solar derived fuels, *Am. Sci.,* **64**, 509–514 (1976b).

Poole, N. J. and A. Smith, The potential of low technology, Octagon Papers No. 3, University of Manchester, U.K. (1976), p. 85.

Povich, M. J., Some limitations of fuel farming, *AIChE Symp. Ser. 158,* **72,** 11–16 (1976).

Povich, M. J., Fuel farming—water and nutrient limitations, *AIChE Symp. Ser. 181,* **74,** 1–5 (1978).

Pratt, H. T., letter in *Chem. Engr. News* (May 1, 1978), p. 64.

Prescott, S. C. and C. Dunn, *Industrial Microbiology,* 3rd ed., McGraw-Hill, New York (1959).

Preston, G. T., Resource recovery and flash pyrolysis of municipal refuse, paper presented at IGT Symposium, Orlando, Fla. (1976).

Prince, I. G. and D. McCann, The continuous fermentation of starches and sugars to ethyl alcohol, *Alcohol Fuels,* **9–11,** 8.17–8.24 (1978).

Pye, E. K., Thermophilic degradation of cellulose for production of liquid fuels, NTIS, CONF-7806107-P2 (1978).

Pye, E. K. and A. Humphrey, University of Pennsylvania interim report to U.S. Department of Energy (August 31, 1979).

Reddy, C. A., Symposium on microbial degradation of lignin, *Dev. Ind. Microbiol.,* **19,** 23–70 (1978).

Redhead, K. and S. J. L. Wright, Isolation and properties of fungae that lyse blue–green algae, *Appl. and Environ. Microbiol.,* **35,** 962–969 (1978).

Reed, T. B., A survey of biomass gasification, Vol. 1, NTIS, SERI/TR-33-239 (1979a).

Reed, T. B., Ed., A survey of biomass gasification, II: Principles of gasification, NTIS, SERI/TR-33-239 (1979b).

Reed, T. and B. Bryant, Densified biomass: A new form of solid fuel, NTIS, SERI Report 35 (1978).

Reed, T. B. and D. Jantzen, Generator gas: The Swedish experience from 1939 to 1945, translation, NTIS, SERI/SP-33-140 (1979).

Reese, K. M., Zooplankton—prolific producers of wax esters, *Chem. Eng. News* (September 20, 1976), p. 84.

Reilly, P. J., Economics and energy requirements of ethanol production, Departmental report, Iowa State University (1978).

Reilly, P. J., Report of the Alcohol Fuels Policy Review, Raw Materials Availability Reports, NTIS, DOE/ET-0114/1 (1979); DOE/PE-0012 (1979).

Rolz, C., S. DeCabrera, and R. Garcia, Ethanol from sugar cane: EX-FERM concept, *Biotechnol. Bioeng.,* **21,** 2347–2349 (1979).

Ruby, E. G. and K. Nealson, A luminous bacterium that emits yellow light, *Science,* **196,** 432–433 (1977).

Rudolph, K., R. Owsianowski and W. Tentscher, Producing ethanol directly from sugar cane, paper presented at Alcohol Conference, Wolfsburg, W. Germany (1977).

Ruehle, J. L. and D. Marx, Fiber, food, fuel, and fungal symbionts, *Science,* **206,** 419–422 (1979).

Russell-Hunter, W. D., *Aquatic Productivity: An Introduction to Some Basic Aspects of Biological Oceanography and Limnology,* Macmillan, New York (1970).

Ryther, J. H., Cultivation of macroscopic marine algae and freshwater aquatic weeds, NTIS, COO/2948-2 (1978).

Ryther, J. H., Remarks at Bio-energy '80, Atlanta, Ga. (1980).

Ryther, J. H., L. Williams, M. Hanisak, R. Stenberg, and T. DeBusk, Freshwater and marine plants for biomass production, Proceedings of 3rd Annual Biomass Energy Systems Conference, Golden, Col. NTIS SERI/TP-33-285 (1979), pp. 13–24.

Ryu, D., R. Andreotti, M. Mandels, B. Gallo, and E. Reese, Studies on quantitative physiology of *Trichoderma reesei* with two-stage continuous culture for cellulase production, *Biotechnol. Bioeng.,* **21,** 1887–1903 (1979).

Sachdev, R. K. and T. Ghose, Immobilization of 1-4β glucosidase on solid supports and its application to hydrolysis of cellobiose to glucose, in T. K. Ghose, Ed., *Bioconversion of Cellulosic Substances into Energy Chemicals and Microbial Protein,* IIT, New Delhi (1978), pp. 181–194.

Saeman, J. F., Kinetics of wood saccharification, *Ind. Eng. Chem.,* **37,** 43 (1945).

Salo, D. J., R. Inman, B. McGurk, and J. Verhoeff, *Silvicultural Biomass Farms,* Vol. 3, Land suitability and availability, NTIS, MTR-7347-3 (1977).

Sanderson, J. E., D. Wise, and D. Augenstein, Organic chemicals and liquid fuels from algal biomass, *Biotechnol. Bioeng. Symp.,* **8,** 131–151 (1978).

Sasaki, T., T. Tanaka, N. Nanbu, Y. Sato, and K. Kainuma, Correlation between X-ray diffraction measurements of cellulose crystalline structure and the susceptibility to microbial cellulase, *Biotechnol. Bioeng.,* **21,** 1031–1042 (1979).

Saterson, K. A., M. K. Luppold, K. M. Scow, and R. E. Lee, Herbaceous species screening program, NTIS, COO-5035-3 (1979).

Savarese, J. J. and S. Young, Combined enzyme hydrolysis of cellulose and yeast fermentation, *Biotechnol. Bioeng.,* **20,** 1291–1293 (1978).

Sayigh, A. A. M., *Solar Energy Engineering,* Academic, New York (1977).

Schaleger, L. L., N. Yaghoubzadeh, and S. Ergun, Pretreatment of biomass prior to liquefaction, NTIS, SERI/TP-33-285 (1979), pp. 119–121.

Scheller, W. A., Energy requirements for grain alcohol production, paper presented at Am. Chem. Soc. national meeting, Miami Beach, Fla. (1978).

Schurz, J., How to make native lignocellulosic material accessible to chemical and microbial attack, in T. K. Ghose, Ed., *Bioconversion of Cellulosic Substrates into Energy Chemicals and Microbial Protein,* ITT, New Delhi (1978), pp. 37–58.

Schwab, C., Legal considerations in the development and implementation of biomass energy technologies, NTIS, SERI/TR-62-265 (1979).

Seth, M. and S. Ergun, The potential for biomass liquefaction, NTIS, SERI/TP-33-285 (1979), pp. 131–137.

Shapiro, I. S., Future sources of organic raw materials, *Science,* **202,** 287–289 (1978).

Sheppard, W. J., Ethanol and furfural from corn, NTIS, CONF-770368 (1978).

Shnaier, E. E., M. Shpultova, and S. Chepigo, A combined method of hydrolysis of corn stalks with concentrated sulfuric acid, *Gidroliznaya: Lesokhimicheckaya Promyahlenost,* **15,** 1–4 (1960).

Shrader, W. D., Effect of removal of crop residues on soil productivity, NTIS, CONF-770368 (1978), pp. 49–74.

Siebert, M., J. Connolly, T. Milne, and T. Reed, Biological and chemical conversion of solar energy at SERI, paper presented at AIChE national meeting, Atlanta, Ga. (1978).

Siemon, J. R., The production of solar ethanol from Australian forests, SES Report 75/5, Commonwealth Scientific and Industrial Research Organization, Australia (1975).

Silverman, H. P., Conversion of natural products by biofuel cells, *AIChE Symp. Ser.,* **158,** 49–51 (1976).

Sitton, O. C., G. Foutch, N. Book, and J. Gaddy, Ethanol from agricultural residue, *Chem. Eng. Progr.,* **75,** 52–57 (1979).

Skinner, K. J., Nitrogen fixation, *Chem. Eng. News* (October 4, 1976), pp. 22–35.

Skrinde, R. T., Review of international biogas programs, NTIS, CONF-7806107-Pl (1978), pp. 823–858.

Smith, F. J., letter in *Chem. Eng. News* (December 19, 1977), p. 4.

Smith, J. E., Cellulose as a fungal substrate in mushroom cultivation, Octagon Papers No. 3, University of Manchester, U.K. (1976), p. 77.

Smith, M. H. and M. Gold, *Phanerochaete chrysosporium* beta-glucosidases: Induction, cellular localization, and physical characterization, *Appl. Environ. Microbiol.,* **37,** 938–942 (1979).

Smith, M. R. and R. Mah, Growth and methanogenesis by *Methanosarcina* strain 227 on acetate and methanol, *Appl. Environ. Microbiol.,* **36,** 870–879 (1978).

Snyder, F. H., Production of sugars from wood product, U.S. Patent 2835611 (1958).

Spano, L. A., Enzymatic hydrolysis of cellulose to glucose, U.S. Army Natick Laboratories quarterly report to U.S. Department of Energy (February 1976).

SRI International, Preliminary economic evaluation of a process for the production of fuel grade ethanol by enzymatic hydrolysis of an agricultural waste, NTIS, HCP/T3891-1 (1978).

Stewart, L. L., Chemical utilization of Douglas fir bark, in J. Mater and M. Mater, Eds., *Technology of Using Barks and Residues as an Energy and Chemical Resource,* Forest Prod. Res. Soc., Madison, Wis. (1976), pp. 110–113.

Stinson, S. C., Methanol primed for future energy role, *Chem. Eng. News* (April 2, 1979), pp. 28–30.

Suzuki, S., I. Karube, and T. Matsunaga, Application of biochemical fuel cell to wastewaters, *Biotechnol. Bioeng. Symp.,* **8,** 501–511 (1978).

Swings, J. and J. DeLey, The biology of *Zymomonas, Bacteriol. Rev.,* **41,** 1–46 (1977).

Takagi, M., S. Abe, S. Suzuki, G. Emert, and N. Yata, A method for production of alcohol directly from cellulose using cellulase and yeast, in T. K. Ghose, Ed., *Bioconversion of Cellulosic Substances into Energy Chemicals and Microbial Protein,* IIT, New Delhi (1978), pp. 551–571.

Takeda, S., report in *Inside R and D* (August 22, 1979).

Taylor, E., Biomass energy conversion workshop for industrial executives, NTIS, SERI/TP-62-299 (1979).

Tilby, S. E., Method and apparatus for separating components of sugarcane, U.S. Patent 3567511 (1971).

Tillman, D. A., Ed., *Wood as an Energy Resource,* Academic, New York (1978).

Tillman, D. A., K. Sarkanen, and L. Anderson, Eds., *Fuels and Energy from Renewable Resources,* Academic, New York (1977).

Timell, T. E., Ed., *Proceedings of the Eighth Cellulose Conference,* Interscience, New York (1976).

Tong, G. E., Fermentation routes to C_3 and C_4 chemicals, *Chem. Eng. Progr.,* **74,** 70–74 (1978).

Torpy, M., S. Barisas, L. Habegger, and S. Chiu, Potential water quality impacts from large scale crop residue harvesting, NTIS, SERI/TP-35-285, (1979), pp. 225–228.

Tosa, T., T. Sato, T. Mori, K. Yamomoto, I. Takata, Y. Nishida, and I. Chibata, Immobilization of enzymes and microbial cells using carrageenan as matrix, *Biotechnol. Bioeng.,* **21,** 1697–1709 (1979).

Tsao, G. T., Fermentable sugars from cellulosic wastes as a natural resource, paper presented at U.S.-Republic of China Seminar, Philadelphia (1978).

Tsao, G. T., C. Gong, M. Chang, and M. Ladisch, A fundamental study of the mechanism and kinetics of cellulose hydrolysis by acids and enzymes, NTIS, COO-2755-4 (1979).

Turbak, A. F., Ed., *Cellulose Technology Research,* ACS Symposium Series, American Chemical Society, Washington, D.C. (1975).

Turnacliff, W., M. Custer, and M. Veatch, Design and evaluation of a methane gas system for a hog farm, NTIS, SERI/TP-35-285, (1979), pp. 401–410.

Tyreus, B. D. and W. Luyben, Controlling heat integrated distillation columns, *Chem. Eng. Progr.,* **72,** 59–66 (1976).

Underkofler, L. A. and R. Hickey, *Industrial Fermentations,* Chemical Publishing Company, New York (1954).

van Bavel, C. H. M., Sun and oil, editorial, *Science,* **197,** 213 (1977).

Voltz, S. E. and J. Wise, Development studies on conversion of methanol and related oxygenates to gasoline, Report for U.S. Department of Energy, Contract E-49-18-1773 (1976).

Wallace, C. J. and B. Stokes, Assessment of Nif-derepressed microorganisms for commercial nitrogen fixation, *Biotechnol. Bioeng. Symp.,* **8,** 153–174 (1978).

Walsh, T. J. and H. Bungay, Shallow-depth sedimentation of yeast cells, *Biotechnol. Bioeng.,* **21,** 1081–1084 (1979).

Wang, D. I. C., C. Cooney, A. Demain, R. Gomez, and A. Sinskey, Degradation of cellulosic biomass and its subsequent utilization for the production of chemical feedstocks, NTIS, COO-4198-7 (1978a).

Wang, D. I. C., R. Fleishchaker, and G. Wang, A novel route to the production of acetic acid by fermentation, *AIChE Symp. Ser. 181*, **74**, 105–110 (1978b).

Wang, D. I. C., I. Biocic, H. Fang, and S. Wang, Direct microbiological conversion of biomass to ethanol and chemicals, NTIS, SERI/TP-35-285, (1979), pp. 61–67.

Ward, D. M., Thermophilic methanogenesis in a hot spring algal–bacterial mat (71 to 30°), *Appl. Environ. Microbiol.*, **35**, 1019–1026 (1978).

Wayman, M. and J. Lora, Aspen autohydrolysis. The effects of 2-naphthol and other aromatic compounds, *TAPPI*, **61**, 50–57 (1978).

Weaver, P., S. Lien, and M. Siebert, Photobiological production of hydrogen—a solar energy conversion option, NTIS, SERI/TR-33-122 (1979).

Weetall, H. H. and L. Krampitz, Two stage bioproduction of hydrogen. *Biotechnol. Bioeng.*, in press.

Weiss, D. E., Energy research interests of the Division of Chemical Technology, CSIRO Research Review (1976).

Weissman, J. C. and J. Benemann, Biomass recycling and species competition in continuous culture, *Biotechnol. Bioeng.*, **21**, 627–648 (1979).

Weisz, P. B., Energy and society, paper presented at Am. Chem. Soc. Symposium, Colorado Springs, Col. (1978).

Wilcox, H. A., Expected yields and optimal harvesting strategies for future oceanic kelp farms, *Bioresour. Dig.*, **1**, 103–114 (1979).

Wilke, C. R., Ed., *Cellulose as a Chemical and Energy Resource*, Biotechnol. Bioeng. Symposium Series No. 5. Interscience, New York (1975).

Wilke, C. R. and H. Blanch, Process development studies on the bioconversion of cellulose and production of ethanol, Lawrence Berkeley Laboratory Report, LBL-8658-UC-61 (1978).

Wilke, C. R. and H. Blanch, Process development studies on the bioconversion of cellulose and production of ethanol, Quarterly report to U.S. Department of Energy (1979a).

Wilke, C. R. and H. Blanch, Process development studies on the bioconversion of cellulose and production of ethanol, Lawrence Berkeley Laboratory Report, LBL-9220-UC-61 (1979b).

Wilke, C. R. and H. Blanch, Process development studies on the bioconversion of cellulose and production of ethanol, Lawrence Berkeley Laboratory Report, LBL-9909-UC-61 (1979c).

Williams, L. A., E. Foo, A. Foo, I. Kühn, and G. -C. Hedén, Solar bioconversion systems based on algal glycerol production, *Biotechnol. Bioeng. Symp.*, **8**, 115–130 (1978).

Wise, D. L., E. Ashare, and R. Wentworth, Fuel gas production from animal and agricultural residues and biomass, NTIS, COO-5099-4 (1979).

Wise, D. L., C. Cooney, and D. Augenstein, Biomethanation: anaerobic fermentation of CO_2, H_2, and CO to methane, *Biotechnol. Bioeng.*, **20**, 1153–1172 (1978).

Woodwell, G. M., Recycling sewage through plant communities, *Am. Sci.*, **65**, 556–562 (1977).

Yand, V. and S. Trindade, Brazil's gasohol program, *Chem. Eng. Progr.*, **75**, 11–19 (1979).

Yokoyama, H., E. Hayman, W. Hsu, S. Poling, and A. Bauman, Chemical bioinduction of rubber in guayule plant, *Science*, **197**, 1076–1078 (1977).

Zehnder, A. J. B., in R. Mitchell, Ed., *Ecology of Methane Formation in Water Pollution Microbiology*, Vol. 2, Wiley-Interscience (1978), pp. 349–376.

Zeikus, J. G., The biology of methanogenic bacteria, *Bacteriol. Rev.*, 514–541 (1977).

Zelitch, I., Photosynthesis and plant productivity, *Chem. Eng. News* (February 5, 1979), pp. 28–48.

Zurrer, H. and R. Bachofen, Hydrogen production by the photosynthetic bacterium *Rhodospirillum rubrum*, *Appl. Environ. Microbiol.*, **37**, 789–793 (1979).

Index